Learning and Teaching

Measurement

Learning and Teaching

Measurement

2003
Yearbook

Douglas H. Clements
2003 Yearbook Editor
University at Buffalo, State University of New York
Buffalo, New York

George Bright
General Yearbook Editor
University of North Carolina at Greensboro
Greensboro, North Carolina

1906 Association Drive, Reston, VA 20191-1502
(703) 620-9840; (800) 235-7566; www.nctm.org

ISBN 0-87353-539-1

The publications of the National Council of Teachers of Mathematics present a variety of viewpoints. The views expressed or implied in this publication, unless otherwise noted, should not be interpreted as official positions of the Council.

Printed in the United States of America

Contents

Preface

Measurement is an important mathematical topic, for several reasons. It has its roots, both historically and in individual development, in significant everyday activity. Thus, it can develop in the earliest years from children's experience, and it readily lends itself to real-world application. Further, it spans and connects mathematics and the other sciences and thus can ideally integrate subject matter areas. Finally, it can serve as a foundation for the development of other topics within mathematics. "The study of measurement also offers an opportunity for learning and applying other mathematics, including number operations, geometric ideas, statistical concepts, and notions of function" (National Council of Teachers of Mathematics 2000, p. 44). The goal of the 2003 Yearbook is to present current thinking about the learning and teaching of measurement, including students' understanding, the mathematics of measurement, estimation and approximation, connections, and pedagogy. In doing so, this volume focuses on research and practice, as well as the integration of the two.

The chapters in this volume illustrate all these aspects of measurement's importance. Through the authors, we see the young child's desire to understand and use ideas of measurement. We get to share classroom experiences that show that the integration of topics, especially geometry, number and operations, and measurement, enhances the learning of them all. Throughout, we see that measurement goes far beyond the domain of skills. Students can be guided to think deeply about spatial extent, time, approximation, number, and myriad other ideas to understand measurement and use it appropriately and skillfully.

We need to help all students have such experiences. U.S. students study geometric measurement less than those in most other countries (National Center for Education Statistics 1996). U.S. students do not perform well on assessments of measurement. Many students use measurement instruments or count units in a rote fashion and apply formulas to attain answers without meaning (Clements and Battista 1992). Less than 50 percent of seventh graders can determine the length of a line segment when the beginning of the ruler is not aligned at the beginning of the line segment. In international comparisons, U.S. students' performance in geometry and measurement is lower than in any other topic (National Center for Education Statistics 1996).

These results, as disturbing as they are, are brought home to us more forcefully when we face students' work on measurement problems. The following is a striking personal experience I had conducting initial field tests of a new curriculum. I presented third graders with the challenge of making a map of their classroom (Clements 1999, p. 5).

> They wished to begin by measuring the room. Pleased, I passed out meter sticks. They began laying these down but soon stopped, puzzled.
>
> "We need more."
>
> "More meter sticks?" I inquired.
>
> "Yeah. There's not enough."
>
> "Maybe you could work together and solve that."
>
> "No. Even all of 'em wouldn't reach."
>
> "I mean is there a way you could measure with just the meter sticks you have?"
>
> After several minutes of futile attempts and useless hints, I believed I was miscommunicating. This led me to a demonstration.
>
> "How about this? Can you lay a meter stick down, mark the end with your finger, and then move it?"
>
> "Wow! Good idea!"
>
> Their surprise and enthusiasm were delightful, but all the way home I thought, "How could this be new to them?" Familiar as I was with students' low performance and misconceptions in mathematics, this left a strong impression.

This volume is divided into two grade-level sections. The first section includes thirteen chapters that consider measurement in the early childhood and elementary years, prekindergarten to grade 5. The second section contains eleven chapters—eight that delve into measurement in middle and high school and three that discuss different issues in measurement education for adults.

The issues of metric measures that wind their way through all these sections reminded me of another experience, in my earliest years in mathematics education: Telling everyone during metric workshops that the United States would be metric in a "few years." I believe we are still suffering, in mathematics education and economically, because we have not made that change. Perhaps it is time to work toward this state on multiple fronts.

The production of this yearbook was made possible by a large amount of work by numerous people. Most especially, my appreciation goes out to my Editorial Panel, as follows:

Veronica Meeks	Western Hills High School, Fort Worth, Texas
Joanne Mendiola	Herman Badillo Bilingual Academy 76, Buffalo, New York
Michelle Stephan	Purdue University Calumet, Hammond, Indiana
Robert F. Wheeler	Northern Illinois University, DeKalb, Illinois
George Bright	University of North Carolina at Greensboro, Greensboro, North Carolina; General Yearbook Editor

The members of the Editorial Panel gave many days of selfless work to this project, always striving to increase the quality and usefulness of the resulting publications. I also wish to acknowledge the assistance of Charles Clements, David Webb, and the production staff at the NCTM Headquarters Office in Reston, Virginia, for their contributions to the editing and production of the manuscript—another enormous task. Finally, I would like to express special gratitude to the authors who responded to the first call for manuscripts. Many of these authors wrote excellent pieces that for a variety of reasons, especially space limitations, could not be included. The fifty-five authors writing twenty-four chapters in the main volume responded promptly and cheerfully to all requests for changes. All these professionals show the dedication to mathematics education that has made the 2003 Yearbook, *Learning and Teaching Measurement*, a unique contribution.

Douglas H. Clements
2003 Yearbook Editor

REFERENCES

Clements, Douglas H. "Teaching Length Measurement: Research Challenges." *School Science and Mathematics* 99 (January 1999): 5–11.

Clements, Douglas H., and Michael T. Battista. "Geometry and Spatial Reasoning." In *Handbook of Research on Mathematics Teaching and Learning*, edited by Douglas A. Grouws, pp. 420–64. New York: Macmillan, 1992.

National Center for Education Statistics. "Pursuing Excellence." Initial findings from the Third International Mathematics and Science Study: NCES 97-198. Washington, D.C.: U.S. Government Printing Office, 1996. www.ed.gov/NCES/timss.

National Council of Teachers of Mathematics (NCTM). *Principles and Standards for School Mathematics*. Reston, Va.: NCTM, 2000.

Introduction

Throughout the chapters in the elementary school section, the authors suggest that educators first need to understand better what learners of mathematics understand about measurement if they are going to design effective measurement instruction. As a consequence, you will find that most of the chapters in this section center on students' conceptual understanding of measurement. A few others illustrate instructional and assessment techniques that can be used to help students develop the sophisticated concepts involved.

The chapters in the elementary school section can be arranged into two groups: lower and upper elementary school measurement topics. The first collection of chapters provides examples of students' thinking as it relates to elementary measurement ideas such as linear and area measurement and estimation. The second collection of chapters addresses notions of how students reason about volume, surface area, angle and time measurement, concepts that start to develop in lower elementary grades but are fully explored in the upper grades.

The articles in the lower elementary school collection begin with a comprehensive literature review by Stephan and Clements. In their review, they discuss current research findings on students' understanding of linear and area measurement. In the second chapter, Barrett, Jones, Thornton, and Dickson illustrate students' understanding of length as they discuss measuring the perimeters of objects. They also provide a helpful table that teachers may use to profile the different ways students are reasoning. Both Yelland's chapter and Grant's and Cline's describe students' thinking about length measurement. The instruction in Yelland's chapter incorporates the use of a computer program, GeoLogo. Joram discusses the importance of implementing instruction that calls for students to develop a repertoire of benchmarks that can help them get a feel for the sizes (lengths, weights, and so on) of everyday objects. Assessment plays an important role in Clarke, Cheeseman, McDonough, and Clark's chapter as they develop a series of student growth points and interviews, which both describe and assess what a student may be able to do. Outhred, Mitchelmore, McPhail, and Gould round out the collection with a chapter that elaborates a measurement curriculum called "Count Me into Mathematics." They also include an assessment of the curriculum from both teachers and students who used it in their classrooms.

The upper elementary school collection begins with an overview of students' reasoning in linear, area, and volume measurement written by Lehrer, Jaslow, and Curtis. This then leads nicely into a more in-depth look by Battista of students' understanding of volume. In addition, Schifter and Szymaszek's chapter investigates students' strategies for structuring arrays for measuring area and describes how teachers can better learn to assess their students' thinking through writing about it. Next, Bonotto investigates students' thinking about surface area, and Kamii and Long examine students' understanding of time measurement. Finally, the elementary school section ends with a chapter by Lipka, Shockey, and Adams, who explain how teachers can use the students' cultural backgrounds to design measurement instruction. For instance, they describe a classroom that integrated the measurement practices of the Alaskan Yup'ik culture into measurement instruction.

Whether your focus is assessment, research, or unique, real-life measurement ideas we believe this section will serve to inspire you to explore and contemplate students' understanding of measurement further at the elementary grades. You may pleasantly find yourself re-examining your own thinking about measurement! Enjoy!

Michelle Stephan and Joanne Mendiola

1

Linear and Area Measurement in Prekindergarten to Grade 2

Michelle Stephan

Douglas H. Clements

RESULTS from the NAEP international assessments indicate that students' understanding of measurement lags behind all other mathematics topics (National Center for Education Statistics 1996). In addition, several research reports reveal that even college-level students have difficulty with certain measurement concepts, in particular area measurement (cf., Baturo and Nason 1996; Simon and Blume 1994). "Something is clearly wrong with [measurement] instruction" (Kamii and Clark 1997) because it tends to focus on the procedures of measuring rather than the concepts underlying them. The reason for this procedural focus probably lies in the facts that measuring is a highly physical activity and that assessing the conceptual or mental understanding that accompanies such physical motion is not as clear to teachers as assessing students' skills. As a consequence, the goal of most instruction is to teach students how to measure, for example, with a conventional ruler. Most researchers agree, however, that the marks on a ruler and the procedures for measuring can mask the very conceptual activity that underlie the tool and the physical activity. Therefore, this article will examine the conceptual underpinnings of measuring rather than measuring skills. To this end, we explore length and area measurement and document both the conceptual building blocks of measuring and accompanying instructional ideas.

LENGTH MEASUREMENT

Length is a characteristic of an object and can be found by quantifying how far it is between the endpoints of the object. Distance refers to the empty space between two points. Measuring consists of two aspects: (1) identifying a unit of measure and *subdividing* (mentally and physically) the object by that unit and (2) placing that unit end to end (*iterating*) alongside the object being measured. The hash marks and numerals on a ruler, therefore, represent the result of iterating 12 inch-sized units.

Most researchers understand subdividing and unit iteration to be complex mental accomplishments that are largely downplayed in typical measurement teaching. As a consequence, rather than focus on only the physical act of measuring, much of the literature investigates students' understandings of measuring as *space covering*. (This is true for area as well; the space is one-dimensional for length and two-dimensional for area.) Students' early measuring experiences generally arise from counting: they count the number of times they iterate a unit. However, measuring is more complex than students' first counting experiences because the "objects" students count when measuring are continuous units (e.g., the length of a rug) rather than discrete units (e.g., fingers, blocks).

Important Concepts in Linear Measurement

There are several important concepts, or big ideas, that underpin much of learning to measure. It is important to understand these concepts so we can use them to understand how students are thinking about space as they go through the physical activity of measuring. These concepts are: (1) partitioning, (2) unit iteration, (3) transitivity, (4) conservation, (5) accumulation of distance, and (6) relation to number.

> **Partitioning** is the mental activity of slicing up the length of an object into the same-sized units.

The idea of partitioning a unit into smaller pieces is nontrivial for students and involves mentally seeing the length of the object as something that can be partitioned (or "cut up") before even physically measuring. Lehrer (in press) suggests that asking students to make their own rulers can reveal how they understand partitioning length. Some students, for instance, may draw hashmarks at uneven intervals, which indicates that they do not partition space into equal-sized units. As students come to understand that units are partitionable, they come to grips with the idea that length is continuous (e.g., any unit can itself be further partitioned).

> **Unit iteration** is the ability to think of the length of a small block as part of the length of the object being measured and to place the smaller block repeatedly along the length of the larger object (Kamii and Clark 1997).

Lehrer (in press) found that initially students may iterate a unit leaving gaps between subsequent units or overlapping adjacent units. For these students, iterating is a physical activity of placing units end-to-end in some manner, not an activity of covering the space or length of the object without gaps. Lehrer goes on to say that when students count each unit iteration, teachers should focus students' conversations on what each number word refers to. For example, if a student iterates a unit five times, the "five" represents five units of space. For some students "five" signifies the *hash mark* next to the numeral five instead of the amount of *space* covered by five units (see also Stephan et al. 2001). Additionally, many students see no problem mixing units (e.g., using both paper clips and pen tops) or using different-sized units (e.g., small and large paper clips) as long as they covered the entire length of the object in some way (Clements, Battista, and Sarama 1998; Lehrer in press). Furthermore, many studies, as well as national assessments, report that students begin counting at the numeral one on a ruler (i.e., 1 as the zero point; Lehrer in press) or, when counting paces heel-to-toe, start their count with the movement of the first foot (i.e., they miss the first foot and count the "second" foot as one from an adult perspective; Lehrer in press; Stephan et al. 2001). The researchers' explanation for this is that these students are not thinking about measuring as covering space. Rather, the numerals on a ruler (or the placement of a foot) signify when to start counting, not an amount of space that has already been covered (i.e., "one" is the space from the beginning of the ruler to the hash mark, not the hash mark itself). In this way, the marks on a ruler "mask" the intended conceptual understanding involved in measurement. A final issue related to unit iteration comes from Stephan et al. (2001) who found that many students initially find it necessary to iterate a unit until it "fills up" the length of the object and will not extend the unit past the endpoint of the object.

Transitivity is the understanding that:

(*a*) if the length of object 1 is equal to the length of object 2 and object 2 is the same length as object 3, then object 1 is the same length as object 3;

(*b*) if the length of object 1 is greater than the length of object 2 and object 2 is longer than object 3, then object 1 is longer than object 3; and

(*c*) if the length of object 1 is less than the length of object 2 and object 2 is shorter than object 3, then object 1 is shorter than object 3.

Being able to reason transitively is crucial for measurement and involves taking a stick, for instance, and using it as an instrument to judge whether two immovable towers are the same size. A child who can reason in this manner can take a third or middle item (the stick) as a referent by which to compare the heights or lengths of other objects.

Conservation of length is the understanding that as an object is moved, its length does not change.

To assess whether students at various ages could conserve length, one group of researchers showed them two strips of paper as illustrated in figure 1.1a. Most students agreed that the two strips were equal in length. However, when the interviewer moved the bottom strip forward a few centimeters, students who could not conserve length answered that the two strips were no longer equal (Piaget, Inhelder, and Szeminska 1960).

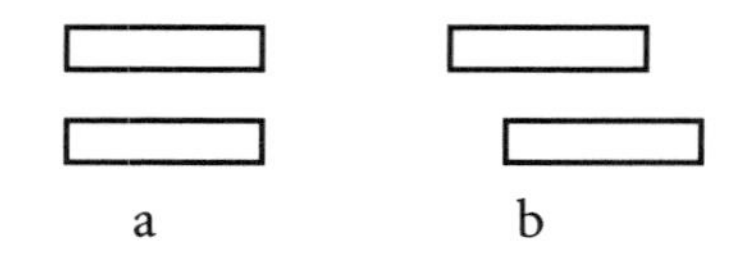

Fig. 1.1. Strips used in a conservation task

Over the past 40 years, researchers have debated both the ages and order in which conservation, transitivity, and measurement, in general, are developed. Many researchers agree that conservation is essential for, but not equivalent to, a full conception of measurement. Furthermore, Piaget, Inhelder, and Szeminska (1960) argued that transitivity is impossible for students who do not conserve lengths because once they move a unit, it is possible, in the student's view, for the length of the unit to change. Some researchers argue that students must reason transitively before they can understand measurement (Boulton-Lewis 1987; Kamii and Clark 1997). Therefore, these researchers go on to conclude that the ruler is useless as a measuring tool before a student can reason transitively (Kamii and Clark 1997). Although researchers agree that conservation is essential for a complete understanding of measurement, several articles caution that students do not necessarily need to develop transitivity and conservation before they can learn some measurement ideas. In fact, Clements (1999) argues that the only two that do seem to require conservation and transitivity are: (*a*) the inverse relation between the size of the unit and the number of those units and (*b*) the need to use equal length units when measuring. Although researchers do not agree on the order in which certain measurement ideas develop for students, we argue that children must develop each of these ideas to reach a full understanding of measurement regardless of the order of development.

The accumulation of distance means that the result of iterating a unit signifies, for students, the distance from the beginning of the first iteration to the end of the last. Furthermore, Piaget, Inhelder, and Szeminska (1960) characterized students' measuring activity as an accumulation of distance when the result of iterating forms nesting relationships to each other.

In Stephan et al. (2001), students measured the lengths of objects by pacing heel to toe and counting their steps. As one student paced the length of a rug, the teacher stopped her mid-measure and asked what she meant by "8." Some students claimed that 8 signified the space covered by the eighth foot, whereas others argued that it was the space covered from the beginning of the first foot to the end of the eighth. These latter students were measuring by accumulating distances. Piaget, Inhelder, and Szeminska (1960) contend that an accumulation of distance interpretation indicates that a student has constructed a complete understanding of linear measurement. Most researchers have observed this type of interpretation in nine- to ten-year-olds (Clements 1999; Kamii and Clark 1997; Piaget, Inhelder, and Szeminska 1960). However, Stephan et al. (2001) showed that, with meaningful instruction, children as young as six years old construct an accumulation of distance interpretation.

> **Relation between number and measurement**—Measuring is related to number in that measuring is simply a case of counting. However, measuring is conceptually more advanced since students must reorganize their understanding of the very objects they're counting (discrete versus continuous units).

Many researchers have investigated the role that counting plays in students' development of measuring conceptions. Inhelder, Sinclair, and Bovet (1974) revealed that students make measurement judgments based on counting ideas. For example, they showed students two equal-length rows of matches (see fig. 1.2).

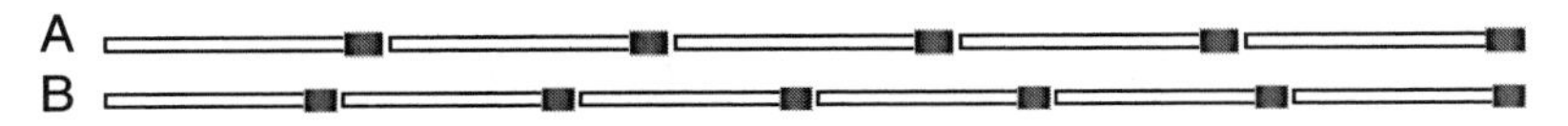

Fig. 1.2. Two rows of matches used in an interview

Although, from our perspective, the lengths of the rows were the same, many children argued that the row with six matches was longer because it had more matches. Other studies have also found that children draw on their counting experiences to interpret their measuring activity. Anyone who has taught measurement knows that students often start measuring with the numeral "1" as the starting point instead of 0. After all, when we measure, the first number word we say is "1." Lehrer (in press) argues that measurement assumes a "zero point," a point from which a measurement begins. The zero point need not be 0, but if students understand measuring only as "reading the ruler," then they will not understand this idea. Lubinski and Thiessan (1996) found that with meaningful instruction, students were able to use flexible starting points on a ruler to indicate measures successfully.

In summary, we have elaborated six important concepts that form the foundation for a full understanding of linear measurement: partitioning, unit iteration, transitivity, conservation, the accumulation of distance, and relation to number. Although researchers debate the order of the development of these concepts and the ages at which they are developed, they agree that these ideas form the foundation for measurement and should be considered during any measurement instruction. When a teacher has these ideas in mind during instruction, she is better able to ask questions that will lead them to construct these ideas. It is clear, however, that traditional measurement instruction is insufficient for helping students build these conceptions.

Instructional Activities That Build the Concepts

Traditionally, the goal of most measurement instruction has been to help students learn the skills necessary to use a conventional ruler. Although researchers agree that students ought to learn to use various measuring devices correctly, there has been a shift in instructional goals toward teaching students to develop the conceptual building blocks that lead to using rulers in meaningful ways. Such an emphasis has found its way into many new curricula in the form of starting measurement instruction with nonstandard units and moving towards the conventional ruler. However, there is some debate in recent literature as to the merits of starting instruction with only nonstandard units. We will not advocate one approach over another but rather emphasize that whichever instructional route is taken, the teacher should focus conversations and thoughts on the meaning that students' measuring activity has for them.

Nonstandard to Standard Devices

Most curricula advise a sequence of instruction in which students compare lengths, measure with nonstandard units, incorporate the use of manipulative standard units, and measure with a ruler (Clements 1999; Kamii and Clark 1997). The basis for this sequence is the developmental theory of measurement outlined by Piaget, Inhelder, and Szeminska (1960). Curriculum developers appear to assume that this approach motivates students to call for a standard measuring unit. Other researchers advocate this approach as well and argue that, when classroom discussions focus on students' meaning during measuring, they are able to construct sophisticated understanding (Lehrer in press; Lubinski and Thiessan 1996; Stephan et al. 2001).

Kamii and Clark (1997) stress that comparing lengths is at the heart of developing the notions of conservation, transitivity, and unit iteration, but most textbooks do not include these types of tasks. Textbooks tend to ask questions such as "How many paper clips does the pencil measure?" rather than "How much longer is the blue pencil than the red pencil?" Although Kamii and Clark advocate beginning instruction by comparing lengths with nonstandard or standard units (not a ruler), they caution that such an activity is often taught only as a procedure. Instead, teachers should focus students on the mental activity of transitive reasoning and accumulating distances. One type of task that involves indirect comparisons is to ask students if the doorway is wide enough for a table to go through. This involves an indirect comparison (and transitive reasoning) and therefore de-emphasizes physical measurement procedures.

Lehrer (in press) describes a sequence of instruction that a second-grade teacher, Ms. Clements, used. Her students began with pacing from one point to another. As students discussed their measuring activity, ideas concerning unit iteration and identical units emerged. Students progressed from counting paces to constructing a "footstrip" consisting of traces of their feet glued to a roll of adding-machine tape.

In a different classroom, the students wrestled with the idea of expressing their result in different-sized units (e.g., 15 paces or three footstrips, each of which had five paces). They also discussed how to deal with leftover space, to count it as a whole unit or as part of a unit. Measuring with these footstrips helped students think about length as a composition of these units. Furthermore, it provided the basis for constructing rulers (Stephan et al. 2001).

Beginning with Standard Measuring Devices

Other researchers encourage using an approach that is a bit different (Boulton-Lewis 1987; Clements 1999; Nunes, Light, and Mason 1993). They agree that very young children should have a variety of experiences comparing the size of objects in various dimensions; for example, finding all the objects in the room that are as long as their forearm. Later, however, rather than measuring initially only with nonstandard units, they argue that students benefit from using numerical measuring devices, either conventional rulers or nonstandard ones. In their study, Nunes, Light, and Mason (1993) found that children were successful on measurement tasks when they used a ruler, and this suggests that the numerical representation provided by rulers is not more difficult for children than starting with nonstandard units.

Clements (1999) and Boulton-Lewis (1987) suggest that using manipulative standard units or rulers is less demanding and more interesting for students. Clements (1999) suggests the following sequence of instruction. Students should be given a variety of experiences comparing the size of objects. Next, students should engage in experiences that allow them to connect number to length. Teachers should provide students with both conventional rulers and manipulatives units, such as unifix or centimeter cubes. As they explore with these tools the ideas of unit iteration (not leaving space between successive units, for example), correct alignment (with a ruler), and the zero-point concept can be developed. He cautions that teachers should focus on the meaning that the numerals on the ruler have for students, such as enumerating lengths rather than discrete numbers. He goes on to say that it is not until the second and third grade that teachers should introduce the need for standard units and the relation between the size and number of units.

Finally, children can construct measurement sense as they work in a computer microworld called GeoLogo (Clements 1999; Clements, Battista, and Sarama 1998). Clements posed problems asking students to estimate or calculate the length of lines and to draw lines of a given length with the Logo turtle. Some students just guessed without making or marking any units. Others created units they could count by drawing hashmarks, dots, or line segments to partition lengths. For some children, however, the segments they marked off were only equal if they were given units; these children's segments often corresponded to these units accurately. The most sophisticated students did not mark off units but drew proportional figures and visually partitioned line segments to assign them a length measure. According to Steffe (1991), young students can impose such a "conceptual ruler" onto objects and geometric figures.

In a related study, Barrett and Clements (2000) found that introducing perimeter tasks not only teaches that important concept, but also introduces children to the need for coordinating measures of parts of paths with the measure around the entire path. Perimeter tasks also emphasize measurable attributes of units as children examine grids and other ways of partitioning the sides and perimeter of a shape. By setting tasks that require a child to identify measured features, like focusing on the edges of a square tile rather than the entire tile as a unit, children learn to discriminate length from area.

AREA MEASUREMENT

Area is an amount of two-dimensional surface that is contained within a boundary and that can be quantified in some manner (Baturo and Nason 1996). Reynolds and Wheatley (1996) explain that the measure of a region is determined by comparing the region to another smaller unit, usually a square unit. They go on to say that there are at least four assumptions involved when assigning a number to a region: (1) a suitable two-dimensional region is chosen as a unit, (2) congruent regions have equal areas, (3) regions do not overlap, and (4) the area of the union of two regions is the sum of their areas. Therefore, finding the area of a region can be thought of as tiling (partitioning) a region with a two-dimensional unit of measure.

Many difficulties abound for students as they learn to measure area. First, the formal method for figuring rectangular areas is to measure the lengths of two sides and then multiply these one-dimensional units to construct a two-dimensional measure. Such a mental activity is sophisticated even for intermediate and middle-school students. Second, the result of iterating a unit along a rectangular region creates an array of units. Although it is obvious to adults that an array is created, the structure of an array is conceptually complex for students. Third, interpreting the area formula for rectangular regions requires reasoning multiplicatively about the product of two lengths. Research has found that reasoning multiplicatively with respect to area is not trivial (Simon and Blume 1994). Finally, often the tools and procedures used in measuring area mask the intended conceptual aspects that underlie area measurement.

Important Concepts in Area Measurement

There are at least four foundational concepts that are involved in learning to measure area: (1) partitioning, (2) unit iteration, (3) conservation, and

(4) structuring an array. As with linear measurement, *partitioning* is the mental act of cutting two-dimensional space with a two-dimensional unit. Teachers often assume that the product of two lengths structures a region into an area of two-dimensional units for students. However, the literature suggests that the construction of a two-dimensional array from linear units is nontrivial. Lehrer (in press) explains that students' first experiences with area might include tiling a region with a two-dimensional unit of choice and, in the process, discuss issues of leftover spaces, overlapping units, and precision to name a few. Discussions of these ideas lead students to partition a region mentally into subregions that can be counted.

Unit iteration is another important concept that students construct as they cover regions with area units. There should be no gaps or overlapping of units. Students also tend to fill in the region with units, but do not extend units over the boundaries of the bigger region (cf., Stephan et al. 2001). Furthermore, when students are given a choice, they choose units that physically resemble the region they are covering. For instance, Nunes, Light, and Mason (1993) found that children chose brick manipulatives to cover a rectangular region, whereas Lehrer's (in press) students used beans to cover an outline of their hands. However, they also reported that students had no problem mixing rectangular and triangular shapes to cover the same region. Finally, the literature stresses that the result of iterating area units ought to signify an array structure for students. Although the structure of an array is conceptually difficult, Outhred and Mitchelmore (2000) suggest that second-grade students are capable of constructing this relationship.

Similar to linear measurement, *conservation* of area is an important idea that is often neglected in instruction. Students have difficulty accepting that when they cut a given region and rearrange its parts to form another shape, the area remains the same (Lehrer, in press). Students should explore and discuss the consequences of folding or rearranging pieces to establish that one region, cut and reassembled, covers the same space. Related research shows that young children use different strategies to make judgments of area. For example, four- and five-year-olds use height + width rules to make area judgments (Cuneo 1980). Children from six to eight years use a linear extent rule, such as the diagonal of a rectangle. Only after this age do most children move to multiplicative rules. This leads to our next concept.

As we have stressed earlier, *structuring an array* is an extremely sophisticated process for students, particularly in the early grades. Battista et al. (1998) argue that students must learn such structuring to understand area. They report that children develop through a series of levels in this learning.

- **Level 1**: No use of a row or column of squares as a composite unit (a "line" of squares thought of as a group). Students at this level have difficulty visualizing the location of squares in an array and counting square tiles that cover the interior of a rectangle.

- **Level 2**: Partial row or column structuring. Some students, for example, make two rows but no more.
- **Level 3A**: Structuring an array as a set of row- or column-composites. Students at this level see the rectangle as covered by copies of composite units (rows or columns) but cannot coordinate those with the other dimension.
- **Level 3B**: Visual row- or column-iteration. These students can iterate a row (e.g., count by fours) if they can see those rows.
- **Level 3C**: Interiorized row- or column-iteration. These students can iterate a row using the number of squares in a column. Only at this level is the usual "formula" method of determining area going to have a firm conceptual basis for most students.

One way to encourage students to construct arrays of units is to have them tile a rectangular region and keep count. However, Outhred and Mitchelmore (2000) caution that using wooden or plastic tiles is too easy a task. These manipulatives mask the structure of an array because students cannot overlap the tiles, for instance. Instead, as students create arrays (with tiles), they should also be encouraged to draw the results of their covering (cf., Reynolds and Wheatley 1996). Drawing the tiles that cover a region leads to surprising pictures showing that the array structure was not as apparent to children as to adults. For example, some students draw a series of square tiles within the region they were measuring, yet there are gaps between tiles. Other students draw arrays that have unequal number of units in each row. Students need to be provided tasks and instruction that leads them through the levels of learning this structuring (Akers et al. 1997; Battista et al. 1998).

Instructional Approaches

What kind of activities help students learn initial area concepts, structure arrays, and finally learn all five concepts to form a complete foundation for measuring area meaningfully? First, students should investigate covering regions with a unit of measure. They should realize that there are to be no gaps or overlapping and that the entire region should be covered. Second, they should learn how to structure arrays. This is a long-term process that can be started in primary grades. Figuring out how many squares in pictures of arrays, with less and less graphic information of clues, is an excellent task (see Akers et al. 1997; Battista et al. 1998). Third, students should learn that the length of the sides of a rectangle can determine the number of units in each row and the number of rows in the array. Fourth, often in the intermediate grades, students who can structure an array can meaningfully learn to multiply the two dimensions as a shortcut for determining the total number of squares. However, if students do not understand array structures, they will have difficulty fully understanding multiplicative formulas for area.

This sequence of conceptual development is similar to the instructional approach suggested by Lehrer, Jacobson, et al. (1998). They suggest that students' development should proceed from informal measurement to more formal procedures. They, along with Nunes, Light, and Mason (1993), caution that teachers should not begin area instruction with rulers. Nunes, Light, and Mason report that students in their study failed to solve area problems when they used a ruler but were able to devise multiplicative solutions when given a chance to cover with a unit. If instruction begins with a ruler, one of the most common mistakes is for children to measure the length of each side and add the two linear measures together. Therefore, Lehrer, Jacobson, et al. (1998) suggests engaging students in tasks requiring them to find the area of an irregular surface with a unit of their choice. For example, their students were asked to trace their hands and find their area using a variety of manipulatives (e.g., centimeter cubes, beans). Although most children chose objects that physically resemble the shape of their hands (i.e., beans), this task provided the opportunity to discuss how to deal with leftover space that was uncovered. Because the students were unsure how to solve this problem, the teacher introduced a square grid as a measurement device. They gradually accepted this notation and used it to estimate and combine partial units.

As a follow-up task students can be asked to draw and measure islands with their newly constructed square grids. This type of task gives students more opportunity to measure with square units and to combine parts of units together to form whole units. It is important to note, say Lehrer, Jacobson, et al., that the teacher must not focus on the calculational processes students develop but rather on the meaning that their procedures have for them. Students may be moved towards building arrays with tasks such as finding the area of zoo cages. Lehrer's students were given a set of various polygonal outlines that represented the floor plan of different zoo cages. Students were provided with rulers if they found them necessary. Although some students measured the lengths of each side of a rectangle, they incorrectly argued that the resulting area would be 40 inches. Other students partitioned the rectangular cages into array structures and argued that they really meant 40 square units. In this way, students were provided a chance to relate the familiar array structure to ideas of length. The reader is invited to see the following references for more detail concerning these types of tasks (Akers et al. 1997; Lehrer, Jenkins, and Osana 1998; Lehrer in press).

In summary, the too-frequent practice of simple counting of units to find area (achievable by preschoolers) leading directly to teaching formulas underemphasizes the conceptual basis of area measurement. Instead, educators should build on young children's initial spatial intuitions and appreciate the need for students to (*a*) construct the idea of measurement units (including measurement sense for standard units); (*b*) have many experiences covering quantities with appropriate measurement units and counting those units; (*c*) structure spatially the object they are to measure; and (*d*) construct the inverse relationship between the size of a unit and the number of units used in a measurement.

CONCLUSION

We began this paper by claiming that students do not develop sophisticated understandings of measurement. We believe the reason that students' understanding of measurement lags behind most other mathematical ideas is that measurement instruction tends to focus on learning the *procedures* for measuring rather than *big ideas* about linear and two-dimensional space. The research indicated that developing measurement sense is more complex than learning the skills or procedures for determining a measure. Rather, researchers emphasize the importance of partitioning space into equal-sized units (either linear units or arrays of two-dimensional units) and counting the iterations of that unit. In order to develop these big ideas teachers must create classroom environments in which students engage in multiple measuring situations that encourage students to measure with standard and nonstandard units. Important to all measurement instruction, classroom conversations should be saturated with talk about the meaning that students' measuring has for them, not merely students' explanations of their procedures.

REFERENCES

Akers, Joan, Michael Battista, Anne Goodrow, and Julie Sarama. *Shapes, Halves, and Symmetry: Geometry and Fractions.* Palo Alto, Calif.: Dale Seymour Publications, 1997.

Barrett, Jeffrey, and Douglas Clements. "Quantifying Length: Children's Developing Abstractions for Measures of Linear Quantity in One-Dimensional and Two-Dimensional Contexts." *Cognition and Instruction* (in press).

Battista, Michael T., Douglas H. Clements, Judy Arnoff, Kathryn Battista, and Caroline Van Auken Borrow. "Students' Spatial Structuring of 2D Arrays of Squares." *Journal for Research in Mathematics Education* 29 (November 1998): 503–32.

Baturo, Annette, and Rod Nason. "Student Teachers' Subject Matter Knowledge within the Domain of Area Measurement." *Educational Studies in Mathematics* 31 (1996): 235–68.

Boulton-Lewis, Gillian. "Recent Cognitive Theories Applied to Sequential Length Measuring Knowledge in Young Children." *British Journal of Educational Psychology* 57 (1987): 330–42.

Clements, Douglas. "Teaching Length Measurement: Research Challenges." *School Science and Mathematics* 99 (1999): 5–11.

Clements, Douglas, Michael Battista, and Julie Sarama. "Development of Geometric and Measurement Ideas." In *Designing Learning Environments for Developing Understanding of Geometry and Space*, edited by Richard Lehrer and Daniel Chazan, pp. 201–25. Mahwah, N.J.: Lawrence Erlbaum Associates, 1998.

Cuneo, Diane. "A General Strategy for Quantity Judgments: The Height + Width Rule." *Child Development* 51 (1980): 299–301.

Inhelder, Barbel, Hermine Sinclair, and Magali Bovet. *Learning and the Development of Cognition*, pp. 131–66. Cambridge, Mass.: Harvard University Press, 1974.

Kamii, Constance, and Faye Clark. "Measurement of Length: The Need for a Better Approach to Teaching." *School Science and Mathematics* 97 (1997): 116–21.

Lehrer, Richard. "Developing Understanding of Measurement." In *A Research Companion to Principles and Standards for School Mathematics*, edited by Jeremy Kilpatrick, W. Gary Martin, and Deborah Schifter. Reston, Va.: National Council of Teachers of Mathematics, in press.

Lehrer, Richard, Cathy Jacobson, Greg Thoyre, Vera Kemeny, Dolores Strom, Jeffrey Horvath, Stephan Gance, and Matthew Koehler. "Developing Understanding of Geometry and Space in the Primary Grades." In *Designing Learning Environments for Developing Understanding of Geometry and Space*, edited by Richard Lehrer and Daniel Chazan, pp. 169–200. Mahwah, N.J.: Lawrence Erlbaum Associates, 1998.

Lehrer, Richard, Michael Jenkins, and Helen Osana. "Longitudinal Study of Children's Reasoning about Space and Geometry." In *Designing Learning Environments for Developing Understanding of Geometry and Space*, edited by Richard Lehrer and Daniel Chazan, pp. 137–67. Mahwah, N.J.: Lawrence Erlbaum Associates, 1998.

Lubinski, Cheryl A., and Diane Thiessan. "Exploring Measurement through Literature." *Teaching Children Mathematics* 2 (January 1996): 260–63.

National Center for Education Statistics. "Pursuing Excellence." Initial Findings from the Third International Mathematics and Science Study. NCES 97-198. Washington, D.C.: U.S. Government Printing Office, 1996. www.ed.gov/NCES/timss.

Nunes, Terezinha, Paul Light, and John Mason. "Tools for Thought: The Measurement of Length and Area." *Learning and Instruction* 3 (1993): 39–54.

Outhred, Lynne N., and Michael C. Mitchelmore. "Young Children's Intuitive Understanding of Rectangular Area Measurement." *Journal for Research in Mathematics Education* 31 (March 2000): 144–67.

Piaget, Jean, Barbel Inhelder, and Alina Szeminska. *The Child's Conception of Geometry*. New York: Basic Books, 1960.

Reynolds, Anne, and Grayson H. Wheatley. "Elementary Students' Construction and Coordination of Units in an Area Setting." *Journal for Research in Mathematics Education* 27 (November 1996): 564–81.

Simon, Martin A., and Glendon W. Blume. "Building and Understanding Multiplicative Relationships: A Study of Prospective Elementary Teachers." *Journal for*

Research in Mathematics Education 25 (November 1994): 472–94.

Steffe, Leslie. "Operations That Generate Quantity." *Learning and Individual Differences* 3 (1991): 61–82.

Stephan, Michelle, Paul Cobb, Koeno Gravemeijer, and Beth Estes. "The Role of Tools in Supporting Students' Development of Measurement Conceptions." In *The Roles of Representation in School Mathematics*, 2001 Yearbook of the National Council of Teachers of Mathematics (NCTM), edited by Al Cuoco, pp. 63–76. Reston, Va.: NCTM, 2001.

2

Understanding Children's Developing Strategies and Concepts for Length

Jeffrey E. Barrett

Graham Jones

Carol Thornton

Sandra Dickson

MR. SMITH and Mrs. Davenport were working together on a measurement unit with Mrs. Davenport's second-grade class. Mr. Smith watched the children use rulers to find the perimeter of their desks. As they finished their work, Mr. Smith posed a new task. He broke off about 2 inches from each end of a ruler and then asked the students how they would use this broken ruler to find the perimeter of a rectangular piece of cardboard (5 inches by 7 inches) that he held in front of them. After the class had thought about the problem, Mr. Smith handed the cardboard rectangle and the broken ruler to Carl and asked him to find the perimeter. Carl responded in the following way:

Carl: [*placing the broken ruler along the longer side*] I put the broken end [*showing the marking of 2*] on this corner and the other corner is at 9. It's 9 inches. [*Moving the broken ruler to the next side*] It's at 7. That's 7 inches.

Mr. Smith: Can you tell us how far it is around the rectangle?

Carl: It's 9 and 7, that's 16 inches. Nine more makes, [*pause*] 25 and then 7 more makes 32. Thirty-two inches?

The work described in this report was funded in part by the National Science Foundation, grant number ESI-9911754. Opinions expressed are those of the authors and not necessarily those of the Foundation.

Carl using a broken ruler

Mr. Smith asked the children to discuss Carl's solution; in particular, he asked the children to talk about whether or not the rectangle measured 9 inches by 7 inches. At the same time, Mr. Smith handed a regular ruler and the original cardboard rectangle to Anna, Jennifer, and Martin. He asked this group to check the perimeter of the cardboard rectangle.

Some children reported they would try another way of measuring, but they said they were not sure how. Most children accepted Carl's measures for the sides of 9 inches and 7 inches. Mr. Smith turned to Anna's group and asked them what they had found using the regular ruler.

Anna: *[pointing to her measure on the regular ruler]* I think there's something wrong with this ruler!

Mr. Smith: Would you like another one? *[handing her a second regular ruler]*

Anna: We came up with 7 inches and 5 inches: that's a perimeter of 24!

Martin: There's something wrong here! Jennifer got 24 with the other ruler, too.

Anna: I still think Carl's right, though.

Like the children in Anna's group, many children in the class still thought Carl's solution (of 32 inches) was correct, rather than the measurement with the regular ruler (24 inches). Concerned, Mrs. Davenport worked to resolve the conflict. She directed their attention to the 3 label and the 4 label on a ruler. She asked them, "How long is the part between the 3 and the 4?" The children all seemed to agree that it was one inch. Then Mrs. Davenport asked why they did not say 4 inches, "... since it ends on the four?" Some children said it might be 4 inches, but others said it had to be just one since it is so short. Dana said, "You have to subtract three from four, and you get one inch." Then, Steven called out, "Oh! We forgot to subtract the two inches that got broken off [from the broken ruler]." Conchita immediately suggested, "We just need to subtract two from every measurement." Anna remarked, "Something really strange is going on here!" Recognizing that not all the children had grasped what Steven and Conchita were seeing, Mrs. Davenport assured them that they would revisit this problem the following day.

The vignette suggests that children's strategies for length measurement do not always coincide with those of their teacher. Why did Carl have difficulty using the broken ruler? Why weren't the other children able to resolve the subsequent dilemma? Why did they doubt their own valid strategy? For Mrs. Davenport, this was an introduction to the struggles children have as they learn to measure length and perimeter. In fact, after this vignette Mrs. Davenport remarked to Mr. Smith, "I really liked how the questions today concentrated on children's thinking. That's something I'd like to work on; helping children develop their own mathematical reasoning rather than just giving them steps to follow." In this article we follow Mrs. Davenport's progress as she tries to guide children's strategy development in the measurement of path length.

MEASURING LENGTH: PRIMARY CONCEPTS

Before proceeding to children's strategy development, we focus briefly on the primary mathematical concepts involved in the measurement of length. In a mathematical sense, measuring length is encouraged by the need to compare lengths. Comparing the heights of two accessible objects, for example, two block towers standing on a child's desk, is a relatively simple process. You simply arrange the objects side by side so that you can compare them visually. However, it is more complex to compare two objects that cannot be brought together such as the height of a chalkboard on one wall and the height of a window on the opposite wall. For this kind of task, measurement is needed. When we measure we associate the length of an object with a number.

Two of the pivotal ideas in length measurement are unitization and iteration. Although these two ideas are interdependent, we examine each idea separately. *Unitization* occurs when we bring in a shorter object or mentally create a shorter object and compare its length to the length of other objects. For example, in comparing the height of the window and the height of the chalkboard we might establish a one-foot long stick as our unit of length. Such a unit has a length that can be compared with the height of the window as well as the height of the chalkboard. By establishing a unit of length, we anticipate iteration.

Iteration is the process of finding how many units would match the length of an object. We take the unit and determine how many units placed end to end are needed to traverse the length of the object we are measuring. In moving the unit along the object, we ensure that we do not leave gaps or create overlaps. We know that we have enough repetitions of the unit along an object when the farthest repetition brings us as close as possible to the end of the object. During this process we also keep track of the number of unit-sized pieces needed to match the entire length of the object we are measuring. This process is called *unit iteration*, and the number of repetitions or iterations is the measured value for the object's length.

In our comparison of the height of the window to the height of the chalkboard, we would iterate the stick along the window to determine the number of repetitions needed. For example, if the height of the window is 3 feet, we would make three iterations to traverse the side exactly and would record 3 feet as its measure. However, if the height is 3 feet 4 inches, three iterations would not traverse the side completely and four iterations would go too far. We either approximate the length as 3 feet because three iterations is closer than four iterations, or we introduce a smaller unit such as an inch and the process begins again. Iteration is the basis for constructing measurement tools such as rulers that can then be aligned with the objects we want to measure.

By describing the processes of unitization and iteration, we have described length measurement both conceptually and procedurally. In our example, we compared the quantity of units needed to match the height of the window and then the chalkboard. If the chalkboard took four iterations of the stick unit, it would be longer than the window that took only three. The larger quantity indicates the longer object. However, as children learn to measure length their conceptions and actions do not always correspond directly to the process described in this section.

CHILDREN'S CONCEPTIONS AND STRATEGIES FOR LENGTH

As a result of our observations and those of other researchers, children's thinking about length measurement has been characterized as having three hierarchical profiles (Barrett and Clements in press; Clements et al. 1997). In the first profile, children compare objects visually but use *no units* (Profile 1); in the second they use addition strategies and show some evidence of using *units* (Profile 2a or Profile 2b); and in the third they may employ multiplication strategies and count *coordinated units* (Profile 3). Although all students in the second profile use units, we use the symbols *a* and *b* to distinguish between those who use inexact units (Profile 2a) and those who use exact units (Profile 2b).

Table 2.1
Profiles characterizing students' thinking on length measurement

Process	Profile 1: No Units	Profile 2a: Inexact Units	Profile 2b: Exact Units	Profile 3: Coordinated Units
Drawing	Draws her own idiosyncratic rectangle [it may be just the image of the card].	Draws a rectangle that incorporates hash marks along each side. These marks may be unevenly spaced and may vary from side to side.	Draws a rectangle that incorporates hash marks along each side. These marks may vary from side to side but are uniform along any one side.	Produces a rectangle without hash marks and assigns a number to each side. If asked, she can identify uniform-sized units along each of the sides among all sides.
Counting	Guesses to assign numbers for the length of the sides of her rectangle or counts finger-taps while moving along a side.	Counts the hash marks as a basis for establishing her units. Identifies dots, hash marks, or numbers when asked to explain her counting.	Counts spaces between hash marks to represent the length of individual sides. Speaks of counting spaces along a given side, but not along the entire perimeter.	Anticipates a number combination that will produce a perimeter of 24 (e.g., 7 and 5). Counting is coordinated with the iteration of a unit or a collection of units to ensure the number combination is achieved. Speaks of counting spaces along the perimeter.
Proportional Reasoning	Assigns a number to each side but is not able to justify how the numbers correspond to the length of the sides except by finger-tap counting.	Assigns numbers based on her units. May even reassign her units to target a sum of 24. Matches larger numbers with longer sides and smaller numbers with shorter sides.	Assigns numbers based on her units and may even reassign her units to target a sum of 24. Matches larger numbers with longer sides and smaller numbers with shorter sides.	Preserves length units when dealing with the various sides of her rectangle. That is, keeps the sides in proportion by coordinating each length number on a side with the drawing of that side.
Inference	Guesses that the perimeter may be 24, whether or not her side measures add up to 24.	Infers that the perimeter must be 24. Sees no inconsistency in dealing with units that are not uniform among sides.	Checks the sum of the sides to see if the perimeter is 24. If not, she changes the size of the units and tries again to get a perimeter of 24.	Infers an appropriate number combination that sets into play a sequence of number compensations leading to further cases of rectangles with perimeter measuring 24. Often infers several rectangles to yield perimeter of 24.
		← Additive Operations →		← Multiplicative Operations →

In outlining each of these profiles in greater detail, we present the general characteristics and also illustrate each profile (see table 2.1) by describing students' responses to the perimeter task. For older children the task was stated as follows: "Draw a rectangle with a perimeter that measures 24 centimeters." For younger children, the teacher held a rectangular-shaped index card, pointed to the perimeter, and said, "Draw a rectangle that is 24 centimeters around its outer edge." In describing the profiles we examine students' thinking on four processes associated with this task: drawing, counting, proportional reasoning (correspondence between number and space), and inference (cf., Lehrer, Jenkins, and Osana 1998)

Profile 1

Children using a "no units" strategy typically range in age from 3 to 7 years (up to grade 2). These children often use a number to tell how long an object is but do not relate their selection of the number to iterations of units. They may even count as they work, but they do not establish units.

For example, in responding to the perimeter task a typical Profile 1 student (see table 2.1) makes a drawing of a rectangle in the hope that its perimeter will measure 24. When probed, some students guess to assign a number for length. Others use nonuniform, finger-tap counting, a strategy that is both arbitrary and unreliable, to assign a number to each side of the rectangle. Profile 1 students often will not even attempt to find perimeter. If they do, they may simply assert that it is 24 without justification, or start a new round of finger-tap counting as they move along a part of the figure.

Profile 2a

Students who use inexact units typically range from 7 to 11 years (up to grade 6). As noted in table 2.1, when these students are confronted with the perimeter task, they make progressive marks along each side of their rectangle and they report their count of marks as the length measure. However, their marks are not always uniform within the one side and may vary between sides. Nevertheless, they usually match larger numbers with longer sides and smaller number with shorter sides.

In dealing with perimeter these students add the numbers they have assigned to the sides. If these numbers do not sum to 24, they modify the numbers they have assigned to the sides by inserting or deleting hash marks. Even though their units are not evenly spaced throughout the rectangle, these students do not see any inconsistency in inferring that the perimeter is 24. Further, they often fail to associate the initial hash mark with zero, counting it as one instead and thereby overstating the value. Alternatively, they may fail to take the final hash mark along an object as the actual value, falling short by one from the number of iterated units.

Profile 2b

Students who use exact units are generally 8 to 13 years of age (up to grade 8). After drawing their rectangle in the perimeter task, students exhibiting Profile 2b draw progressive marks along each side of their rectangle in a manner similar to Profile 2a students; that is, the marks may vary between sides. However, unlike those characterized as Profile 2a, they use uniformly spaced marks along any given side. Profile 2b students count spaces rather than marks as they assign length. In dealing with perimeter they change their units rather than simply add hash marks in attempting to produce a perimeter measure of 24 (see table 2.1, Proportional Reasoning and Inference). Students exhibiting either Profile 2a or Profile 2b tend to rely on addition strategies for combining units along a side and for finding perimeter.

Profile 3

Students who use coordinated units are generally 10 years or older. As illustrated in table 2.1 with respect to the perimeter task, they draw a rectangle without recording hash marks and assign a number to each side. Probing reveals that these students have a length unit in mind that they use consistently around all sides of the rectangle. In fact they often use a helpful aggregate such as 5 units to determine side lengths. In doing this, they maintain proportionality among side lengths as they draw a rectangle. They are able to generate many cases of rectangles having the given perimeter of 24. In generating these cases, they often shift a unit from width to length (or vice versa) so as to maintain the perimeter. For example, they might first think of a rectangle that is 10 long and 2 wide and then create another that is 11 long and 1 wide by visualizing the movement of one unit. Furthermore, in communicating their strategy, they verbalize the process of shifting the units from side to side. As a consequence they generate several rectangles with perimeter 24, all of which are based on the same unit. Hence they exhibit proportional reasoning by preserving unit length across different rectangle drawings.

These broad profiles provide a developmental account of the prior knowledge that teachers might expect students to possess concerning measurement of length. In the next section we look at how Mrs. Davenport improved her instruction by reflecting on these profiles of students' strategies (see table 2.1). Our description encompasses a period of two years through which Mrs. Davenport worked to extend her students' conceptions of length measure by understanding their present knowledge of length and by devising activities and problems to promote more sophisticated ways of understanding units and iteration (Jacobson and Lehrer 2000).

Building Children's Length Concepts

Mrs. Davenport is an experienced second-grade teacher who considers social studies and literature-based teaching her strengths. The broken-ruler incident (described at the beginning of this article) occurred toward the end of a school year and prompted her to look for new ways to teach path length and perimeter. Because her students seemed unable to employ the regular measuring procedures she had wanted them to use, she began to look at how her students were thinking about length measurement.

At the beginning of the next school year, she began participating in a lesson study (Lewis 2000) with other teachers and the first author. In accord with the notion of a lesson study, this team planned, observed, and discussed lessons in order to improve the way they taught elementary children to measure path length and perimeter. Mrs. Davenport recounted her concerns at the time:

> I didn't like geometry or measurement. So I really only taught it by showing the kids what to do. When they didn't get it, I had to figure they would get it later; that it must be a developmental thing. But when I saw that my kids really didn't get what I was showing them as a procedure for finding perimeter, it made me ask more questions. I always want to find a better way to teach when I find that my children are not able to use a strategy I give them.

The team focused on three themes: (*a*) identifying the children's own strategies for comparing and measuring length, (*b*) emphasizing ways of picking a unit and describing it, and (*c*) describing measuring as an iteration process (Stephan et al. 1998). In order to link iteration and length measurement, Mrs. Davenport focused on children's explanations of their counting patterns and asked them to show how their counting helped them measure the length of a path.

Markers or Spaces?

During one lesson, Mrs. Davenport distributed metric rulers to the students (her students were unfamiliar with metric units) and asked them to find how many millimeters there were in a single centimeter. One of the students, Meghan, immediately asked her small group whether the little marks (between the two centimeter marks) were millimeters. Her group determined they were and she later reported that her group decided there were "eight millimeters in a centimeter." Mrs. Davenport asked Meghan to show the class why her group came up with 8 millimeters. Meghan used the portion of a metric ruler that Mrs. Davenport had drawn on the board (see fig. 2.1a). Meghan pointed to the smallest hash marks and counted them in the manner shown in figure 2.1b. That is, she excluded the two end hash marks and the middle hash mark in coming up with 8 millimeters to the centimeter. When Mrs. Davenport saw her strategy, she asked her, "What about the ends?" Meghan responded by not only including the end hash marks, but also including the middle hash mark. Consequently, she came up with 11 millimeters to the centimeter. Mrs. Davenport asked, "How does this link up with the spaces between the hash marks?" In response, Meghan counted the number of spaces and came up with 10 millimeters. The dialogue continued with Mrs. Davenport asking the class which was right, 10 millimeters to the centimeter, 11 millimeters to the centimeter, or 8 millimeters to the centimeter. Subsequently, Mrs. Davenport took several sessions to resolve this issue.

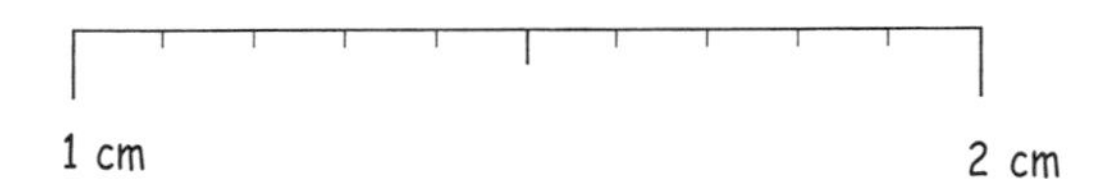

Fig. 2.1a. The portion of a metric ruler drawn by Mrs. Davenport

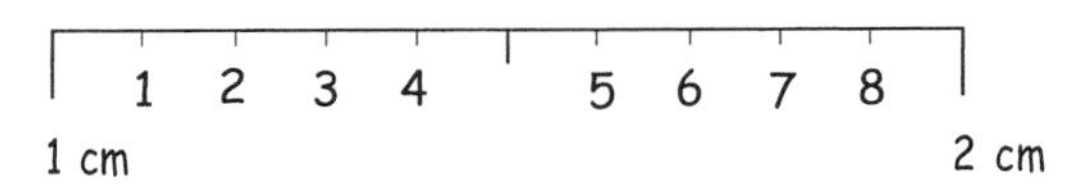

Fig. 2.1b. The hash marks Meghan picked as she counted along the drawing

Meghan's thinking seems to exemplify Profile 2a. Although she could be at Profile 1 because she is counting hash marks that are already present, we had already seen her construct and count her own hash marks as a way of finding the length of an object. For Meghan, the correspondence between the hash marks and the length is intuitive; her thinking has not reached the stage where she sees length as an aggregation of unitary spaces or segments (Profile 2b). Because she works intuitively, Meghan sees no inconsistency in leaving out the hash marks at the end or in the middle or, for that matter, putting these hash marks back in response to Mrs. Davenport's question. At first, Mrs. Davenport tried to overcome Meghan's struggle by focusing her attention on the end points. When this didn't work, she redirected her attention to spaces rather than marks along the ruler. Reflecting on this interaction, Mrs. Davenport claimed that children like Meghan need to carry out extensive iterating movements with units to unravel the meaning of a ruler. She resolved to focus on units and iteration in a different way next year.

Extending Students' Thinking: Building Units to Iterate

When Mrs. Davenport returned to this set of lessons the following year, she decided to have the children employ a wider range of measuring tools to prompt them to examine units and iteration. She gave them rulers without number labels, broken sections from rulers, and most important, she gave them each a 1-inch-long strip of cardboard to use as they measured several paths and polygon perimeters. She began listening closely to determine whether her students were counting spaces, counting hash marks, or counting number labels. Whenever she noticed her students focusing on hash marks or number counting with rulers, that is, reflecting Profile 2ạ, she asked them, "How would you label this point [*putting her finger on the initial endpoint*]?" Alternatively, she gave them a task that prompted them to specify the meaning of that initial hash mark. The following incident illustrates the task and how she used it.

Extending Students' Thinking: Beginning with Zero and One

Mrs. Davenport reminded them about finding and recording their own heights. She asked, "If you were measuring the height of this stick person that I have drawn on the board, and the person is standing against a wall with marks like this (see the large stick person in fig. 2.2a), how would you find the stick person's height?"

Figure 2.2a. Two stick figures next to a set of hash marks on a wall.

In the discussion that followed, some students said it would be 5 feet, and others said 6 feet. Mrs. Davenport asked each group in turn to explain how they determined whether it was 5 feet or 6 feet. The group arguing for 5 feet said they counted up five spaces. The group arguing for 6 feet said they counted the six marks, starting with the floor. On the one hand, Mrs. Davenport recognized that the first group was counting spaces and hence exhibiting at least Profile 2b. On the other hand, the second group was exhibiting Profile 2a, because they were counting all the marks, including the marks at both end points. In essence, the second group was not coordinating marks with spaces. In view of this difficulty, Mrs. Davenport looked for a question to help the latter group focus on the iteration of spaces. She added a shorter stick person to fit in the first space (see fig. 2.2a) and asked the group, "How tall is this one?" Not surprisingly, some students thought it was 2 feet, whereas others thought it was 1 foot. Mrs. Davenport challenged the group who said 2 feet to come to the board and draw a stick person that would be 1 foot tall.

Disagreement followed, and no student volunteered to draw such a stick person. At the same time, those students who were counting spaces claimed that Mrs. Davenport's shorter stick person was already 1 foot tall. Lindsay said, "It's got to be, because that is a foot." Mrs. Davenport said, "Lindsay, show me where you would put zero and the other numbers on our chalkboard scale." Lindsay responded by labeling the scale as shown at the left in figure 2.2b.

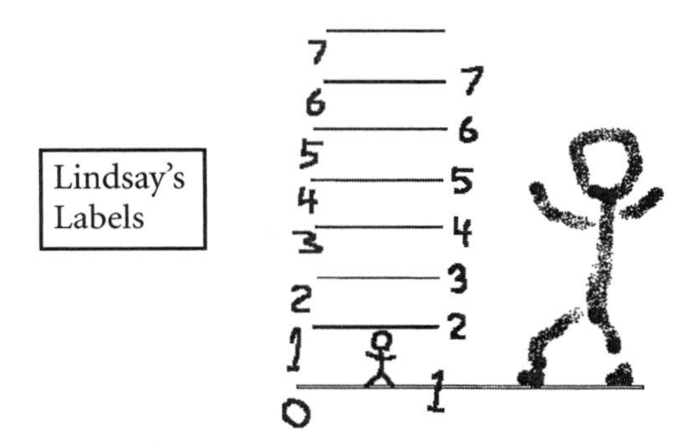

Figure 2.1b. The Hash Marks Meghan picked as she counted along the drawing

Next, Mrs. Davenport asked the class, "Have Lindsay's labels helped us resolve our disagreement? If we count the hash marks like some of you were doing, we don't seem to be able to draw a one-foot-tall stick person. On the other hand, by counting spaces and labeling zero at the floor, the first space shows the height of a one-foot tall stick person." This situation began to resolve the conflict experienced by students who were counting hash marks including the floor. They began to accept the reasonableness of starting with zero and using spaces to measure paths (cf., Stephan et al. 1998).

By requiring the children to decide what the unit is, where to label the zero value, and how to iterate this unit, Mrs. Davenport used her knowledge of the profiles to set up a task that produced conflict and eventually led some students to more sophisticated strategies for dealing with path measurement. For her, the notion of using a task to challenge and extend children's thinking had become a central element of her mathematics teaching.

Mrs. Davenport's Developing Model of Her Students' Thinking and Strategies

Mrs. Davenport reflected on her progress in understanding children's thinking about measurement of length:

> Now, I really teach measurement.... These kids cannot say why they don't get the concept. So I have to find out what it is they don't understand. But now I look to see what they are doing, and why. I try to show it to them another way.

Since beginning the lesson study, Mrs. Davenport recognized that she had learned to distinguish between children who used coordinated units of length (Profile 3), those who used exact (Profile 2b) or inexact units (Profile 2a), and those who only talked about units as labels for their own guesses or estimates (Profile 1). Rather than merely showing them a procedure for reading from a standard ruler, she encouraged the children to use a unitary object like a 1-inch strip of cardboard to traverse the length of a ruler in order to help them understand the way a ruler is constructed. In addition, she prompted them to describe and explain how they used units to find length.

Mrs. Davenport also focused on making drawings from scratch, instead of always measuring objects that are already segmented. Moreover, she asked the children how the numbers they reported for their measures would be related to the objects they measured: "What does it mean when you say you found 4?" Perhaps most important of all are questions she asks herself while she teaches:

- Do the children count intervals, or do they count hash marks?
- Do they connect their counting with how many spaces they have traversed along the path?
- Do they match the beginning of the first interval with the number zero?

IMPLICATIONS FOR IMPROVING INSTRUCTION ON MEASURING PATH LENGTH

Principles and Standards for School Mathematics (National Council of Teachers of Mathematics [NCTM] 2000) recommends that instructional programs in measurement for young children should enable all students to compare and order objects according to attributes like length. Children's strategies for comparing and ordering objects according to length develop over time; development is reflected in the sequence of profiles. Understanding these profiles of children's thinking can help improve instruction on length measurement.

Instruction becomes more effective when it focuses on children's thinking processes as they measure path lengths. Having more mature thinking and strategies positions children to employ both formal and informal measuring procedures in meaningful and appropriate ways. As the profiles indicate children do not always understand units and iteration in ways that their teachers expect. Their counting may be linked to markers rather than units, and their concept of zero in measurement may be problematic. Recognizing that students bring this kind of thinking to the classroom helps teachers when they are designing, implementing, and assessing instruction in measurement.

Instruction has greater impact when teachers design tasks and questions that recognize where students are in their thinking and then move them to more sophisticated strategies. In particular, it is useful to determine whether students are guessing without units, using inexact units, using exact units, or are able to coordinate collections of units. Having recognized the range of thinking strategies that may exist in their classrooms, teachers are in a favorable position to use tasks like the broken-ruler task or the one-foot stick person task to tease out children's thinking and provoke conflicts about important concepts like unit definition, the placement of zero, and the linking of number to path length.

Instruction needs to help children make connections among their drawings of paths and perimeters, their counting movements along these paths and perimeters, and their verbal descriptions of what they are doing. In essence, children need to connect their counting movements with the number of spaces they traverse along a path. In describing these actions, students need to convince both their teacher and classmates that their counting movements are coordinated with unit iteration. That is, the number that a student uses to describe the length of a path must correspond precisely with her aggregation of units along the path.

Developing children's understanding of length measurement presents special challenges. As teachers attend to the mathematical importance of units and iteration, and become more familiar with the thinking profiles children bring to the classroom, they are in a stronger position to build on "students' intuitive understandings and informal measurement experiences" (NCTM 2000, p. 103).

REFERENCES

Barrett, Jeffrey E., and Douglas H. Clements. "Quantifying Length: Children's Developing Abstractions for Measures of Linear Quantity in One-Dimensional and Two-Dimensional Contexts." *Cognition and Instruction* (in press).

Clements, Douglas H., Michael T. Battista, Julie Sarama, Sudha Swaminathan, and Sue McMillen. "Students' Development of Length Measurement Concepts in a Logo-Based Unit on Geometric Paths." *Journal for Research in Mathematics Education* 28 (January 1997): 49–70.

Jacobson, Cathy, and Richard Lehrer. "Teacher Appropriation and Student Learning of Geometry through Design." *Journal for Research in Mathematics Education* 31 (January 2000): 71–88.

Lehrer, Richard, Michael Jenkins, and Helen Osana. "Longitudinal Study of Children's Reasoning about Space and Geometry." In *Designing Learning Environments for Developing Understanding of Geometry and Space*, edited by Richard Lehrer and Daniel Chazan, pp. 137–67. Mahwah, N.J.: Lawrence Erlbaum Associates, 1998.

Lewis, Catherine. "Lesson Study: The Core of Japanese Professional Development." Paper presented at the Special Interest Group on Research in Mathematics Education of the American Educational Research Association, New Orleans, 2000.

National Council of Teachers of Mathematics (NCTM). *Principles and Standards for School Mathematics.* Reston, Va.: NCTM, 2000.

Stephan, Michelle, Paul Cobb, Koeno Gravemeijer, and Kay McClain. "Reconceptualizing Measurement Investigations: Supporting Students' Learning in Social Context." Paper presented at the Annual Meeting of the American Educational Research Association, San Diego, April 1998.

3

Making Connections with Powerful Ideas in the Measurement of Length

Nicola Yelland

THIS article presents and discusses examples of children engaging with measurement concepts in a curriculum that encouraged an investigative approach to mathematical learning with computer-based tasks embedded in the sequence of activities. The Investigations in Number, Data, and Space (Seymour 1998) curriculum has a variety of software included in it as essential resources. In the Year 2 measurement of length unit, *How Long? How Far?* (Goodrow et al. 1997), tasks using software called Geo Logo (© Douglas H. Clements 1994) accompany activities based in the real world and paper-based work examples. Geo Logo is a computer environment in which commands can be given to a turtle so that it is instructed to perform various actions (Clements et al. 1997). The turtle is directed with distance moves, forward and back, with numeric inputs of measure that are called turtle steps (e.g., forward 30). The turtle can also be directed to turn right or left as determined by a numeric input measured by degrees (e.g., right 90). There are a number of tools built into the environment that were designed to assist children in planning what they wanted to do with the turtle and what they were required to do in the various tasks that form part of the curriculum unit. These include an on-screen ruler to determine distances and the ability to show lengths and angle size with a label. Additionally, features built into Geo Logo assist children's thinking about a wide range of mathematical ideas. For example, the turtle turns slowly with rays shown for each 30-degree segment to facilitate the visualization of angles and turns, and a step procedure allows the interrogation of code in the problem-solving process.

Geo Logo is one example of a number of Logo languages that have developed since the original was produced in the late part of the 1960s. Not all versions of Logo contain the same elements. Geo Logo is particularly suited to the exploration of mathematical ideas because of the built-in features, such as those just mentioned above.

The tasks in *How Long? How Far?* were designed to enable children to explore nonstandard units to measure length and then to work on tasks that require them to determine the ways in which a variety of standard units of length can be combined over specified distances of length. There are opportunities for connections to be made between the different-sized units of measures that represent the more sophisticated levels of thinking.

The content of the unit is related to the exploration of linear measurement and includes the processes of determining, analyzing, and comparing both arbitrary and standard units of length. Traditionally the concept of length is introduced with the use of arbitrary units of measure so that children can explore the relationship between distance and the number of units needed to traverse it.

In *How Long? How Far?* a number of initial activities are suggested in which the children use nonstandard units of length to compare and classify objects in their environment. They record this in a variety of ways on paper and discuss their findings with the whole group. In the first computer-based task, called Steps, pairs of children can take turns directing their individual turtle to a randomly generated item with steps of the same size. In the next activity, Giant Steps, although the items are placed at an equal distance from each turtle, the children have to use different-sized steps in order to reach the item. Next there is a maze task, in which the children are required to collaborate to take a mother turtle to her baby, located at the end of the maze. The final two activities of the unit are completed first off the computer and then on it. They involve directing an ant and a turtle, respectively, around a make-believe town in order to visit various locations. An integral part of the task is that the turtle has to be "recharged" at a battery stop after ten distance moves, since this determines the paths that can be taken. Visits to the battery for recharging therefore had to be incorporated into the planning process, otherwise the turtle could not resume the journey. The sequence of the tasks was designed so that children could explore the ways in which paths are created. They also provide opportunities for children to discover and articulate different lengths of paths for discussion, as well as for descriptive and comparative purposes. This enables them to make connections between the lengths of the paths or journeys taken and subsequently the numbers used to quantify them. The data from three tasks, Steps, Giant Steps, and Maze, will form the basis of the discussion in this article.

The Context

The children who participated in this study were in the second year of elementary school (average age 7 years 3 months) in an Australian state that does not provide a compulsory kindergarten year at the beginning of school. The sequence of activities in the curriculum unit *How Long? How Far?* aligned with the scope and sequence of local curriculum documents. However, *How Long? How Far?* goes beyond traditional sequences and activities by including tasks that facilitate comparisons of length and further opportunities for comparisons between the different number of units needed. Additionally, it includes computer-based tasks, which enable the children to engage with length concepts in new ways that were not possible without the technology.

In both on- and off-computer tasks the children worked in pairs that remained constant throughout the unit of work. The pairs were selected by the teacher, who wanted children of similar ability working together in this instance. The children were familiar with working in pairs and small groups. Previous work with the teacher had indicated that a useful form of classroom organization included time for initial class discussions, followed by group work before coming together again to discuss findings and share strategies. This is particularly significant in the Investigations curriculum because many of the tasks not only have different paths to solution but also have different solutions, which the children always found interesting to share.

Previous work with the Investigations curriculum had revealed the importance of scaffolding children's learning so that they were able to demonstrate higher levels of thinking and conceptual understandings. The various processes of scaffolding used have been outlined elsewhere (Yelland 1999), but in the context of *How Long? How Far?* the main form of scaffolding was questioning the children regarding the commands that they gave to the turtle. For example: how they knew which number to match to the distance traveled and the ways in which they were able to relate the number chosen to the size of the steps that they were using. Additionally, diagrams and tables were provided for recording findings so that connections between number and size could be discussed and analyzed more effectively. Calculators were also provided so that discoveries about the relative comparison of numbers could be achieved more easily. The scaffolding techniques used supported the children's learning by—

- extending their existing understandings to create new ones at a more sophisticated level;
- supporting their explorations so that they could take risks and make connections about various aspects of the process during their investigations.

Understanding Length in *How Long? How Far?*

Throughout the unit *How Long? How Far?* it was evident that the environment and tasks were interesting to the children and that they constituted a useful context for the exploration of concepts in ways that were not possible without the technology. This was an important aspect of the work because this engagement and interest in mathematical ideas, in a context that afforded the opportunity for active exploration of length concepts, was a salient contrast to that experienced in traditional mathematics classrooms. The research project took place in a classroom, and videotapes of the children were made and analyzed. In order to depict the main findings, some vignettes from the transcripts are provided to illustrate salient issues throughout the text.

THE TASKS

Steps

In Steps (see fig. 3.1) the children were instructed to take their turtle to a toy that was placed directly in front of it. Each time the activity is generated, the toys are placed at a random distance from each turtle. The children in the study often made the activity into a game to see who could get their turtle to the toy first.

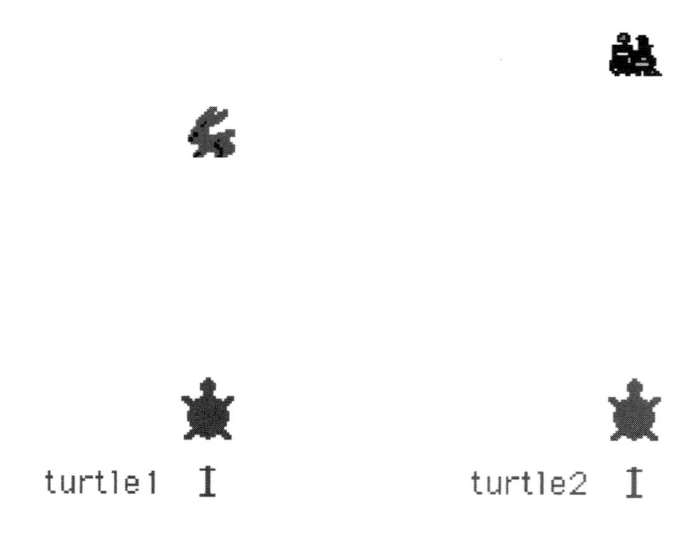

Fig. 3.1. Steps

A feature of the activity was that if you attended to the number of steps taken by your partner, you could use it as the basis for a good estimation of the distance needed to be traveled by your own turtle. Only a few children did this the first time they played it, and, of course, they were more successful as a result. This is clearly shown in the vignette below, which captures the interactions between the pair and the teacher who is questioning the children in an attempt to help them to make the connection between distance and number. In early interactions this technique was not always successful. In this vignette the teacher intervenes to scaffold the pairs with questions that provided opportunities for the children to think about the numbers that they were using to move the turtle forward to their toy.

Teacher: Altogether ...?
Boy: I don't know.
Teacher: Altogether you had 17, so you'll need more than 17, won't you? So what number do you think you'll try first?
Boy: 9.
Teacher: Is 9 more than 17?
Boy: Ummm, no.
Teacher: No?
Boy: 16?
Teacher: Is 16 more than 17?
Boy: Ummm.
Girl: No.
Boy: 18?
Teacher: Well, 18 is more. Do you what to try 18?
Boy: Yep.
Teacher: OK.

Prior to this interaction, it was evident from the spontaneous comments of the children that they could understand the notion that larger numbers had to be used when the toy was a greater distance from the turtle. However, this vignette shows that their numerical knowledge concerning the relative value of numbers was often limited in its application. The scaffolding of this pair continued, and by the fourth turn on the game they were making comments to indicate that they were using an input number based on a consideration of the estimated distance in relation to previous experiences as well as in relation to the amount entered by their partner.

The tendency to be conservative in their estimates of distance and number was also apparent in a previous study (Yelland 2001) when the children completed the tasks without adult intervention. This was interesting because it demonstrated that other classroom factors influenced the ways in which the children made decisions about inputs for measurement in these tasks. The children were reluctant to use numbers greater than 20 because they "had not done them yet!" When we advised the children that they could use any numbers that they wanted, it was apparent that they were considering the distances from the turtle to the toy in new ways. For example, instead of traveling a distance of 30 steps by edging up to it in increments—say, 10, 10, and 10—they would view it as one distance of 30. In this way they were playing with combinations of numbers and realizing that there were a variety of ways in which the final destination could be reached.

When the children attempted the tasks on a second occasion some weeks later, they were not only more accurate, as would be expected with a practice effect, but they were also able to articulate the relationship between the number of steps made—the distance to be covered—and demonstrate that they were using their partner's move as a basis for their own decision making. The following vignette provides an example of a child using her partner's move as a reference point. It is of the same pair of children, but it is now a couple of weeks later. The children have had experience with a number of similar games, and they are returning to play the Steps game again. This is their fourth turn in the session, and the girl has made the move to nearly reach her turtle with 21 steps (it was 24 steps away). The boy's object, a helicopter, is closer, at 18 steps.

Boy: You should have gone more! Mine is not so far... so it will be less.

Girl: It's nearly there, though,.... I need only a few more ... 2 or 3 to get it!

Boy: So I have to do less ... how many? ... only a few less....[*He puts his fingers on the screen to try and measure how many.*]

Boy: You did 21! I will do 18. [*He types and the turtle moves to the toy.*] I got it in one try! Cool!

Teacher: Good try! That was a good try!

Giant Steps

In Giant Steps (see fig. 3.2) the randomly generated toy appeared the same distance away from each turtle but the turtles had different-sized steps.

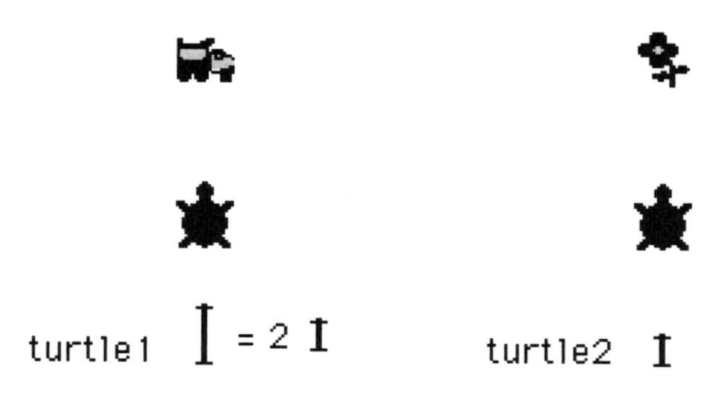

Fig. 3.2. Giant Steps

Thus, the number needed to reach the toy was not the same. In this way making connections between the number needed to get the length accurate, and the size of the steps, was important for success in the task. Again, in initial experiences with the task the children mainly used trial-and-error strategies to direct the turtle the length to the toy. In the first game they were often confused when they used the same number as their partner only to find the turtle ended up in a different location. When the relative sizes of the steps were highlighted to the children, they were able to recognize that they needed different inputs to reach the toy. Some started to make the connection between the relative sizes, which we restricted in the first instance on the basis of results obtained in a previous study (Yelland in press). This study revealed that when a turtle with step size 2 was presented with another that had step size 1 in the first instance, the children were more able to use the doubling process to determine the number needed rather than to work in the reverse, where the concept of halving seemed to confuse them. The language inherent to this process appeared to be distracting the children. For example:

T: This step size is half the size, so…

B: So we have to half…. What is half of 5? [*looks puzzled*]

T: No, if it is half the size, we have to use twice as many steps.

B: Oh, that is easy—double 5 is 10! Why is it half? It's two times bigger, you said….

Of course when the smaller step size turtle was moved first, it was more difficult to scaffold the children's choice of number with appropriate questions, since halving numbers is complex and also because the concept of division is not introduced until later in the curriculum. Thus, the children often reverted back to the strategy of edging up to the turtle with a visual approximation of the distance needed. However, on revisiting the task the second time around—that is, after the children had completed Steps, Giant Steps, Maze, and Steps again, when the children were scaffolded with questioning techniques and provided with explanations about halving with the use of a calculator—the notion of dividing by 2 was more easily understood. It was then adopted as a strategy for determining not only the exact number needed for a move (relative to a partner's) but also whether the children were able to state the relationship between the two numbers. It should also be noted at this stage that the children soon realized that they had a better chance of reaching the toy first if they went second and used their partner's move as a basis for their own decision making. For example:

G: Yours was 16, and you had step 1..., and you need to go a bit more. Do you think 18 ... mine is 2, so I need half ...

G2: What is half of 18? Get the calculator. You have to do that one [*points to calculator buttons*] and 2 then equals ...

G1: [*On the calculator presses 18 divided by 2 equals*] It's 9. Half of 18 is 9! Forward space 9 [*she watches the turtle move*] Yesssssss! [*claps and smiles*] I got it!

As previously stated, we could expect a practice effect in relation to the efficiency of moves made to achieve the target on the second attempt with the task. However, it was the children's ability to demonstrate their understanding by articulating the relationships between the numbers that we were concerned with, and this was only evident after scaffolding them through questioning and with the use of aids such as the calculator. Their comments demonstrated that they realized the distance was constant but that the different-sized steps meant that they needed to use different numbers, which were connected in a relationship determined by the step size. In this way they were able to show that they understood this relationship and could determine the correct number of steps by performing the appropriate mathematical operation, which may have been doubling, halving, or multiplying by 3.

Maze

In the Maze activity (see fig. 3.3), the children are told that they have to take the "mother" turtle to the baby at the end of the maze because the baby is lost and the mother needs to get there as quickly as possible. In this activity turns are introduced for the first time, as well as an "energy meter," which depletes each time a move is made on the screen. The children are told that each time you tell the mother turtle to make a move, she uses up some energy. In this way they are advised that they have to make as few moves as possible and take the shortest route to the baby so that the baby does not get worried and so that the mother does not run out of energy on her journey through the maze. Each time the activity is played, a different step size is generated, and the size is shown visually. Previously the size was indicated with a single measure and a numeral next to it.

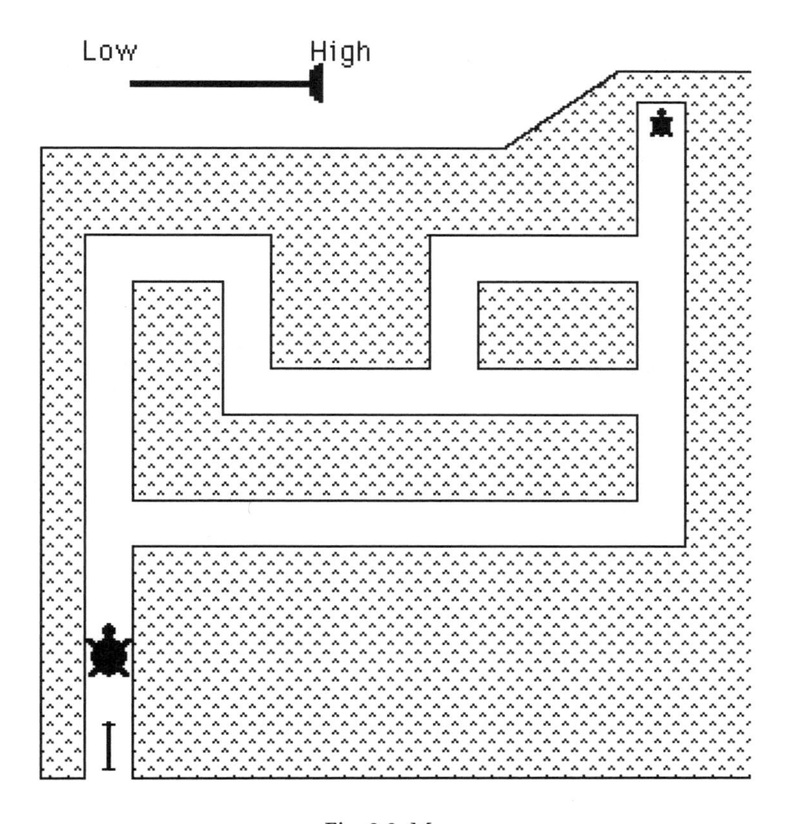

Fig. 3.3. Maze

As with Steps and Giant Steps, the pairs of children completed the task twice. The first time occurred after Steps and Giant Steps, and then again after these tasks were completed a second time. In a previous study (Yelland in press) the children were left alone to direct the mother through the maze. They did so using a variety of strategies, but generally following a trial-and-error technique. When questioned the pairs gave no indication that they were using the step size as a referent, instead of just guessing a number and not once making any comments about the relative size and number of the step sizes either spontaneously or through questioning. The step sizes ranged from 1 to 3. Although the size of the steps was randomly generated, we used only step 1 and 2 for the first attempt at Maze and then, when the pairs returned to the task after completing Steps and Giant Steps, we introduced step size 3. We were interested in seeing if this would elicit statements that indicated that the children were making connections between the numbers. It also seemed important to establish a connection between step size 1 and 2—before the pairs could go on to make comparisons between step size 1 and 3.

Further, we incorporated the use of a matrix, shown in figure 3.4, in order to record the number of steps that were made in each section of the maze, to help the children make connections between the numbers that were used for the different step sizes. Children could record the number of steps that they were using each time, and we asked them questions about the relationships between the numbers.

The Maze **Name**__________

Section	Step Size 1	Step Size 3
1	6	2
2	24	8
3	18	6

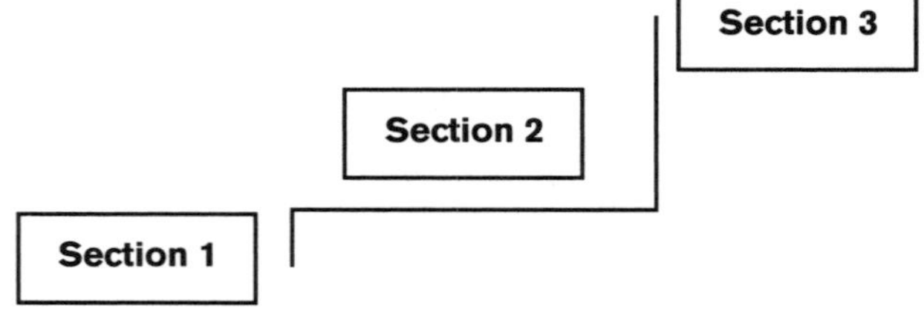

Fig. 3.4. The matrix scaffold for Maze

In the following vignette of two children playing this game for the first time, you will note that they are making guesses about the number of steps needed in the various sections and in every instance underestimate the number. It was decided that by using the table as a form of scaffolding the relationship between the numbers would be made more explicit and could be supported with strategic questions. Further, recording the numbers in the table seemed to facilitate this process and provide a focus for the children responding to the questions. In this example, with a boy-girl (BG1) pair, both called Jesse, the researcher asks questions to elicit responses pertaining to the nature of the relationship between the numbers, in order to encourage the children to think about the step sizes. At first the children are not able to make the connections:

Teacher: If you took 3 in step size 2, how many do you think you need in step size 1?

Girl: Ummmm …

Boy: 4? or 3?

Girl: 5.

Boy: OK…. Well, I'll take … 3….

Teacher: Now, just have a think. If this is only half the size, do you think 3 will get you there?

Girl: [*shakes her head*] 5.

Boy: What's half of 3?

Girl: Half of 3?

Teacher: Well, it's half the size. That means for every one of these steps you need two steps.

Boy: Oh, 2 … 3 … half of 3? That'd be….

Girl: Half of 3?

Teacher: OK, let me draw you a picture…. OK, if this is the first road here, we took 1, 2, 3 steps but that was step size 2…. Step size 1 is this big, so you'll need two of them, for just one of the steps. So how many will you need altogether?

Boy: 2!

Girl: 2.

Teacher: That's only got you that far. It hasn't got you up to there [*pointing to diagram*].

Boy: Hmm.

Girl: 6.

Teacher: You think 6.

Boy: OK, then. [*He types in 3.*]

Teacher: Hang on.

Boy: Oh, 6. [*He deletes 3 and types in 6, and the turtle moves forward.*]

Boy: Yes! Good, Jesse, good!

Teacher: So write 6 there. Well done!

By the time the children return to this task two weeks later, the boy is able to state clearly the way in which he is going to determine the number of steps needed to get down the various sections of the path. He spontaneously offers:

B: When it was, if it was step size 3, it would take 3… actually, … ummm, 3 more. So if we had 1 step, we would count that as 3. Then, so what we'd do, we'd make the 3s that we'd count from little steps into 1 whole big 3 size step size and see how many it takes.

T: I see. All right….

When the pair complete the task with step size 3 and then with step size 1, they not only recall their previous experience but also calculate the amount needed as they go. The boy used a personally developed system based on his fingers, and the girl checked the number by multiplying on the calculator.

Teacher: Our 2 became 6.

Boy: Wait, there. 1, 2, 3, …, 4, 5, 6 … [*counts with his fingers in groups of three*]

Girl: 16…. 16.

Boy: 7, 8, 9, …, 10, 11, 12, … [*then keeps counting on fingers*]

Boy: I reckon it's 24.

Teacher: 24.

Girl: I bet it's 16.

Boy: I reckon it's 24.

Teacher: Jesse, just before you go, do you understand how Jesse got that? [*Girl shakes her head.*]

Boy: Because, see, every step is 3 times bigger, and it equals 3. So and ours is 3 of little ones, that's how … that's how I was working it out.

Teacher: So for every step …

Boy: 8 times 3.

Teacher: So for every step, there's 3 little ones in it.

Boy: So I reckon it's 28.

Teacher: So how could you do it on the calculator?

Boy: 3 times 8.

Teacher: OK. Do you want to try that, Jesse?

Girl: 3 times 8 … 24 … he said 28.

On both occasions all the children were able to get the mother turtle to her baby without running out of energy. The video data showed that some pairs from both groups were monitoring their progress along the way by making comments such as, "We are going OK," "We have enough [energy] left to get to the baby …" and "We are nearly there, and the mummy is pleased." It was also apparent, however, that the energy meter was not the main focus of the task after the first four or five moves had been made.

DEFINING LEVELS OF PERFORMANCE

When the children were observed completing the tasks, it was evident that the performance of the pairs was qualitatively different. In the teaching of mathematics we often have as our goal the use of higher-order thinking skills to promote more effective problem solving. It is useful not only to distinguish between performances that are qualitatively different but also to describe what we mean in task-specific examples of what constitutes higher-order thinking skills. In this way we are able to guide children toward their effective use in problem-solving contexts with appropriate scaffolding techniques.

In this study the children's performance was described as being either *naïve* or *knowledgeable*. There were also children who were characterized as being in a transitional phase in which they exhibited the use of both naïve and knowledgeable strategies but were inconsistent in their applications of either. The naïve strategies included making random guesses, moving the turtle along gradually until they were happy with its position and location, and just choosing a number because they "liked it." In contrast a knowledgeable performance was characterized by the use of higher-order thinking skills. These included suggesting what could be done before the task was started, monitoring progress and evaluating moves as they were made in order to determine if they were successful in light of reaching the goal, using processes such as comparing, calculating, and hypothesizing in order to generate strategies, and drawing on existing mathematical knowledge for problem solving (e.g., number sense and operations on numbers) in order to calculate the precise size of move needed. A summary of the types of strategies used and examples from the data are provided in table 3.1.

In this study, the use of scaffolding techniques resulted in more examples of knowledgeable performances being observed. This is important because it illustrates the fact that teachers need to work with children to elicit the higher-order thinking skills so that they become part of the children's repertoire of skills. Further, when the pairs made a second attempt at the task, returning to it after completing the Maze task, they were observed to be using knowledgeable strategies on more occasions than in previous sessions. It was during performances characterized as knowledgeable that the children clearly articulated that they had made mathematical connections to complete the task, as opposed to arbitrary ones that had the same end product but were more a result of trial and error.

Summarizing Statements

This study explored the strategies that young children employed while they were completing computer-based tasks related to the concept of length. Throughout the study it was apparent that the children were interested in, and engaged with, the ideas that were embedded in the tasks, and their enthusiasm remained high throughout the time period under review. It was also apparent that the children were working mathematically and using the ideas inherent to the concept in different ways. Some were able to examine

Table 3.1
Performance Levels in How Long? How Far?

Steps/Giant Steps–Year 2

Knowledgeable	**Naïve**
Used step size as a point of reference to determine the number of steps needed to reach the toy	Made random guesses in order to direct turtle to toy
Reflected on prior moves to determine size and extent of move	Showed no evidence of reflection on previous moves
Compared numbers to determine size of move	Were intuitive in their use of number ("I will use 2 'cause it's curly!")
Articulated connections in number between the different sized steps	Were not able to articulate number connections
BG2: *G*: [*counting up the screen with finger*] Your pinky is the pace!	**BG3:** *T*: Will you need more or less than 17? *B*: More. *T*: So how many? *B*: 8 …
BG2: *G*: So I just have to count … 1, 2, 1, 2 like that (pointing to each of her partner's steps). It's double!	**BB1:** *T*: What can you tell me about this game and the size of the steps? *B1*: I lose, and he wins, and it goes phrrrrt! *B2*: I win, and he lost three times

Note: BG denotes a boy-girl pairing, BB is a boy-boy pair.

their actions in detail and develop skills in higher-order thinking during their problem solving.

The data from the study showed that the children were capable of high levels of task participation when scaffolded, as demonstrated by more pairs performing at a knowledgeable level. Additionally, on revisiting tasks after additional experiences in the computer context, the children were able to articulate the mathematical connections spontaneously as well as when answering questions posed by the teacher related to the relationships between various numbers.

The study also revealed that when young children are provided with complex and challenging mathematical tasks, they are able to approach them with enthusiasm and interact with powerful mathematical ideas. In this unit of work the Year 2 children were using concepts such as multiplication and division with products over 20, which are not usually encountered in the curriculum until later years. They were also able to relate these aspects of number to the computer tasks that they were working on in computer-based learning contexts.

Thus, it is apparent that when computer activity is embedded in a curriculum that is characterized by active exploration, inquiry, and problem solving, children maintain interest and are able to engage in high levels of mathematical thinking. Of importance here is that teachers provide support so that the children are allowed the opportunity to articulate the ways in which they are using mathematical ideas to connect with concepts. It is through such interactions that higher levels of thinking and performance can be achieved with well-designed activities. It was evident that computer-based tasks such as the ones described here provide a context in which this was possible, because they were dynamic and afforded the opportunity for the children to play with ideas in an active way. However, the role of the teacher as facilitator was crucial. The scaffolding techniques used in this study provided contexts in which children were able to discuss and question the ways in which they attempted to solve problems, and they described the ways in which children used processes. This was achieved in a classroom using small-group–based discussions incorporating an active learning inquiry method and problem-solving situations. They illustrate that the mathematics that was discovered and displayed not only was more interesting for young children, as evidenced by their spontaneous comments, but also afforded them the opportunity to link the conceptual areas of the subject in meaningful ways to create contexts for their own understandings to develop.

REFERENCES

Clements, Douglas, Michael Battista, Julie Sarama, Sudha Swaminathan, and Sue McMillen. "Students' Development of Length in a Logo-Based Unit on Geometric Paths." *Journal for Research in Mathematics Education* 28 (January 1997): 70–95.

Goodrow, Anne, Douglas Clements, Michael Battista, Julie Sarama, and Joan Akers. *How Long? How Far?* Palo Alto, Calif.: Dale Seymour Publications, 1997.

Seymour, Dale. Investigations in Number, Data, and Space. A curriculum. Palo Alto, Calif.: Dale Seymour Publications, 1998.

Yelland, Nicola. "Reconceptualising Schooling with Technology for the Twenty-first Century: Images and Reflections." *Information Technology in Childhood Education Annual* 1 (1999): 39–59.

———. "Girls, Mathematics, and Technology." In *Sociocultural Foundations of Mathematics Education,* edited by Bill Atweh, Helen Forgaz, and Ben Nebres, pp. 393–409. Hillsdale, N.J.: Lawrence Erlbaum Associates, 2001.

———. "Creating Microworlds for Exploration of Mathematical Concepts." *Journal of Educational Computing Research,* in press.

4

Developing the Building Blocks of Measurement with Young Children

Theresa J. Grant

Kate Kline

TRADITIONALLY measurement instruction has focused on *how* to measure, rather than on what it *means* to measure. Mathematics educators have pointed to results of the National Assessment of Educational Progress (NAEP) over the last two decades to support this sentiment. An often-cited example is one requiring students to determine the length of an object pictured above a ruler but not aligned with the end of the ruler. On the 1991 NAEP, only 24 percent of fourth-grade students and 62 percent of eighth-grade students got this item correct (Kenney and Kouba 1997). Research on children's learning of measurement has provided insights into both the complexities involved in understanding what it means to measure and the kinds of activities that serve to focus on specific aspects of measuring. This article discusses some important measurement ideas and illustrates ways in which these ideas can be grappled with in a first-grade classroom.

DEVELOPING IMPORTANT IDEAS IN MEASUREMENT

Comparisons are one of the earliest activities that help children understand the idea that they are measuring specific attributes of objects. Direct comparisons involve physically lining up two items to determine which is the longest, whereas indirect comparisons involve the process of comparing two items that cannot be physically juxtaposed. For example, if one wanted to compare the height of a doorway to the length of a teacher's desk, one could directly compare by physically placing the desk adjacent to the doorway, or indirectly compare by using a piece of string as an intermediary device. Engaging in both direct- and indirect-comparison activities is

This article is based on work supported by the National Science Foundation under grant no. ESI-9819364. The opinions expressed are those of the authors and do not necessarily reflect the views of the National Science Foundation.

important because they do not require students to deal with numbers or units and they thus allow students to focus on understanding measurable attributes and the basic processes of measuring.

Activities involving measuring with units involve many important issues for young students to consider. First, they need to determine an appropriate unit for measuring the attribute in question. For example, using one's foot to measure the length of a room would be more reasonable than using a paper clip. Once a unit is chosen, the student must also decide how to use that unit to measure, iterating the units without allowing either gaps or overlapping units. The issue of how to iterate is both procedural (how does one physically measure) and conceptual (what it means to measure with a unit, and thus, how it makes sense to iterate the unit). Another important issue arises when measurements do not work out evenly, such as when the room is more than 15 paces long and yet not quite 16 paces long. Finally, there is an inverse relationship between the size of the unit and the number of units in the measure; that is, as the unit gets larger, the measure gets smaller. All these issues should be brought into play when designing lessons to enhance children's understanding of the process of measuring.

Evidence suggests that not all children will be able to resolve all these issues by the end of first grade (Wright, Mokros, and Russell 1998). However, it is valuable for students to begin to grapple with these ideas at a young age, since it necessitates that they think more deeply about what they are doing when they measure and enables them to begin to formulate conjectures about these complex issues.

A Look in the Classroom

Although having a variety of rich activities and investigations related to the ideas discussed previously is essential, one must also consider the pedagogical aspects involved in supporting young children as they explore measurement. It is crucial to think carefully about the questions that are posed to children in the context of measurement activities. The questions should make children think carefully about the decisions they need to make when they are about to measure, when they are in the process of measuring, and when they are interpreting what they've found. In order to get a sense of what kinds of questions might be appropriate at these times and how young children might respond to them, several episodes taken from one lesson from a first-grade classroom are described below.

Although it is important to use real conversations from real classrooms, one caveat is that no single episode or lesson is beyond reproach. The examples that follow contain some missed opportunities for probing thinking or for pursuing ideas more completely. Reality dictates that teachers, who make decisions minute by minute and second by second in many instances, will necessarily miss some opportunities. The realness of the classroom episodes does provide, however, an authentic illustration of some good questioning as well as a genuine context for thinking about additional possibilities.

Choosing an Appropriate Unit

The discussion that follows occurred in the beginning of a lesson in the middle of a unit on measurement from the Investigations in Number, Data, and Space curriculum. Parts of the discussion referring to additional suggestions about units have been omitted, but what remains provides an essence of the type of discussion the students were having about appropriate units to measure length. The students had already done a variety of comparison activities related to length, and they were ready to grapple with more complex measurement ideas.

Teacher: We already did a little bit of measuring, and it was talking about *longer than* and *shorter than*. What if we wanted to know how long something was? What could we use to measure something and find out how long it was? Audrey? Think hard about it. What could we use to measure something to find out how long it is?

Ryan: Use your arm.

Teacher: What do you mean?

Ryan: You could lay your arm against it until you touched the swing.

Teacher: So, lay your arm down against the swing and kind of mark on your arm where it was?

Ryan: Yeah.

Teacher: What if it was more than one arm?

Ryan: Um, do another arm.

Teacher: Do another arm? So it would be one arm and then another arm. It would be a pretty high swing, wouldn't it? Yeah, you could use your arms. Okay, what if instead of seeing how high the swing was off the ground, we wanted to see how far it was from here to the door. Could we do it with Ryan's arms again?

Students: No, no, no.

Tessa: We don't have that long of arms.

Jamie: But we could just keep using arms and another arm and another arm, but, um, that seems too far.

Teacher: So what else could we do?

Tessa: Um, a person could lay down here, and then another person could get their, like, head in front of them so if they have to peek or, like, or you could take your shoes off, and you could go all the way to the door …

Teacher: So, you're finding how many bodies it was from here to the door? So do one body and then another body and then another body to see how many kids it was from here to the door? So see how many kids it was from here to the door. That might work. Matthew?

Matthew: Um, you could, um, use mark, um, use, um, markers, um, to see how much colors it takes to get to the door.

Teacher: How would we do that?

Matthew: Like, take the one color and put it, um, on one side, and then take another one, and put it right on top of that, and keep doing that until it gets to the door.

Teacher: Okay. Tessa?

Tessa: Um, you put one foot down and count the number, and you could count and, like, if you put one foot down and count, like, number 1, and number 2, and number 3 of each steps.

In this episode, the students are asked to think about what kinds of nonstandard units they could use to measure length. Notice how the teacher encourages them to think about different possibilities and has them explain how they envision using the unit they suggest. She also asks them to consider different lengths to think about whether a particular unit is appropriate for a variety of situations. For example, after an arm is suggested as a unit to measure the height of a swing, she asks if it would be an appropriate unit for measuring the length of the classroom, to which many responded no. It is interesting that the students did not react to the suggestion of using markers in the same way. After all, markers are smaller than arm lengths, and it would certainly require a large number to measure the length of the classroom. It may have been the appeal of using markers, which would be expected of most first graders, that hindered any thoughts about the appropriateness of doing so. It may also have been that the teacher could have pursued this issue further but missed an opportunity to encourage the students to think about unit size compared to the size of the object being measured.

The Inverse Relationship between the Size of the Unit and the Number of Units

The second exchange that follows involves the issue of the inverse relationship between the size of units and the number of units required to measure. In other words, as a unit increases in size (going from a baby's handprint to an adult's handprint), fewer units are required to measure the same object. The following episode took place after the discussion about units described above. The teacher asked the students to measure a strip of tape placed on the floor with their feet. One student (Jillian) had already walked along the strip to figure out that it took her 14 footsteps to cover the strip. The teacher then asks the class to think about how many feet it would take A.J. (who has bigger feet than Jillian), and the following discussion ensues.

Teacher: A.J., let's see your feet. Hold a foot up in the air. OK. It might be a little bigger than yours [*Jillian*], right? All right, so if his foot is a little bit bigger, would he need more? Would we have more feet going down the row or less feet going down?

All: More! ... Less! [*Students yell out both answers.*]

Teacher: Let me see a thumbs up for more. Let me see a thumbs up for less. Where are the mores? Josh, why do you say more?

Josh: Because his feet are bigger.

Teacher: His feet are bigger so he would need more feet to go across? OK. Where is someone that says less? Brittany, why less?

Brittany: Because, um, Jillian's feet are smaller so, um, it'll take less because it's like going back a little, and his feet are bigger so it's going forward more.

Teacher: Hmmm.

Student: I agree.

Teacher: Interesting. Did those arguments change anyone's mind? It changed your mind, Matthew? Well, goodness, pretty good arguments then. Let's see it again. Give me a thumbs up if you think there's gonna take, if A.J. is gonna need more steps to go across. Give me a thumbs up if it will take A.J. less steps to get across. [*Most students vote for less.*] Well, Brittany, you are pretty convincing.

[*A.J. walks along the strip and counts to 13. This is recorded on the board.*]

Devon: My feet are smaller.

Teacher: Thank you. So, your feet are smaller than Jillian's? So, would you be more than 14 or less than 14?

All: More ... less ... more ... no, less ... more ... mine are smaller. [*Students yell out different answers.*]

Teacher: Okay. Hold on. Let's think. Devon says his feet are smaller than Jillian's.

Student: And A.J.'s.

Teacher: If A.J.'s feet were big—not too big, just bigger than Jillian's. A.J.'s feet were bigger than Jillian's, and he had less footsteps. Jillian's feet were smaller than A.J.'s, so she needed more footsteps than A.J.'s. So would your number be higher than 14 or less than 14?

All: More ... less, less, less ... more ... more ... less ... more! [*Students yell out different answers.*]

Teacher: You say more. What do you think, Devon?

Devon: More.

Teacher: Why more?

Devon: Because my feet are smaller and, um, and 'cause her feet are bigger, and it's just like, ... like, A.J.'s feet are bigger than hers, and he got one less.

Teacher: Are you convinced? [*Said to the entire class.*]

All: Yes ... No ... Yes ... I don't get it ... I do ... I do! [*Students yell out different answers.*]

Teacher: Well, should we have him do it and see? Maybe that will show you.

Devon steps along the tape and ends up with a measurement between 15 and 16 whole steps, and this leads to disagreement among the students about the measure. This exchange will be discussed in the next section.

What is powerful about the preceding episode is the way in which the teacher, through her questioning, encouraged the students to grapple with the mathematical ideas themselves. She engaged all students in this task by encouraging them to listen to one another's ideas and decide for themselves whether the ideas were correct or not. For example, she first asks the students to vote on whether A.J.'s measure is going to be less than or more than Jillian's measure. Then she asks for an explanation from both sides of the argument. Some may think that this kind of discussion, where young children are asked to provide evidence for their position, is too difficult. However, it is apparent that the students in this classroom are both willing and able to participate in such a discussion.

One example is Brittany's argument for why A.J.'s measure would be less than Jillian's. She says, "Because, um, Jillian's feet are smaller so, um, it'll take less because it's like going back a little, and his feet are bigger so it's going forward more." Parts of this explanation are difficult to interpret precisely. Does the phrase "it'll take less" refer to A.J.'s measure? Does the phrase "it's like going back" refer to Jillian's feet? Although some may interpret these phrases differently (and some might change those interpretations on seeing and hearing Brittany rather than just reading her words), most would agree that Brittany is genuinely thinking about this situation and has made a reasonable attempt at explaining her thinking. It is important to provide this opportunity to enable young students to clarify their thinking, even though they will be challenged to do so clearly.

Most of the students were persuaded by Brittany's argument and believed that it would require fewer of A.J.'s feet to measure the tape. However, the fragility and developing nature of this understanding was apparent when Devon entered the conversation by questioning what would happen with his smaller feet. Recall that a significant number of students thought smaller feet would lead to a smaller measurement. Even after the measurements were accomplished and the students saw that A.J. required less feet and Devon required more feet than Jillian, some students still did not understand why this happened. One may question why the teacher did not step in at this point and resolve the issue for the students. Knowing that an exploratory phase followed this discussion in this particular lesson and that they were only midway through the unit helped the teacher realize that there would be several opportunities later to arrive at a resolution. She also wanted to make certain that some of the students had more time to explore this issue during the investigations that were to follow in future lessons, to give them the opportunity to think more about this complex issue. But it is noteworthy that timing these expectations for resolutions is perhaps one of the most challenging aspects of teaching for understanding.

Finally, one other technique used by the teacher that made this a particularly effective exchange was that she did not condone the correctness of the ideas presented by the students. Rather, she encouraged them to decide and empowered them by giving them the authority to make sense of the mathematics they were learning. This transmission of authority from resting solely with the teacher to the entire classroom community is difficult to make happen. But several questions that the teacher asked, such as "Did those arguments change your mind?" and "Are you convinced?" contributed to this empowering atmosphere in the classroom.

Whole Units and Part of Another Unit

This brings us to the disagreement students were having about Devon's measure. This was a perfect context to discuss the issue of what happens when you have part of a unit and how to represent this measure with words and symbols. The episode below contains the discussion that occurred about this issue.

Teacher: [*To Devon*] All right, start right there on the end [and count as you go].

Students: 1, 2, 3, 4, 5, 6, 7, 8, 9, 10, 11, 12, 13, 14, 15,...

[*Teacher and students pause, because there is not enough room for another whole foot. Then, as Devon places another foot and it hangs over the edge of the strip, the students call out 16 and 16 1/2.*]

Teacher: Oh? How many?

Students: 16 … 15 … 16 1/2! [*Students shout out different answers again.*]

Teacher: Hold on. Did he get a whole sixteenth foot on there?

Students: No.

Teacher: So could it be 16 1/2? Could it be more than 16?

Student: No, it has to be 16 1/2.

Teacher: Well, let's think about that. Fifteen was his last whole foot, and then he had a little bit more. So he had 15, and then a little bit more. So, is that 15 or 16?

Students: 16 1/2 … 15 1/2! [*Students call out answers.*]

Teacher: Let's turn around and face the white board. Let's see: 14 was Jillian, 13 was A.J., and 15 was Devon. But it wasn't just 15, was it? His fifteenth foot was … here is our strip of paper, and then he had … [*the teacher draws a whole foot next to the strip with a little piece of the strip remaining*]. So here is number 15, and you had this one here [*the teacher draws another whole foot going past the edge of the strip*]. If he got all the way here, it would be 16. But, instead, it is just a little bit more than 15. So what is that? What is that, Scott?

Scott: 15 1/2.

Teacher: Why is it called 15 1/2?

Scott: 'Cause, um, 'cause there was a half left, and half of it got taken off because there wasn't enough.

Teacher: So, half of his foot got used, and half got taken off, so there is a half there [*pointing to the strip*] and there is a half there [*pointing to the part hanging over the strip*]. So we had 15 whole feet and a half of a foot. Does that make sense, Jillian?

Jillian: No.

Teacher: Hmmm, well, did he use 16 whole feet?

Jillian: No.

Teacher: Did he use that whole foot? This is our tape [*points to the strip on the white board*].

Jillian: No.

Teacher: No, so he didn't get to use that whole foot. He used 15 whole feet and then half of another foot. So it's 15 and a half. So instead of counting this as 16, we count 15 and then a half. So, it is 15 1/2.

Although the teacher is more directive in this episode than in previous episodes, it is still impressive that she makes an effort to pursue a challenging mathematical issue as it arose in the context of their discussion. When considering what the students were thinking in this episode, an important distinction needs to be made between their understanding or recognition of partial units and their decisions about how to represent them using words or symbols. The teacher attempts to do this by focusing first on the recognition aspect by asking the students if Devon got "a whole sixteenth foot on there." It was evident that most of the students understood or recognized that there was not another whole unit when Devon was measuring. However, they were challenged by having to think about ways to represent the answer with words or symbols. To help the students think about how they might represent the measure, the teacher then uses language that might be easier for them to understand, saying the measure was "a little bit more than 15," while still allowing them to pursue the idea of halves.

One may argue that the teacher's efforts would have produced better results if she had been less directive in this exchange. Although she attempted to get students to think with her questioning, she was often narrowing the range of possibilities through pointed and leading questions. Imagine, conversely, that she had asked the students to provide an explanation or argument for their measure (15, 16, 16 1/2) directly after those measures were given. As in the discussion about A.J.'s measure, this would have given students the opportunity to think more and attempt to develop an argument and clear explanation. It would also have given the teacher the opportunity to understand better how to pursue this issue of representing the measure. This highlights the challenge of teaching for understanding and consistently asking questions that get students to think rather than doing the thinking for them. It is important to realize that there can be a real tendency to fall back on more leading questions, particularly when the issue being discussed is difficult, and that a greater vigilance should be brought to bear at these times.

After this last episode in the whole-group discussion, the students were allowed to spend the remainder of the time working with a partner to measure other lengths in the room and record their findings on paper. This allowed the students to investigate these ideas on their own and continue to formulate their conjectures about how to measure. During this exploration time, the teacher continued to pursue the measurement ideas discussed previously. She asked the students questions about their choice of units (they were asked to use footprints or handprints), why partners may have gotten different measures (different foot or hand sizes), whether they could predict if one partner would have a larger or smaller measure than the other and why, and how they would represent a part of a unit.

It is important to note that many of the students were in what could be called the "developing stage" of thinking about some of these ideas. They continued to struggle with how to represent part of a unit. For example, one pair counted out their footsteps on a short strip of tape as 1, 2, 3, 4, 5, 6, 7 1/2, just as in the episode above. After a discussion with them about their answer and prodding them with skillful questioning to think harder about their answer, the teacher realized that they understood they had a part of a unit but did not yet understand how to represent it. They seemed to want to indicate somehow that the measure was almost or close to 7, and 7 1/2 seemed to make more sense than 6 1/2. The teacher then asked them to record their answer as 7 1/2 and place it in their math folders. Accepting their solution at this time is essential for allowing children to learn with meaning and develop at their own pace. With continued exposure to these types of investigations and rich discussions about them, the students will find ways to represent their solutions more accurately.

In watching and discussing this entire episode (the whole-group discussion and small-group exchanges) with other teachers, it is interesting to note that they often suggest that manipulatives should have been used to represent the units. For example, they believe that if you have 15 footprints cut out of paper that you can line up on the strip with a half of another footprint, that students will say 15 1/2 when they go back and count up the group of footprints. This suggestion is puzzling given the exchange that occurred in the whole-group discussion. When the students were asked if there was another whole foot after 15, they all said no. Many who realized that there was part of a unit and 15 whole units still chose to represent it as 16 1/2 units. Given the fact that this also occurred during the partner work in the context of using a much smaller number of units (6 1/2), this suggests that some of the students are struggling with a complex idea that most likely would not be overcome simply with the use of manipulatives.

CONCLUSION

There are many issues embedded in beginning to understand what it means to measure. The first graders described in this article were grappling with three of these issues: choosing an appropriate unit for measuring a particular attribute, exploring the relationship between the size of a unit and the number of units required to measure, and dealing with measurements that have whole units and parts of a unit. Struggling with these concepts is a crucial component of understanding the measuring process and preventing the kinds of misconceptions evident in the NAEP data discussed earlier. Although the classroom episodes demonstrated that young children can think about these ideas, they also made clear the pedagogical changes necessary to facilitate this process. The teacher in these episodes engaged her students in thinking about measurement by working to create a classroom environment in which the correctness of solutions was determined through discussion among students. Through her questioning and her orchestration of classroom discourse, the teacher succeeded in making measurement a thought-provoking activity rather than a set of procedures to follow. This not only served to develop students' understanding of important mathematical ideas related to measurement but also enhanced their beliefs about what it means to understand mathematics in general.

REFERENCES

Kenney, Patricia Ann, and Vicky L. Kouba. "What Do Students Know about Measurement?" In *Results from the Sixth Mathematics Assessment of the National Assessment of Educational Progress*, edited by Patricia Ann Kenney and Edward A. Silver, pp. 141–63. Reston, Va.: National Council of Teachers of Mathematics, 1997.

Wright, Tracey, Jan Mokros, and Susan Jo Russell. *Bigger, Taller, Heavier, Smaller*. A grade 1 unit in Investigations in Number, Data, and Space. Palo Alto, Calif.: Dale Seymour Publications, 1998.

5

Benchmarks as Tools for Developing Measurement Sense

Elana Joram

"MEASUREMENT sense" (Hope 1989) can be thought of as including, but not being limited to, having a "feel" for units of measurement and possessing a set of meaningful reference points or benchmarks for these units. For example, when we asked thirty-six undergraduates the question "When you think about a mile, what do you think of?" fully three-quarters of them gave responses like "4 times around my old high school track" (Joram et al. 1996). These individuals thought of a mile as a distance that was personally known to them, and this translation process apparently helped them represent the magnitude of the mile unit. Because of the capacity of benchmarks to enhance the meaningfulness of standard units of measure, they are often recommended by mathematics educators as an important component of instruction on measurement and measurement estimation (e.g., Bright 1976; Carter 1986; Markovits, Hershkowitz, and Bruckheimer 1989; McIntosh, Reys, and Reys 1992; Tierney et al. 1998).

Although numerate adults, like the undergraduates above, may have a repertoire of benchmarks readily available for representing measurement units, school-aged children may not be as fortunate. Markovits, Hershkowitz, and Bruckheimer (1989) report that few of the sixth and seventh graders in their study had knowledge of the measurements of everyday objects, such as the height of an eight-story building, that might have allowed these objects to serve as benchmarks. Further, when asked to estimate the measurements of related objects, they found that students' answers were internally inconsistent. For example, students gave estimates indicating that the width of a car was greater than the width of a road. The results of this study punctuate the need to help students develop a repertoire of benchmarks that can be easily repre-

The preparation of this article was supported by a research grant from the First in the Nation in Education (FINE) Foundation. The ideas described in this article were developed while the author was on a University of Northern Iowa professional development leave. I would like to thank Myrna Bertheau, Bonnie Smith, and the students at Cedar Heights Elementary School for their assistance.

sented and used for estimation. These meaningful representations of units should serve to increase students' understanding of measurement as well as improve their ability to estimate measurements.

Benchmarks are usually thought of as tools for enhancing students' representations of measurement units; however, they can also be used to teach students about measurement systems and principles. In this article, the multiple instructional uses of benchmarks are examined, with an eye to explaining how benchmarks can help students develop greater measurement sense. For clarification, examples are given for linear measurement only; however, the topics under discussion could be extended to other attributes such as weight.

RATIONALE FOR USING BENCHMARKS

Benchmarks typically consist of nonstandard units whose lengths are used to represent the lengths of standard units (e.g., the length of the top half of my thumb equals about one inch in length) or multiples of units (e.g., the height of an average man is about six feet). Benchmarks should not be confused with manipulatives that denote objects that students handle; although benchmarks may start off as such objects, they typically evolve into mentally represented objects, whereas manipulatives do not. For example, students may use a piece of bubble gum to represent an inch; to be considered a measurement benchmark, students would need to be able to represent the bubble gum mentally when estimating.

To an individual estimator, a particular measurement attribute will be salient with respect to any given benchmark. For example, when thinking about the piece of gum described above as a benchmark, it is the length of the gum that is most significant to the estimator, rather than its weight or other attribute.

Benchmarks are sometimes referred to as reference points because they provide individuals with a standard to refer to when thinking about measurements. For example, if you know the height of an average man, you can estimate that a shorter-than-average man might be about five and a half feet tall, or that a building about the height of three men stacked on top of one another would be approximately 20 feet tall. Just as landmarks in a city allow one to navigate through unfamiliar territory, measurement benchmarks allow one to generate estimates about unfamiliar quantities.

Hope (1989) notes, "A knowledge of a wide variety of everyday measurement references, such as that doorways are about two meters in height, is the foundation of good measurement sense as well as good number sense" (p. 15). Such a repertoire, according to Hope, is necessary both for producing estimates and for judging the reasonableness of an estimate. For example, students who have learned appropriate referents for numbers in everyday life will realize that it is not reasonable for a child to be 10 meters tall (Thompson and Rathmell 1989).

Benchmarks are thought to be more meaningful to individuals than standard units because they typically consist of familiar objects, which may be more easily represented and consequently more accurate. An additional benefit of benchmarks is that in the context of estimating, a benchmark that consists of multiple units (e.g., a person known to be six feet tall) reduces the number of units that must be mentally concatenated. For example, to estimate the length of a room, using a person's body as a benchmark, one need only imagine two person lengths rather than twelve individual foot-long units lined up. This lessens the cognitive load faced by the estimator by reducing the number of units that must be tallied and thereby the likelihood of error arising from using unequal-sized units or from leaving spaces between units.

What Kind of Benchmarks?

Although any object could potentially serve as a benchmark, there are important considerations when deciding which benchmarks students should use. First, although tempting because of their availability, body parts may not make ideal benchmarks for children because their size changes as the children grow (Bright 1976). However, body parts may be entirely appropriate when instructing adults, for example, on the metric system. Knowing that a meter is approximately the distance from one's nose to the end of one's outstretched arm can be enormously helpful for remembering an element of what has remained an elusive measurement system for many Americans. Second, many objects that are readily available in classrooms such as paper clips and Unifix cubes may not make the best benchmarks, or at least should not be the only benchmarks that students use. We have argued that such objects, although easily obtained by teachers, may not be readily represented by children, nor may they be particularly memorable for them. Instead, we have suggested that, when possible, children generate their own "personal benchmarks" (Joram et al. 1996). These personal benchmarks can be generated for different attributes and units over a period of weeks and recorded on a chart (Tierney et al. 1998). Tierney and colleagues depict a simple chart (p. 28), labeled by individual students as "My Benchmarks," on which objects that are to serve as benchmarks along with their corresponding measurements are drawn (e.g., a crayon with an arrow pointing to the tip and "2 millimeters" written above).

There are obviously times when objects that are readily available in classrooms, such as paper clips, will be advantageous to use for classroom exercises and demonstrations. These objects can be obtained by teachers in quantity and are inexpensive. Bright (1976) points out that having students select objects from home to act as benchmarks, in addition to those present in the classroom, will increase the likelihood that students will use benchmarks to estimate in daily life. Thus, we can conclude that benchmarks in classrooms would best consist of some combination of student-generated and teacher-given objects.

Because benchmarks should ultimately become represented as images for students, there are a number of other important considerations in using them. Repeatedly using the same benchmark for estimating should help students develop and remember mental images for these objects. Thus, it may be helpful for students to work with a small set of personal benchmarks for different units (e.g., one benchmark for an inch, and a different one for a foot) and to practice estimating with this set instead of employing a wide variety of different benchmarks. Students need practice visualizing their benchmarks. Exercises devoted to this—such as imagining the benchmark's magnitude, showing it with their hands, and drawing it—may all help the benchmark evolve from an object to an image.

A final consideration in the use of benchmarks is when they should be related to standard units. Although it is often recommended that nonstandard units be introduced first and followed by standard units, Clements (1999) pointed out that it may be preferable to introduce both nonstandard and standard units early in measurement instruction. He suggested that this will be of greater benefit than the "traditional nonstandard-then-standard-then-ruler sequence" (p. 6). Thus, benchmarks should be initially and continually related to standard units for students.

COMMUNICATING TO STUDENTS WITH BENCHMARKS

As well as helping students develop a set of personal benchmarks, benchmarks can also be used by teachers and parents to communicate the magnitudes of measurements. For example, Carter (1986) pointed out that when a child asks how far it is to the zoo, it is much more meaningful to respond that it is "about as far as the park" than "three-quarters of a kilometer." The familiarity of the distance to the park allows the child to picture the magnitude of the distance, whereas the measurement "three-quarters of a kilometer" will likely be difficult for the child to represent. One could argue that benchmarks in this capacity serve merely as a temporary cognitive crutch that allows the student to bypass the effort of learning to represent the magnitude of measurements; indeed, this is a possibility. Communicating the magnitude of measurements to students through benchmarks, however, may help them to learn to visualize the magnitude of the measurement. Further, some measurements seem intrinsically difficult to represent, particularly long distances, and it may be that such measurements always require a translation process into a familiar object that embodies the measurement.

We can infer that many authors of magazine and newspaper articles know that benchmarks provide a powerful way to think about measurements, because we often find "measurement similes" in everyday texts. For example, a magazine article might state that one should pay attention to skin moles that are more than half a centimeter in diameter, about the size of a pencil eraser. In this example, an object whose size is thought to be known (the pencil eraser) is used to help the reader form a mental "picture" of the measurement "1 cm" in diameter. This journalistic technique seems particularly appropriate when the measurements discussed are thought to be difficult to appreciate, because they are very large or small or when the units are unfamiliar. In one informal experiment I conducted, a group of college students was given texts with measurement similes (e.g., "a spider whose abdomen was 6 cm in diameter, about as wide as a lemon"), whereas another group was given texts with only measurements (e.g., "a spider 6 cm in diameter"). The students were then asked to draw the lengths of various objects that had been embedded in the texts (either with or without similes) and, predictably, the drawings of those who had been given measurement similes were much more accurate in length.

Measurement similes may be useful for helping students to develop their own repertoire of personal benchmarks. The results of the informal study described above suggest that one way to encourage students to construct a set of benchmarks might be to give them measurement similes to complete, for example, "the wave was 20 ft. high, about as high as a/the ________ (e.g., school)." The measurement in the first part of the simile is meant to be unfamiliar to the student or difficult to represent because it is relatively large or small, whereas the magnitude of the object in the second part of the simile (that the student fills in) is meant to be familiar. Students in small groups could each give a response to the same simile, monitoring that the objects generated by their peers were within a reasonable ballpark of the target measurement. Students might go outdoors or walk in the school building to look for objects that would serve as benchmarks to help them relate measurements to real-world experiences. As in the benchmark chart task specified by Tierney et al. (1998), students could also draw pictures of the objects used in their similes in order to help them visualize the magnitude of these objects and their corresponding measurements.

An additional benefit of the exercise described above is that it gives students practice with a second type of estimation, specified by Bright (1976), in which students are given a measurement and have to find an object of that magnitude. He pointed out that most students practice only one type of estimation task, where an object is given and the student is required to give a corresponding measurement (e.g., "How many inches long is this pen?"). Both types of estimation exercises, that is, going from object to measurement and from measurement to object, are important for developing meaningful mediating representations of units.

In summary, benchmarks may be very effective tools for helping students develop meaningful, accurate mental representations of standard measurement units. It has been suggested here that teachers should promote the construction of a repertoire of personal benchmarks that are eventually represented as mental images. In addition to increasing the meaningfulness and accuracy of representations of standard units, benchmarks can also be used to teach important measurement concepts and principles in a way that is both engaging and memorable.

USING BENCHMARKS TO ENHANCE STUDENTS' UNDERSTANDING OF MEASUREMENT SYSTEMS

One potential liability of using benchmarks to teach students about measurement is that they may develop associations between specific measurements and their magnitudes but not understand how benchmarks are related to one another (Joram, Subrahmanyam, and Gelman 1998). In other words, a student may know that one inch is about the magnitude of his or her thumb from the knuckle to the tip, and that six inches is about the length of a dollar bill, but not realize that one and six inches are related. We have suggested that students need to learn to align benchmarks on measurement scales rather than simply build associations for individual units and reference objects (Joram, Subrahmanyam, and Gelman 1998). That is, images for units must take their place within an ordered system of units that is governed by mathematical principles. Gibson and colleagues (Gibson and Bergman 1954; Gibson, Bergman, and Purdy 1955) referred to the mental representation of units on a scale as a *conceptual scale* and noted that such a scale might contain fractions and multiples of the basic unit.

A natural way to help students develop conceptual scales for units is to have them work with visual representations such as an individually constructed or class-constructed chart. Students, or groups of students, can place their benchmarks on a chart, possibly referring to the benchmark chart they have constructed, as described above. Alternatively, or in addition, a class chart can be constructed on which benchmarks are superimposed. This might consist of a long piece of paper taped to a chalkboard, with the number of inches written horizontally along the bottom (e.g., "1 inch; 6 inches") and the corresponding magnitudes (heights) shown vertically along the left side (a yardstick can be taped in a vertical orientation on the left side of the board for this purpose). Students can attach benchmarks of different sizes to this chart with masking tape; for example, at the "1 inch" mark along the bottom, a piece of bubble gum one inch high can be attached; at the "3 inch" mark, a ketchup package; and at the "6 inch" mark, a dollar bill (see fig. 5.1). Because the focus is for students to develop a conceptual scale, students should be encouraged to close their eyes and visualize the chart and benchmarks.

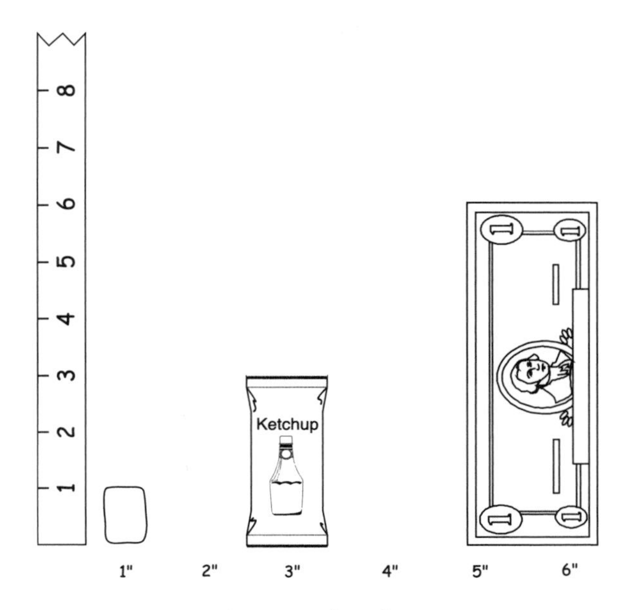

Fig. 5.1. A class chart

A class chart provides the teacher with a powerful tool for illustrating the additive and multiplicative relations among units on the scale using benchmarks. For example, the teacher can show that two ketchup packages stacked one on top of the other, each three inches in height, are equivalent in length to a dollar bill. Students can also see that, for example, a dollar bill and a ketchup package together are about nine inches in height. Further, benchmarks depicted on a chart can be used to help students understand the relations among different units, for example, seeing that three pens (each six inches long) are equivalent in measure to one and a half feet. Thus, students can learn some practical strategies for using benchmarks in an estimation context (i.e., doubling or adding benchmarks together) and can also learn about the relations among units.

USING BENCHMARKS TO TEACH MEASUREMENT PRINCIPLES

Benchmarks provide a convenient way to illustrate important measurement principles such as the "inverse relationship between unit size and number" (the idea that as the size of the unit increases, fewer units are needed to measure a given object), and the "unit covering" principle (the notion that, in linear measurement, the object to be measured or estimated must be completely covered with units). These principles continue to be problematic even for students in middle school. About 40 percent of seventh-grade students in a cross-national study had difficulty with a problem tapping the inverse relationship between unit size and number (Beaton et al. 1996). Hiebert (1981) discovered that only 62 percent of middle school students could find the number of paper clips needed to fully cover the length of a drawn line segment, indicating difficulties understanding the unit-covering principle. More recently, the TIMSS study echoed this finding, when only half of seventh and eighth graders in an international sample were able to correctly determine the length of a pencil that was lined up with part of a ruler (Beaton et al. 1996).

To help students understand the inverse relationship between unit size and number, teachers can ask them to predict which benchmarks (of varying lengths) they will need more of to cover a given extent or distance. Students can then work with their benchmarks to discover the inverse relationship between unit size and number on their own, finding that one needs fewer longer units to measure or estimate a given object than shorter units.

Benchmarks can be used to elicit students' intuitions about measurement principles, which, once articulated, can be discussed and honored when estimating. Even very young children seem to be sensitive to graphically illustrated violations of measurement principles, and presenting these situations may elicit students' correct intuitions about measurement. For example, I recently showed a first-grade class a stuffed lizard that I was going to "measure" using pieces of bubble gum. I lined up the gum alongside the lizard and spread it out so there were spaces between the units (see fig. 5.2a). "Is it okay to measure the lizard like this?" The students emphatically responded "No!" "Why not?" I then asked. They responded that when something is measured, it is not okay to leave spaces between the units. I repeated this series of questions after pushing the gum together but leaving a piece of the lizard's tail uncovered (see fig. 5.2b). Finally, I arranged the pieces of gum in different orientations, again asking if this was "okay," and "why not" (see fig. 5.2c). Students were then asked to measure their own objects the "right" way, by lining up pieces of gum in the same orientation so that the unit sizes were equal and by completely covering the to-be-measured object. In order to encourage students to visualize units, they were eventually asked to estimate objects, keeping in mind the measurement principles they had previously identified.

In each of the three cases described above, the first graders were able to articulate, with reasonable accuracy, why the violation was inappropriate. In doing so, they also articulated a measurement principle. Because the bubble gum was used to show discrete units, it provided a means of illustrating violations of the principles in a way that would be difficult with a ruler.

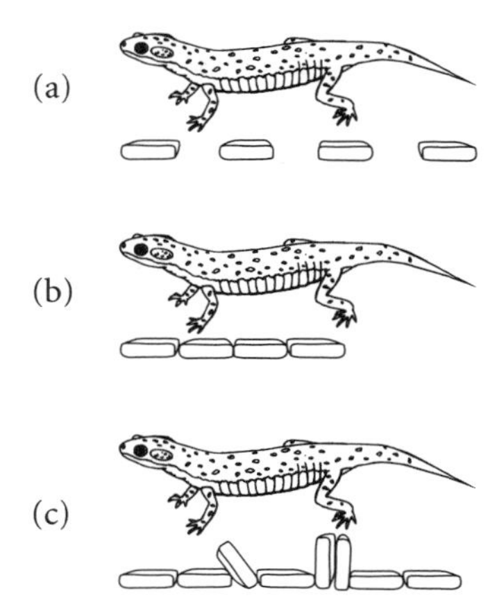

Fig. 5.2. An illustration of violations of measurement principles

Conclusions: Learning the Language of Measurement

Number sense, which is notoriously difficult to define (McIntosh, Reys, and Reys 1992), has been explicated through the metaphor of learning "the lay of the land," with the domain of numbers as a conceptual environment (Greeno 1991). Another apt metaphor for number sense, and its aligned measurement sense, is learning a language. Developing measurement sense can be conceptualized as learning the language of measurement, with attention paid to the semantics, syntax, grammar, and pragmatics of measurement.

There are three levels that characterize the ways students interact with benchmarks. At the first level, benchmarks are merely nonstandard units that students handle; this would be appropriate in the early elementary grades when students' capacity to construct and maintain accurate mental images is not highly developed. Although suitable early in mathematics instruction, it is important that teachers make efforts to take students beyond this level.

At the second level, through deliberate scaffolding by the teacher, benchmarks are represented as visual images: they can then serve as reference points when estimating. We have referred to this elsewhere as developing a "semantics of measurement," where individuals come to understand the meaning or magnitude of a given unit or multiples of that unit (Joram, Subrahmanyam, and Gelman 1998). Above, it is suggested that having students practice visualizing, showing with their hands, and drawing personal benchmarks will promote the development of a semantics of measurement. At this level, however, students may see benchmarks only as individual objects of a given magnitude and may not yet understand how they fit into a measurement system.

At the third level, again through efforts on the part of the teacher, students recognize the place of individual benchmarks within a measurement system. Because this awareness permits students to understand the relationship among units within the system, we have referred to this elsewhere as understanding the "syntax" of measurement (Joram, Subrahmanyam, and Gelman 1998). Working with charts on which benchmarks are placed to point out the relationships among units described above as one way to promote understanding the syntax of measurement. Relating units across scales should also be encouraged, for example, using the image of a six-inch-long dollar bill to find the 1/2 point on a scale for feet. At this level, students also appreciate the "grammatical" rules that apply to measurement: the principles and constraints that govern what one can and cannot do in a measuring context, such as the unit-covering principle.

Finally, an understanding of the "pragmatics" of measurement is needed, as it is with all communication systems. Students may understand much about benchmarks and yet fail to mobilize them in appropriate contexts. It has been suggested that using benchmarks that are personally meaningful may make it more likely that students will use them when appropriate. Students also need to understand the situational constraints on measurements—for example, that a cook may measure ingredients in pinches when cooking but that this would be inappropriate for a pharmacist when measuring medicine (Hope 1989).

In this paper, benchmarks are presented as a powerful vehicle for learning the language of measurement and for developing measurement sense. The metaphor of language learning points to the need to attend to multiple aspects of use and not to restrict benchmarks to serving only as mnemonics for single units.

REFERENCES

Beaton, Albert E., Ina V. S. Mullis, Michael O. Martin, Eugenio J. Gonzalez, Dana L. Kelly, and Teresa A. Smith. *Mathematics Achievement in the Middle School Years: IEA's Third International Mathematics and Science Study*. Boston: TIMSS International Study Center, 1996.

Bright, George W. "Estimation as Part of Learning to Measure." In *Measurement in School Mathematics*, 1976 Yearbook of the National Council of Teachers of Mathematics (NCTM), edited by Doyal Nelson, pp. 87–104. Reston, Va.: NCTM, 1976.

Carter, Heather L. "Linking Estimation to Psychological Variables in the Early Years." In *Estimation and Mental Computation*, 1986 Yearbook of the National Council of Teachers of Mathematics (NCTM), edited by Harold L. Schoen, pp. 74–81. Reston, Va.: NCTM, 1986.

Clements, Douglas H. "Teaching Length Measurement: Research Challenges." *School Science and Mathematics* 99 (January 1999): 5–11.

Gibson, Eleanor J., and Richard Bergman. "The Effect of Training on Absolute Estimation of Distance over the Ground." *Journal of Experimental Psychology* 48 (1954): 473–82.

Gibson, Eleanor J., Richard Bergman, and Jean Purdy. "The Effect of Prior Training with a Scale of Distance on Absolute and Relative Judgments of Distance over Ground." *Journal of Experimental Psychology* 50 (1955): 97–105.

Greeno, James G. "Number Sense as Situated Knowing in a Conceptual Domain." *Journal for Research in Mathematics Education* 22 (May 1991): 170–218.

Hiebert, James. "Units of Measure: Results and Implications from National Assessment." *Arithmetic Teacher* 28 (February 1981): 38–43.

Hope, Jack. "Promoting Number Sense in School." *Arithmetic Teacher* 36 (February 1989): 12–16.

Joram, Elana, Anthony J. Gabriele, Rochel Gelman, and Kaveri Subrahmanyam. "Building Meaning for Units of Measurement: A 'Personal Anchors' Approach." Paper presented at the annual meeting of the American Educational Research Association, New York, April 1996.

Joram, Elana, Kaveri Subrahmanyam, and Rochel Gelman. "Measurement Estimation: Learning to Map the Route from Number to Quantity and Back." *Review of Educational Research* 68 (winter 1998): 413–49.

Markovits, Zvia, Rina Hershkowitz, and Maxim Bruckheimer. "Research into Practice: Number Sense and Nonsense." *Arithmetic Teacher* 36 (February 1989): 53–55.

McIntosh, Alistair, Barbara J. Reys, and Robert E. Reys. "A Proposed Framework for Examining Basic Number Sense." *For the Learning of Mathematics* 12 (November 1992): 2–8, 44.

Thompson, Charles S., and Edward C. Rathmell. "By Way of Introduction." *Arithmetic Teacher* 36 (February 1989): 2–3.

Tierney, Cornelia, Margie Singer, Marlene Kliman, and Megan Murray. *Measurement Benchmarks: Estimating and Measuring.* Menlo Park, Calif.: Dale Seymour Publications, 1998.

6

Assessing and Developing Measurement with Young Children

Doug Clarke

Jill Cheeseman

Andrea McDonough

Barbara Clarke

YOUNG children bring a range of knowledge, understanding, and experience about measurement ideas to school, and their enthusiasm and curiosity provide a wonderful opportunity for teachers to build on what they know and can do.

As part of an Australian research and professional development project, teachers used a framework of "growth points" in early mathematics learning and a related, task-based, one-on-one interview to assess children's understanding of important ideas. The growth points offered a series of conceptual landmarks along the way to understanding pivotal mathematical ideas, as well as a lens for teachers through which they could view the things they saw in the classroom and build on their children's current skills and concepts.

In this article, we outline the framework, share examples of interview tasks that provide insight into children's understanding of measurement, describe some examples of creative teachers' use of a children's storybook to develop a range of classroom activities that focused on measurement concepts, and share some insights from the project.

The Early Numeracy Research Project was supported by grants from the Victorian Department of Employment, Education and Training, the Catholic Education Office (Melbourne), and the Association of Independent Schools Victoria. We are grateful to our colleagues in the university team (Glenn Rowley, from Monash University, and Ann Gervasoni, Donna Gronn, Marj Horne, Pam Montgomery, Anne Roche, and Peter Sullivan from Australian Catholic University), and our coresearchers (the teachers in ENRP research schools) for insights that are reflected in this paper.

Background

The Early Numeracy Research Project (ENRP) involved 35 Australian elementary schools, approximately 350 teachers, and 11,000 children in grades K–2 over three years. The research project was a joint collaboration among Australian Catholic University, Monash University, and the three school sectors (public, Catholic, and independent) in the state of Victoria (for details see Clarke 2001 and Sullivan et al. 2000).

In this project, teachers and university researchers were seeking to find the most effective approaches to teaching mathematics in the early years of school. At the beginning of the project, the research team identified the need for developing of a comprehensive and appropriate learning and assessment framework for early mathematics and a tool for assessing young children's mathematical thinking. The inappropriateness of pen-and-paper assessment at these grade levels led to the development of a task-based, one-to-one interview schedule.

The ENRP had a major professional development component, with teachers meeting regularly with project staff for statewide, regional cluster, and local in-service programs. One of the major issues that was addressed over the course of the project was elementary school teachers' own confidence with mathematics, both the mathematics they were wishing to teach and the later mathematics for which they were preparing their students.

Measuring Mathematics Learning

The impetus for the project was a desire to improve children's mathematics learning, and so it was necessary to quantify such improvement. It would not have been adequate to describe, for example, the effectiveness of the professional development according to teachers' professional growth or the children's engagement, or even to produce some success stories. It was decided to create a framework of pivotal growth points in mathematics learning. Growth points can be considered primary stepping stones along the way to understanding important mathematical ideas. Students' growth in understanding could then be quantified by considering the movement of students through these growth points over time.

The project team studied available research on the development of young children's mathematics learning in the mathematical domains of counting, place value, addition and subtraction, multiplication and division (in number); time, length, and mass (in measurement); and properties of shape and visualization and orientation (in geometry). In this article, the focus will be on the measurement domains of length and mass.

Particularly helpful references in developing the measurement growth points included Brown et al. (1995); Dickson, Brown, and Gibson (1984);

Pengelly and Rankin (1985); and Wilson and Rowland (1993). Although the research base in measurement is not nearly as strong as in number, for example, these references provided research evidence on children's developing understanding and insights into the kinds of tasks that would be useful to assess this.

In the development of the framework, it was intended that the growth points would

- reflect the findings of relevant research in mathematics education from around the world;
- emphasize important concepts and skills in early mathematics in a form and language readily understood and, in time, retained by teachers;
- allow the description of the mathematical knowledge and understanding of individuals and groups;
- provide a basis for task construction for interviews, and the recording and coding process that would follow;
- allow the identification and description of improvement where it exists;
- form the basis of planning and teaching;
- assist in identifying those students who may benefit from additional assistance;
- have a sufficient "ceiling" to describe the knowledge and understanding of all children in the first three years of school.

These intentions guided the process of developing and refining the framework as outlined in the next section.

The Development of the Framework

Within each mathematical domain, growth points were developed with brief descriptors in each instance. There were typically five or six growth points in each domain. The growth point descriptors for measurement are given in a generic form in figure 6.1.

Within the project, this generic form was used for length and mass, but we believe that it also applies to developing skills and understandings in capacity and area. One example of the kind of rich discussion that occurred during the professional development program was during a session in the final year of the project focusing on children's developing understanding of capacity. Teachers were able to discuss the strengths and limitations of the framework as it applied to this attribute, with general agreement that comparing capacities is generally more complex than comparing lengths for young children.

1. The child shows awareness of the attribute and its descriptive language.
2. The child compares, orders, and matches objects by the attribute.
3. The child uses uniform units appropriately, assigning number and unit to the measure.
4. The child chooses and uses formal units for estimating and measuring, with accuracy.
5. The child can solve a range of problems involving important concepts and skills.

Fig. 6.1. ENRP generic growth point descriptors for measurement

The growth points provided a sense of the typical order in which important understandings and skills develop, with recognition that there are potentially many components of understanding and skills between any two growth points. For example, once a child is able to compare, order, and match objects by length, there are a number of important measurement principles that need to be grasped before the next growth point is evident. These may involve the child learning that the same unit can be used iteratively to give the measure, the importance of starting at one end of the object to be measured, using the units with no overlap, and that the count, together with the appropriate unit name, gives the measure. An understanding of each of these concepts or principles about measurement is necessary for children to demonstrate the skills involved in the interview task, where children are asked to measure the length of a plastic straw with paper clips.

In a discussion of higher-level growth points in a given domain, the comments of Clements et al. (1999) in a geometrical context are helpful: "the adjective *higher* should be understood as a higher level of abstraction and generality, without implying either inherent superiority or the abandonment of lower levels as a consequence of the development of higher levels of thinking" (p. 208). As children move toward measuring with informal or formal units, they nevertheless still need to be able to compare two objects by length, for example.

It is important to stress that the framework, although based on available research, is not a document written specifically for the research community, but rather for teachers, with the intention of use and eventual "ownership" by them.

THE INTERVIEW

Once the early drafts of the framework were developed, assessment tasks were created to match the framework. A major feature of the project was a one-to-one interview with every child over a thirty- to forty-minute period. Regular classroom teachers were provided with release time to interview all their children in a quiet place away from the busy classroom, at the beginning and end of the school year. The timing enabled the teacher to build up a clear, mathematical picture of each child early in the year and to see growth over the year. The interview was very hands-on, with considerable use of manipulative materials. Small, plastic teddy bears were a major feature of many tasks, and they were appreciated by the children.

Although the full text of the interview involves sixty tasks in the various mathematical domains listed earlier (with several subtasks in many instances), no child attempts all of them. The interview is of the form "choose your own adventure," in that the interviewer makes one of three decisions after each task. Given success with the task, the interviewer continues with the next task in the given mathematical domain, as far as the child can go with success. Given difficulty with the task, the interviewer either abandons that section of the interview and moves on to the next domain or, in the instance of the number section of the interview, moves into a detour designed to elaborate more clearly the difficulty a child might be having.

Of course, decisions on assigning particular growth points to children are based on a single interview on a single day, and a teacher's knowledge of a child's learning is shaped by a wider range of information, including observations during everyday interactions in classrooms. However, teachers agreed that the data from the interviews were revealing of students' mathematical understanding and development in a way that would not be possible without that special opportunity for one-to-one interaction. It appeared that the children also enjoyed that special time having the teacher all to themselves. Teachers reported that children appreciated the opportunity to show what they knew and could do. They enjoyed sharing their strategies. Two of the more interesting responses to the question "How did you work that out?" were "My brain told me," and "God told me!"

Some Examples of Measurement Tasks from the Interview

A range of tasks was used to establish the child's understanding of important ideas of length measurement, for which the following equipment was required: a 25-cm wooden skewer (blunted); 30-cm of string; a 20-cm plastic drinking straw; eight large (5-cm) paper clips; a 30-cm ruler; several 2-cm-wide, straightened-out party streamers about 180 cm long; and several party streamers exactly 93 cm long.

In each instance, the instructions to the teacher are given in italics.

Length Task 1: The String and the Stick

Drop the string and the skewer onto the table.

1. By just looking (*without touching*), which is longer: the string or the stick?
2. How could you check? (*Touching is fine now.*)
3. So, which is longer?

Length Task 2: The Straw and the Paper Clips

Get the straw and show the child the long paper clips.
Here are some paper clips. Here is a straw.

1. Measure how long the straw is with the paper clips. (*If the child hesitates, say, "Use the paper clips to measure the straw."*)
2. What did you find? (*no prompting*)

If the correct number is given (e.g., 4) but no units, ask, "Four what?"

Length Task 3: Using the Ruler

Here is a ruler. (*Place the ruler in the child's hand.*)
Here is a straw (*20 cm*).

1. Please measure the straw with the ruler.
2. What did you find?

If the correct number is given (20) but no units, ask, "Twenty what?"

Length Task 4: Tearing the Streamer

Without referring to it, place a pen on the table in front of the child. Give the child a long piece of reasonably straight streamer (around 180 cm).
Here is a piece of streamer.

1. Please tear off a piece that you believe is about 1 meter long.

Hand the child the prepared 93-cm streamer and the ruler.

2. I tried to tear off a 1-meter piece. Please measure how long my streamer is. (*Allow the child to use the pen to mark the streamer if necessary, without prompting.*) What did you find?
3. How far out was I? (*If the child is unclear, ask, "How far off a meter was I?"*)

RELATING THE TASKS TO THE GROWTH POINTS AND STUDENTS' ACHIEVEMENT

In table 6.1, each of the four length tasks is related to the growth points for which it provides assessment information. The percents of children at each grade level (kindergarten, grade 1, grade 2, grade 3, and grade 4) who were able to complete successfully each task in the last few weeks of the school year in 2001 are given. Although the ENRP was a grades K-2 project, a small extension of the original project found that the interview and framework were useful and relevant also for teachers of grades 3 and 4. Generally, in Australia, most kindergarten children are five years old at the start of the year; grade 1, six years old; and so on, similar to the United States.

It should be noted that the data in table 6.1 are from the control sample—a group of children whose teachers were not participating in the professional development program. This group was chosen because it is our best representation of typical children in Victorian schools.

Not surprisingly, children's proficiency with these tasks increases over the grade levels. We see also that although it is a relatively easy move from understanding the attribute to comparing objects of different lengths, it is a much greater "distance" to the next growth point because of the various related skills that are needed to measure an object, whether with formal or informal units.

The desire to create an interview that would challenge all children has apparently been achieved, given that only one child out of more than 11,000 who have been interviewed at this point has succeeded on all tasks presented to him in number, measurement, and geometry.

TABLE 6.1

Relationship of Tasks to Growth Points with Percent Success on Each Length Task (November 2001)

Growth Point Description	Related Tasks	Percent Success at End of School Year				
		K (n=475)	1 (n=470)	2 (n=395)	3 (n=157)	4 (n=135)
1. Shows awareness of the attribute of length and its descriptive language.	LT1*	96	99	100	100	100
2. Compares, orders, and matches objects by length.	LT1*	94	99	99	99	100
3. Uses uniform units appropriately, assigning number and unit to the measure.	LT2	39	63	84	92	97
4. Chooses and uses formal units for estimating and measuring length, with accuracy.	LT3	2	9	39	69	88
5. Can solve a range of problems involving key concepts of length.	LT4	0	1	6	15	33

*LT1 represents "Length Task 1."

OTHER MEASUREMENT DOMAINS

Teachers have found that the important ideas behind the growth points for length measurement are easily translated to other measurement domains, with the exception of time, for which quite different growth points were created. That is, the growth in children's understanding of mass, capacity, and area can be described in similar words.

Examples of the initial interview tasks on mass are the following:

Mass Task 1: What Do You Notice?

Present the child with a box containing a range of objects: a 1-kg mass, a ball of string, a rock, a piece of foam, a tall and thin plastic bottle, and a short and broad plastic bottle. A balance is placed to one side.

Please take these things out of the box and put them on the table.

1. What do you notice about them?
2. Which things are heavy and which things are light?

Push all items aside, except for the two plastic containers.

3. Take these two plastic containers. (*Place one plastic container in each hand for the child to feel.*) Which do you think is heavier?
4. How could you check?
5. Do you know about balances? (*Allow some time for the child to become familiar with the balance.*) Use the balance to see which container is heavier.

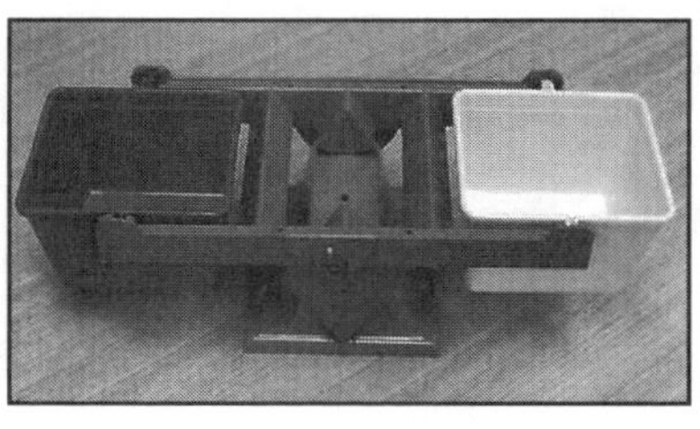

6. Were you right? How did you know?

These tasks enable the teacher to determine whether the child has an understanding of the attribute of mass and its descriptive language and is able to compare masses using the balance.

ASSESSMENT GUIDING INSTRUCTION

Most readers would agree that the major purpose of assessment is to inform those involved, with a particular focus on providing guidance for planning and teaching. With the framework of growth points increasingly "owned" by teachers through its extensive use in their assessment and planning, teachers have told us that their interactions with children—whether individually, in small groups, or as a whole class—are seen through the lens of the growth points. An example of teachers using their knowledge of children's growth to develop worthwhile mathematical experiences now follows.

Teachers in grades K–2 at Mountain Gate Primary School used a storybook, *Alexander's Outing* (Allen 1992), as a springboard for mathematical work. In this story, a little duck, Alexander, lives with his mother and four brothers and sisters in Sydney in a beautiful pond. The ducks head off for a walk one day, but Alexander (who is a bit of a dreamer) lags behind and then falls into a deep hole. With much quacking and excitement, several people try unsuccessfully to help Alexander out of the hole, until a little boy has the idea of tipping water from a nearby fountain into the hole, which continues to the point where Alexander rises to the surface. All the surrounding people join in, and Alexander is soon free to head back home with mother duck and the rest of the family. The teachers saw great potential in this story for mathematical investigations (in counting, ordinal number, capacity, estimation, informal units, and so on) and planned a range of activities for their students, to follow the reading and discussion of the story.

In a kindergarten class, the mathematical focus was on capacity. The children used different-sized containers to fill the jar that contained Alexander, to see how many scoops from each different container were needed to bring Alexander to the surface. The focus here was on informal units (related to Growth Point 3), with a particular focus on the concept of larger containers meaning that fewer scoops were required.

In one grade 1 class, the teacher provided plastic "holes" of different sizes for each group of children. A lit-

tle duck was placed at the bottom of each hole, with a magnet on his stomach. The children were asked to estimate and then check how many paper clips joined together would be needed to help Alexander rise to the top. (Notice the clear relationship between this task and helping children move toward length Growth Point 3.)

In another grade 1 class, the children completed the same activity with a Barrel of Monkeys game, predicting and checking how many linked monkeys were needed to fill each hole.

The teacher then brought the class together and encouraged the children to sort the different holes in a variety of different ways, from smallest to largest depending on the attribute of interest (see Growth Point 2). The problem was then posed about how many monkeys would be needed if all the holes were stacked one on top of another, moving the mathematical focus to strategies for addition and subtraction.

These are just a few examples of worthwhile and exciting mathematical experiences that ENRP teachers created, with a clear mathematical focus and the learning needs of their children in mind.

TEACHERS' STATED PROFESSIONAL GROWTH

During the second year of the project and following the opportunity for teachers to gain experience using the framework and one-to-one interview, teachers were asked to identify changes in their teaching practice, if any. There were several common themes:

- More focused teaching (in relation to growth points)
- Greater use of open-ended questions (in response to the wide range of understandings evident in students' responses to interview tasks)
- Giving children more time to explore concepts
- Providing more chance for children to share strategies used in solving problems

- Offering greater challenges to children, as a consequence of higher expectations
- Greater emphasis on "pulling it together" at the end of a lesson
- More emphasis on links and connections among mathematical ideas and between classroom mathematics and real-life mathematics
- Less emphasis on formal recording and algorithms, allowing a variety of recording styles

Teachers were also asked to comment on aspects of children's growth that they had observed that were not necessarily reflected in movement through the growth points. Common themes were the following:

- Children were better at explaining their reasoning and strategies.
- Children enjoyed mathematics more, looked forward to mathematics time, and expected to be challenged.
- Children showed greater overall persistence on difficult tasks.
- Children were thinking more about what they have learned and are learning.
- All children were experiencing a level of success.

Further information on the professional growth of teachers within this project can be found in Clarke (2001).

CONCLUSION

We have learned many things through the ENRP. The insights gained from one-on-one interviews in mathematics supply important information on what children know and can do and an excellent basis for teachers' planning. One teacher said, "The assessment interview has given focus to my teaching. Constantly at the back of my mind I have the growth points there, and I have a clear idea of where I'm heading and can match activities to the needs of the children. But I also try to make it challenging enough to make them stretch."

The growth points provided a mapping of the mathematics territory for teachers, something they could carry in their head, to structure what they see and hear as they work with children. Teachers found that they were doing more incidental mathematics as it arose in the classroom. For example, reading a story about a six-meter, baby blue whale, a grade 1 teacher quickly asked several children to walk from the chalkboard to a point they believed was six meters away. One child then checked their estimates with a trundle wheel to give a clear image of the length of the whale. The use of storybooks and hands-on activities became more common over the course of the project.

One powerful teaching strategy in meeting the needs of individuals was the use of open questions, where children could respond to the task at their own level, in their own way. One example is the following: "Belinda and Sandra measured the basketball court with metersticks. Belinda said it was 20; Sandra said it was 19 1/2. Why might that be?" This task offers great insight into children's understanding of important measurement principles. One common response from children is that one stick was longer than the other—but who had the longer stick?

Having high expectations of all children and encouraging a feeling that we can all do mathematics were very important and appeared to have very positive effects. Young children continue to excite us with their enthusiasm and their creative strategies for solving problems. Teachers in grades K–2 inspire us with their accumulated wisdom, their openness to new ideas and approaches, and their intense desire for all children to develop their mathematical thinking.

REFERENCES

Allen, Pamela. *Alexander's Outing.* New York: Penguin Putman, 1994.

Brown, Margaret, Ezra Blondel, Shirley Simon, and Paul Black. "Progression in Measuring." *Research Papers in Education* 10 (June 1995): 143–70.

Clarke, Doug M. "Understanding, Assessing and Developing Young Children's Mathematical Thinking: Research as a Powerful Tool for Professional Growth." In *Numeracy and Beyond,* Proceedings of the 24th Annual Conference of the Mathematics Education Research Group of Australasia (MERGA), edited by Janette Bobis, Bob Perry, and Michael Mitchelmore, Vol. 1, pp. 9–26. Sydney: MERGA, 2001.

Clements, Douglas H., Sudha Swaminathan, Mary Ann Z. Hannibal, and Julie Sarama. "Young Children's Conceptions of Space." *Journal for Research in Mathematics Education* 30 (March 1999): 192–212.

Dickson, Linda, Margaret Brown, and Olwen Gibson. *Children Learning Mathematics: A Teacher's Guide to Recent Research.* London: Holt Rinehart Winston, 1984.

Pengelly, Helen, and Leanne Rankin. *Linear Measurement: Children's Developing Thoughts.* Adelaide, South Australia: Department of Education, 1995.

Sullivan, Peter, Jill Cheeseman, Barbara Clarke, Doug Clarke, Ann Gervasoni, Donna Gronn, Marj Horne, Andrea McDonough, and Pam Montgomery. "Using Learning Growth Points to Help Structure Numeracy Teaching." *Australian Primary Mathematics Classroom* 5 (January 2000): 4–8.

Wilson, Patricia S., and Ruth Rowland. "Teaching Measurement." In *Research Ideas for the Classroom: Early Childhood Mathematics*, NCTM Research Interpretation Project, edited by Robert J. Jensen, pp. 171–94. New York: Macmillan, 1993.

7

Count Me into Measurement

A Program for the Early Elementary School

Lynne Outhred
Michael Mitchelmore
Diane McPhail
Peter Gould

What young children learn about length, area, and volume can form a strong foundation for much of their later mathematics learning. We organized a research-based program, Count Me into Measurement, designed to teach measurement of these spatial attributes in the first three years of formal schooling. The program was recently implemented in a sample of New South Wales (NSW) elementary schools. For the sake of brevity, we shall refer to length, area, and volume as *spatial attributes* and their measurement as *spatial measurement.*

Recent Research on Spatial Measurement

In the first years of school, measurement is usually introduced as comparison (e.g., longer or shorter than), then developed and extended in subsequent years to more complex measurement concepts. These early concepts include an understanding of the spatial attribute itself and the use of informal units to measure and compare. Such concepts are crucial, since they provide the basis for the estimation and measurement skills taught in late primary and early secondary school.

Length concepts are particularly important. If students do not have a sound understanding of linear measurement, they are unlikely to succeed in measuring areas and volumes. In addition, spatial measurement supplies the context for much number work (e.g., multiplication and division). Students also need to learn that measurement differs from number concepts in fundamental ways—especially regarding the subdivision of continuous quantities into units.

Although spatial measurement is a major component of the elementary school mathematics curriculum, there is a considerable body of evidence showing that many secondary school students do not have a thorough knowledge of the relevant concepts (Carpenter et al. 1988; Hart 1989; Schwartz 1995; Wilson and Rowland 1993). For example, the results of the most recent international comparative study of mathematics achievement indicate that many Australian secondary school students still have problems with linear measurement (Lokan, Ford, and Greenwood 1996). Commonly reported errors in these studies involve inadequate understanding of the attribute being measured and of the units of measurement.

Several recent studies have focused on the role of the unit in spatial measurement. These studies are not concerned with unit conversion (e.g., "change 2.5 km to meters") but a far more fundamental aspect: students' awareness of the principles that units must be congruent and must fit together without gaps or overlap. These principles determine what we shall call the *unit structure*, that is, the way the units fit together in the process of spatial measurement. Separate studies of length, area, and volume measurement have come up with very similar findings, as we shall now demonstrate.

Length

In a recent study of the understanding of length (Bragg and Outhred 2000a, 2000b), students aged six to ten years were found to follow mainly procedural strategies when using a ruler to measure length. They tended to count units, unit marks, or unit spaces, often making errors by counting marks instead of spaces or by aligning the ruler incorrectly. Even those students who correctly used these strategies did not seem to understand linear measurement in such a way that they could generalize the procedures to practical problems. An explanation for students' reliance on procedures might be that teachers rely largely on worksheets and textbook exercises. Such an emphasis on techniques does not develop more abstract concepts, such as a knowledge of how a scale is constructed. A teaching emphasis on counting units may obscure the linear nature of the unit of measure, since a counting procedure is not easily connected with partitioning length into units. To develop such conceptual knowledge would seem to require practical experiences of marking out linear units rather than simply counting given units.

Area

Students will not be able to subdivide a region into equal parts if they cannot visualize the unit structure. Young children may understand the unit structure for length quite easily, but the unit structure for area—a rectangular array (tessellation) of squares—is much more difficult to understand (Battista et al. 1998; Outhred 1993; Outhred and Mitchelmore 2000). Students need to learn that a rectangle can be partitioned into rows and columns and that there are equal numbers of units in each row and in each column.

Manipulative materials are popular in elementary school. However, the use of rigid materials, such as wooden or cardboard tiles, to teach area may actually mask the unit structure (Doig, Cheeseman, and Lindsay 1995). When students make arrays with such materials, they do not have to attend to the unit structure. However, when they fill a region by repeatedly drawing a single unit, they must solve the problem of how to fit the units together. The leap from making an array using concrete materials to representing it pictorially may thus be a very large one.

Students' first step in learning the unit structure for area seems to be the realization that each row (and column) has the same number of units. They then use a row or column of units as a composite unit, repeating rows or columns instead of drawing each individual unit. Students who have constructed rows and columns as composite units generally represent them using lines. Outhred (1993) found that only students who drew the unit structure using lines were able to find the area of a rectangle from the lengths of its sides. Thus, understanding the unit structure is fundamental to the understanding of area measurement and, in particular, of the link to multiplication.

However, Outhred and McPhail (2000) found that teachers seem to conceive of area measurement as a process of covering and counting rather than subdividing a region. In their descriptions of area measurement, the teachers did not mention any structural features of a covering (e.g., gaps or overlaps, congruent units, rows and columns). An emphasis on counting may be detrimental when students move away from concrete materials to more abstract representations that require the knowledge of how to subdivide a region. Moreover, counting units might suggest that area is a set of discrete points rather than a region. Teachers' conception of area measurement as a counting procedure appears to parallel their conception of linear measurement and may have similar consequences, that is, the area units may not be perceived as a subdivision of a region.

Volume

Research on students' understanding of volume has shown the difficulty of abstracting the structure of a three-dimensional packing of cubes in a cuboid (Battista and Clements 1996; Hart 1989). Volume measurement seems to involve the same process of constructing composite units (in this instance, layers or sections), but the process is more complex because students have to coordinate three dimensions and diagrams cannot show the layer structure clearly. Campbell, Watson, and Collis (1990) found that elementary school students tend to attempt to count the number of individual cubes in diagrams of regular rectangular prisms and that many of them count only the visible cubes. Once again, we see an unhealthy emphasis on counting when attention to the unit structure might be more productive.

Early experiences of volume are often limited to filling containers. (In this instance, volume is often referred to as capacity.) Such an approach completely avoids the structural problems of packing. The two situations appear to be very different, and students may not see that both involve the measurement of volume. Filling may, in fact, appear more similar to length measurement: the units are easily countable, and many containers for measuring liquids (such as beakers and jugs) are marked with a linear scale.

Teaching Early Measurement

Our interpretation of the literature summarized above led us to the conclusion that it was important, in the early stages of teaching spatial measurement, to emphasize unit structure. A search of popular textbooks and curriculum documents suggested that this step is usually omitted or treated only in passing. We were therefore faced with three questions:

- Could we design a spatial measurement curriculum for the first three years of school, which would include an emphasis on unit structure?
- Would teachers be able to teach this curriculum?
- Would the curriculum be effective in improving students' understanding of spatial measurement?

THE COUNT ME INTO MEASUREMENT PROGRAM

A numeracy program, Count Me In Too, was introduced into NSW schools in 1996 with the aim of developing teachers' knowledge of how students learn early number concepts. The program includes a "learning framework in number" that can be used in both teaching and assessment, since it provides a means of documenting students' progress.

In 1999, the program was extended to early measurement with a focus on students of five to seven years of age. Outhred and McPhail developed the Count Me into Measurement (CMIM) program, comprising a learning framework in measurement, accompanied by small-group tasks in which teaching and assessment could be linked. A principal aim of CMIM was to improve teachers' knowledge of general measurement principles by emphasizing similarities across the different spatial attributes (length, area, and volume). In addition, the measurement tasks were designed to be practical and to develop students' awareness of the approximate nature of measurement (Outhred and McPhail 2000).

The Learning Framework in Measurement

We sought to construct a conceptual framework for the early teaching and learning of spatial measurement (that is, before the introduction of formal units), which would include similar stages for length, area, and volume. One stage would focus on unit structure. Our reading of the research literature led us to conclude that this stage should be preceded by two other stages: one in which students identify the spatial attribute and one in which they learn to measure it informally.

- **Stage 1: The identification of the attribute** includes directly comparing quantities, partitioning quantities, and realizing that a quantity is unchanged if it is rearranged (the principle of conservation).
- **Stage 2: Informal measurement** includes finding the number of units to cover, pack, or fill a given quantity without overlapping or leaving gaps; knowing that the number of units used gives a measurement of quantity; and using these measurements to compare quantities.
- **Stage 3: Unit structure** includes replicating a single unit to cover, pack, or fill a given quantity, either by drawing or visualizing the unit structure; and realizing that the larger the unit, the fewer will be needed.

The resulting learning framework in measurement is shown in table 7.1, with each stage divided into two substages. It is conceived as three separate strands developing in a similar way, not as a single, general developmental path. Thus, students are not expected to be at the same stage across the three strands. Many aspects of area and volume measurement depend on a knowledge of length; and although it is difficult to conceive that volume (packing) might be taught before area, volume (filling) does not depend on area.

The framework is deliberately designed to illustrate the similarity of the teaching sequence for each spatial attribute. Once teachers understand this framework, they are less likely to omit crucial stages in students' development and more likely to teach general measurement concepts that might generalize to the measurement of other attributes (mass, time, temperature, and angle) later. At the same time, teachers are likely to differentiate the attribute and the units more clearly in each strand. For example, the common practice of using small, wooden cubes to measure length, area, and volume can be very confusing to students (Outhred and McPhail 2000; Bragg, personal communication).

TABLE 7.1

The Learning Framework in Measurement

	Length	Area	Volume
The identification of the attribute			
Stage 1.1	Make direct comparisons of lengths.	Make direct comparisons of areas.	Make direct comparisons of volumes (pack or fill containers).
Stage 1.2	Order lengths by direct comparison.	Order areas by direct comparison (e.g., by superimposing).	Order volumes by direct comparison (e.g., by packing or filling).
Informal measurement			
Stage 2.1	Find how many identical units fit along a line.	Find how many identical units cover an area.	Find how many identical units pack or fill a container.
Stage 2.2	Use numbers of units to compare lengths.	Use numbers of units to compare areas.	Use numbers of units to compare volumes.
Unit structure			
Stage 3.1	Replicate a given unit to fit along a line.	Replicate a square unit to cover rectangular areas.	Replicate a unit to pack or fill rectangular containers.
Level 3.2	Relate unit size to the number of units used to measure length.	Relate unit size to the number of units used to measure area.	Relate unit size to the number of units used to measure volume.

The CMIM Teaching Materials

Each CMIM teacher is supplied with a single curriculum document that includes, in addition to an introduction to the learning framework in measurement, a range of activities and sample lesson plans. For the six substages in each of the three strands (length, area, and volume), constituent knowledge and skills are listed and a variety of small-group activities are suggested (see table 7.2). Table 7.2 illustrates the format of the knowledge and strategies and activities for Length 2.1. The knowledge and strategies listed for each strand and substage will not always apply to all the activities listed.

Expanded lesson notes (see table 7.3) are provided for one activity at each substage in each strand. As well as showing the steps in a lesson, these notes illustrate questioning techniques that teachers could use to develop measurement concepts. In many of these lessons, students are required to record their findings and report them to the whole class. In the expanded lesson notes in the CMIM document both the knowledge and strategies and the resources are included so that teachers need only to refer to the one page. These aspects have been omitted from table 7.3.

The first two stages, the identification of the spatial attribute and informal measurement, do not differ markedly from the current teaching guidelines (New South Wales Department of Education 1989), except that the importance of emphasizing the attribute being measured is stressed. However, stage 3, unit structure, is new to NSW teachers. The crucial stage 3 focuses on an understanding of the structure of the repeated units, and in particular, that there must be no gaps or overlaps. The knowledge of row and column structure is seen as fundamental to an understanding of area and volume measurement. The stage 3 activities all involve regular shapes, for two reasons: first, to assist students' understanding of unit structure, and second, to avoid the added complexity of fractional units.

The use of appropriate concrete materials is emphasized. For example, rigid tiles (wooden or cardboard) can be used in stages 1 and 2, but the stage 3 activities require students to use drawn or paper units so that the units cannot simply be fitted together or traced.

Table 7.4 shows the activities given for the four sample lessons in stage 3.1. In most of these activities, students have only one unit available. In both the length and the area activities, the teacher is expected to discuss with students how the units fit together. The volume (filling) activity illustrates the similarity between filling and measuring length. The volume (packing) activity is far more complex, since the layer structure is hidden from view; also, we have not found a way of structuring this activity using a single unit.

The teacher's guide gives only rough guidelines for the most appropriate grade level for each activity. Teachers are expected to assess a sample of students in their class and use the results to establish a suitable level for teaching. The activities are designed so that teachers can establish small groups at

Table 7.2

Length 2.1 Teaching Activities: Informal Measurement

Knowledge and Strategies (K & S)

1. Use identical units to measure a length.
2. Align units end-to-end along the given line without overlapping or leaving gaps.
3. State or record that the length is the number and type of units used.
4. Use approximate language to explain parts of units (e.g., "a bit left over").
5. Measure a circumference using string or paper strips, without overlapping ends.

Lesson Notes	Resources	K & S
Choose my unit		
Students choose from a bucket of different units the ones they will use to measure a line (this might be taped on the floor). It is essential that students choose and use a set of identical length units. Record which units were used and how the line was measured.	a selection of units: paper clips, straws, Popsicle sticks, and so on (multiple copies of each unit)	1 2 3 4
Making lengths		
Make a length the same as one made by the teacher using units (e.g., Popsicle sticks, paper clips) and glued onto cardboard. Students can place their units— • in a straight line; • not in a straight line (e.g., in a curved or zigzag line). Students compare different ways of making the same length.	a selection of units: paper clips, straws, Popsicle sticks, matchsticks, and so on	1 2 4
Alternatives		
Use different units to measure the same length, for example, "My desk is 6 straws or 9 Popsicle sticks wide." Record which units were used and how the length was measured.	a selection of units: matchsticks, straws, Popsicle sticks, and so on	1 2 3 4
Who has the biggest head? (See expanded lesson notes.)		
Students measure around their head with string and mark correctly without overlap. Measure the length of the string in units (rods, straws, and so on) to find who has the biggest head in his or her group. Record group measurements and the units used.	string, scissors; units: rods, straws, matchsticks, and so on	1 2 3 4 5

TABLE 7.3

Length 2.1 Expanded Lesson Notes for "Who has the biggest head?"

Students measure around their head with string or paper strips and mark correctly without overlap. Measure the length in units (rods, straws, and so on) to find who has the biggest head in the group. Record group measurements and the units used. This activity may be extended by asking students to work in groups of four and listing the four head measurements in order of size for a group report. All group members will need to use the same unit of measure.

Activity	**Questioning/Comments/Discussion**
Step 1: Whole-class discussion	
• Discuss the measurement of length and the terminology used.	What are we measuring when we use the words *long, short, longer than, shorter, the same length, shorter than?*
• Discuss how to measure head size. Students may suggest and demonstrate some alternative methods, which can be discussed by the class.	How could we measure your heads? What advice can you give to someone who is measuring with units?
• Teacher models how to measure head circumference with string and emphasizes the skills of placing and counting the measuring units (no gaps, no overlaps).	How can we record the head measurements?
• Discuss what to do with fractional units.	
• Display the names of the units for students to use in recording.	
Step 2: Work in pairs or small groups	
• Students help one another to measure around heads.	This may be an opportunity for individual assessment to check that students have—
• Students choose, align, count, and record the number of units used to measure.	• chosen one kind of unit; • aligned and counted units correctly.
Step 3: Whole-class discussion	
Report back to a small group, the teacher, or the whole class.	Which units were good to use, and why? Can we compare head sizes if Casey used blocks for units and Tim used matchsticks?

TABLE 7.4

Sample CMIM Activities at Stage 3.1

Length: Make a ruler	**Area: Which is bigger?**
Students make their own cardboard rulers based on an object (Popsicle stick, paper clip) or a body part (foot, hand span). When an object (foot or hand) is used, students should cut a ribbon or string the same length as the object to use as the unit when marking the scale. Student use their ruler to measure classroom objects.	Given just one unit, students compare the size of two rectangles outlined on the floor (or made of cardboard and fixed to students' desks). The rectangles should not be movable, or the students may simply superimpose them.
Volume (filling): Let's estimate	**Volume (packing): How many will fit?**
Students predict which container will hold 2 cups, 5 cups, and so on. They estimate how many cups it will take to fill a container, when the level for one cup is marked on its side. Students can nominate upper and lower limits (cupfuls) for several containers and justify their measures. They record these limits on a number line, then measure and check their estimates.	Given a few identical blocks (cubes), students work out how many would be needed to fill rectangular containers (e.g., their lunchboxes). How many layers? How many blocks in each layer? Students may also make boxes (e.g., $2 \times 3 \times 3$; $1 \times 3 \times 3$) from nets drawn on grid paper.

different levels in their classroom. Each activity is keyed to a list of knowledge and skills for that substage, thus enabling teachers to assess students during small-group teaching and take whatever corrective action is necessary. Students work through a variety of activities in each substage, moving on to the next substage when the teacher thinks they understand the underlying concepts.

The Implementation of CMIM

The CMIM program was first implemented in a sample of NSW schools in the second half of 2000. In each of the forty school districts, the mathematics consultant was asked to identify one school and one teacher (called a facilitator) in that school who was willing to implement the program. The facilitators, who had no special expertise in mathematics, were required to attend one training day during which they were introduced to the learning framework in measurement and worked through some of the suggested activities.

Each facilitator then worked with a minimum of three other teachers across kindergarten to grade 2 in her school. The program was also implemented with some grade 3 classes.

Each of the CMIM teachers (including the facilitators) taught two of the three strands for five lessons each. The facilitator met with her teachers regularly to discuss students' learning and to plan subsequent lessons. All facilitators and teachers were asked to assess five of their students (one of low ability, three of middle ability, and one of high ability) before and after teaching each strand, as well as to complete a questionnaire on the program and its effectiveness. In all, 154 teachers were involved.

Teachers' Responses to CMIM

Teachers were overwhelmingly positive about CMIM. They reported that the learning framework in measurement had helped them recognize links among topics that may not have been apparent previously, and several aspects of the program appeared to have helped teachers reflect on the way they taught measurement or clarified their own measurement concepts. Typical comments were, "Yes, it gave measurement a sequence—something that I didn't understand" and "I taught concepts I'm not confident in teaching. [It] made me use the correct vocabulary." One facilitator wrote, "It made the class teacher reflect on the present style of teaching concepts and gave them practical examples of how to achieve new concepts in a motivational way." Another wrote, "We have all learned some valuable lessons from this project, e.g., the value of hands-on teaching, the value of recording to organize and consolidate learning, and the need for concise use of language in measurement."

As we suspected, many teachers seemed to have previously used only textbook examples when teaching measurement. A facilitator commented, "At our school we have been using a textbook and working from that. Now I can see the gaps in our textbook and how it just covers an area without teaching the strategies and skills involved. I have found it vital to use a variety of materials in 'hands-on' experiences with the students. I have valued the recording process which is more often than not left out of the textbook situation." Many teachers did not have sufficient resources for practical work, but they obtained these in a variety of ways—by asking parents or staff to collect resources, by putting together a shared box of resources, or by making their own.

Many teachers mentioned the importance of language and commented that the emphasis on explicit language, combined with questioning and discussion, assisted students in understanding the concepts. They did not seem to have been aware of the importance of mathematical language, perhaps because some measurement terms (e.g., *length, area, row,* and *column*) are commonly used in everyday life. Several teachers were surprised that the students did not know the mathematical language associated with measurement; one commented, "The language or lack of it was an eye opener and emphasized the necessity of doing these lessons."

The CMIM activities often require students to write about mathematics. Although asking students to write or draw explanations of their mathematical thinking appeared new to many teachers, many of them commented on its value. One teacher wrote, "I especially liked the idea of students recording what they had learned so (1) they thought about it and (2) I was aware of their strengths and needs." Another wrote, "It made me realize that I take some things for granted. For example, I may have assumed that a particular child knew something that they didn't yet, or I may have underestimated some students' abilities and been surprised with their response or interpretation." Such remarks show that teachers were able to use the framework to assess their students' progress, although some teachers pointed out that it was an onerous task to assess students who were not yet writing independently.

Students' Responses to CMIM

Several teachers sent in samples of their students' work. The texts of three seven-year-old students' work are shown in figure 7.1. They show how recording explanations enables students to indicate their understanding of unit structure for area.

Fraser appears to have a clear idea of rows and columns as composite units. He estimated that each row comprises 9 squares and that there are 4 rows before he covered the rectangle with squares. However, it is not clear from the work sample exactly how he obtained his solution of 32 squares, and his drawing (not shown) is accurate only in its rows—he drew a 4-by-12 array.

Hayley has discovered a profound relationship—that perimeter and area are not related—although she cannot yet explain why every shape with a constant perimeter (12 matchsticks) does not have an area of 8 squares. If the teacher had limited students' experimentation to rectangles, Hayley might have perceived there is a relation between the tessellation of squares and the side lengths of a rectangle. This activity was a teacher's modification of a stage 2 activity, making different rectangles with the same area. If the teacher had not supplied the tiles but had the students imagine the array, the activity would have been suitable for stage 3.

Imogen was asked to find how many squares would cover two rectangles that were drawn on a page. She traced units along two adjacent sides of each rectangle (5-by-2 and 4-by-3) to find how many units would be needed to cover each one. She stated that one has an area of 10 square units, the other 12 square units. Imogen's work sample shows that she can visualize the unit structure in this context because she does not need to draw all the squares.

Fraser

I estimate that 35 squares will fill the Surface area of the rectangle because I looked at the square and Said I think 9 Squares will go across and I think 4 rows down and 9 x 4 = 35. I found out that 32 squares cover the Surface area of the rectangle.

Name Hayley

Stick constructions

We used 12 matches to make a rectangle and 8 tiles covered the area. We were asked to change the shape. We expected to use 8 tiles no matter what the Shape. We found that by making irregular Shapes we always used less than 8 tiles to cover the area. why because the Shapes are different

6

5

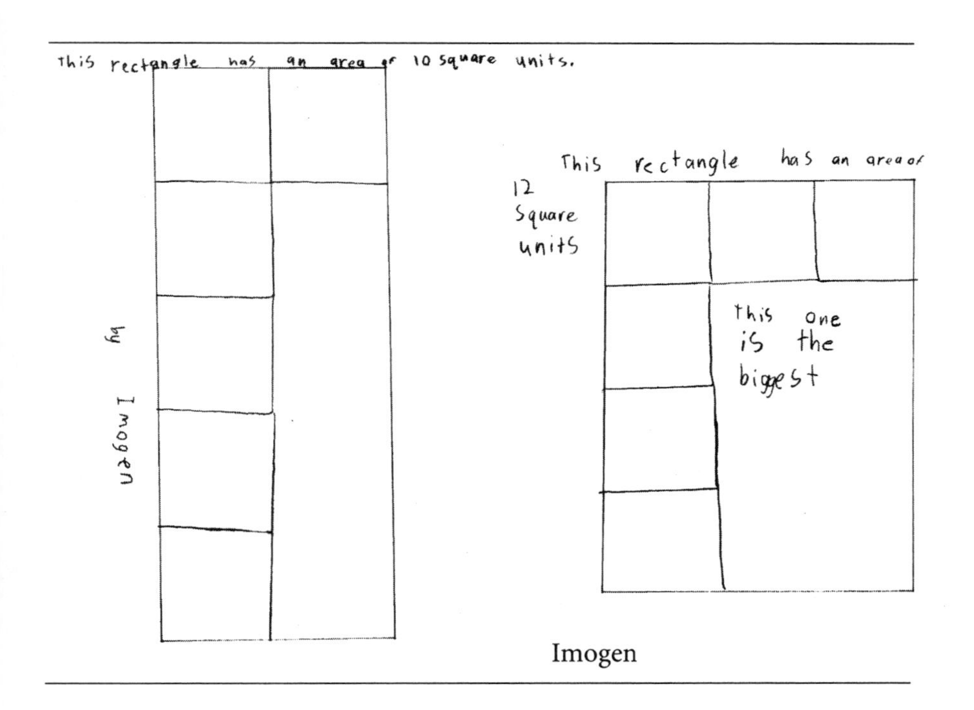

Fig. 7.1. Samples of students' work on the unit structure for area

The distinction between visualizing the squares and tracing them all is an important one. Students may successfully trace a rigid rectangular unit, such as one made from cardboard or plastic, yet they may not have an understanding of array structure because the materials structure the array that they draw. When students cannot trace the unit—for example, they are asked to use a paper square or copy a picture of a unit—they can complete the task successfully only if they have some knowledge of array structure (Outhred 1993).

Figure 7.2 shows some students' drawings of the unit structure for the volume of a rectangular prism, made after they had packed containers with blocks. A grade 3 student, Jackson, drew figure 7.2a. Figures 7.2b and 7.2c were drawn by grade 1 students, and figure 7.2d by a grade 2 student. The drawings indicate that these young students had a good knowledge of the structural features of volume. The students in grades 1 and 2 had previously been involved in a sequence of five lessons on the structural features of an area array, giving them a basis on which to build volume concepts.

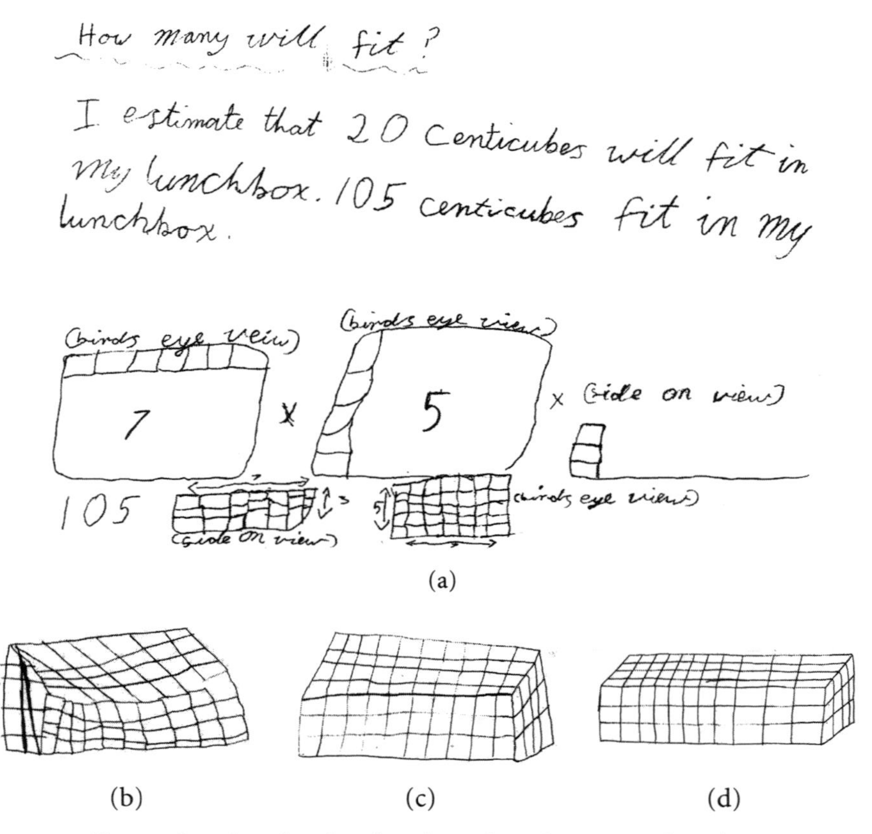

Fig. 7.2. Samples of students' work on the unit structure for volume

Jackson's drawing in figure 7.2a is particularly interesting. It shows that he had a good grasp of the structure of packing a rectangular container with cubes, since the view has been drawn from each perspective (volume stage 3.1). He has clearly shown a link between the number of cubes that fit along each side and multiplication to find the number of cubes that will fit in his lunchbox. Another student, Jake—his work is not shown—said that his box "held 144 blocks, and I had 4 rows and 6 squares and I had 4 layers." Jake's explanation suggests that he has grasped the idea of layers but is having trouble calculating the total number of units.

Students' Learning in CMIM

A student recording sheet, linked directly to the learning framework in measurement, was developed for each strand. An example is shown in figure 7.3. Teachers wrote the names of their five target students in the boxes at the top, entered their assessments (on a 3-point scale) before (marked under "1") and after (marked under "2") teaching the five lessons from the strand they had chosen.

Student Recording Sheet — Length

School:________________________ Class: ______

Assessment 1 (A1) Date: ______ Assessment 2 (A2) Date: ______

Length: knowledge and strategies								
Indicator: achieved =2; partly achieved = 1; not achieved = 0	1	2	1	2	1	2	1	2
L1.1 Use length vocabulary (e.g., *long, high, tall; short, low;* the same)								
L1.1 Put two lengths side-by-side to find if they are the same								
L1.1 Straighten a curved or bent length to check if two lengths are the same								
L1.2 Use comparative language, (e.g., *longer than, higher than, shorter than, the same as, …*)								
L1.2 Align ends for comparison (establish a baseline)								
L1.2 Compare lengths systematically and explain why a length fits into a particular ordering								

Fig. 7.3. Student recording sheet for the Length strand

In general, this form of assessment was seen as both useful and practical. Teachers liked the detailed information given by assessments of individual students even though these were time-consuming to do. Providing individual profiles of students' achievement in measurement also seemed to fit in well with methods of reporting to parents. Several teachers also commented on the value of listening to and observing students instead of simply marking textbook work.

To obtain some idea of how effective CMIM had been in promoting students' learning, all teachers' assessments were collected and analyzed. Not all assessments could be included, since many teachers did not assess common tasks at the beginning and end of the strand. The number of students included in the analysis varied from 800 for Length stage 1.2 down to 40 for Length stage 3.2. Also, more, younger students were included in the stage 1 and 2 results than were in stage 3.

For the Length strand, the results show an extremely large change in students' achievement. The only stage that more than 50 percent of students had mastered at the beginning of the teaching was Length 1.1 (Identify objects of equal length). However, by the end of the teaching, more than 80 percent of students had achieved Length stage 1 and 2; that is, they could compare and order lengths and also use informal units to measure and compare lengths. The results for Length 3.1 and 3.2 also improved dramatically—according to the teachers about 50 percent of the students assessed had achieved these stages.

As expected, the results indicate that area concepts are more difficult than length concepts. At the beginning of the teaching, only about 20 percent of the students assessed had achieved the first two stages. After the CMIM lessons, the assessments again showed dramatically improved performance on all stages. About 60 percent to 80 percent of students had mastered the first two stages (identification and informal measurement of area). Even more encouraging, the teachers' results indicated that about 50 percent had mastered Area stage 3.1 and had developed the idea of structuring a covering by rows and columns, a prerequisite to understanding the formula for the area of a rectangle. One facilitator commented, "We have been most happily surprised with the outstanding learning of our students. We showed them the links between multiplication and area and even our kinder[garten] students developed understandings about rows and columns." However, only about 20 percent of the students assessed on Area stage 3.2 appeared to have an understanding of the relationship between unit size and the number of units required to cover an area.

The teachers' assessments show similar substantial learning in the volume strand. At the beginning of the teaching, students seemed to be more familiar with volume than area. Before beginning teaching, about 60 percent were considered to have mastered stage 1.1, identify volumes, and 40 percent had mastered stage 2.2, use numbers of units to compare volumes, compared with about 20 percent for stages 1.2 and 2.1. However, teachers may have assessed only Volume stage 1.1 and stage 2.2 using activities that involved capacity, that is, filling rather than packing. The marked improvement after the CMIM lessons is consistent with the Length and Area results. Overall, from 60 percent to 90 percent of the students had achieved the first two stages and Volume stage 3.2 by the end of the teaching. The exception, Volume stage 3.1 (about 30 percent of students), was difficult because students had to work out the structure of a three-dimensional packing of unit cubes, given only one cube.

Although these results appear very positive, they must be interpreted with extreme caution, since the teachers were very enthusiastic about the CMIM program and therefore may not have been totally objective in their assessments. They also may have used different activities to assess knowledge and strategies at each stage. It is also possible, as the results from the preassessments suggest, that the teachers had not taught many measurement lessons prior to the program (Outhred and McPhail 2000), so these results may be partly due to teachers spending more time teaching measurement than they had previously. Also, the stage 3 results may be particularly unstable because of the small numbers of students who were assessed. Nevertheless, the results seem to show that students improved markedly in their knowledge of spatial measurement. Considering that the teachers taught only five lessons in each strand, the students' progress appeared substantial.

CONCLUSION

The evaluation of the first implementation of CMIM shows that the answers to all three questions we posed at the beginning of the chapter are yes. It was possible to develop a spatial measurement curriculum that emphasized spatial structure, classroom teachers could implement it, and it was effective.

The provision of a detailed teachers' guide combined with the assistance of a teacher within the school appeared to have been effective in the short term in changing teachers' attitudes toward measurement and their ways of teaching it. Many teachers commented that they enjoyed teaching CMIM and that students enjoyed learning about measurement. It is hoped that their positive responses will result in continued use of CMIM when support is withdrawn. Moreover, the assessments of students attest to the effectiveness of the CMIM program in improving performance on measurement concepts. The improvement seemed to be due to a combination of factors: teachers were given a knowledge of the structure of measurement, teachers were required to consolidate measurement concepts by teaching a sequence of five lessons using small-group practical activities instead of textbook exercises, and students were required to record and explain their findings. Teachers particularly noted the improvement in students' mathematical language.

This project is still in its infancy. It offers a clearer structure for teachers than the current NSW syllabus and incorporates a step—the knowledge of unit structure—that research implies is fundamental to an understanding of spatial measurement. Nevertheless, at present there are insufficient data to show how well teachers promoted links across spatial attributes and whether an emphasis on common processes could have contributed to students' learning. The most exciting aspects of the project were the teachers' positive responses to teaching measurement and the students' progress in their knowledge of measurement concepts.

REFERENCES

Battista, Michael T., and Douglas H. Clements. "Students' Understanding of Three-Dimensional Rectangular Arrays of Cubes." *Journal for Research in Mathematics Education* 27 (May 1996): 258–92.

Battista, Michael T., Douglas H. Clements, Judy Arnoff, Kathryn Battista, and Caroline Van Auken Borrow. "Students' Spatial Structuring of 2D Arrays of Squares." *Journal for Research in Mathematics Education* 29 (November 1998): 503–32.

Bragg, Philippa, and Lynne Outhred. "What Is Taught versus What Is Learnt: The Case of Linear Measurement." In *Mathematics Education Beyond 2000, Proceedings of the 23rd Annual Conference of the Mathematics Education Research Group of Australasia, Fremantle, Australia*, edited by Jack Bana and Anne Chapman, Vol. 1, pp. 112–18. Sydney, Australia: MERGA, 2000a.

———. "Students' Knowledge of Length Units: Do They Know More than Rules about Rulers?" In *Proceedings of the 24th Annual Conference of the International Group for the Psychology of Mathematics Education*, edited by Tadeo Nakahara and Masataka Koyama, Vol. 2, pp. 97–104. Hiroshima, Japan: Program Committee, 2000b.

Campbell, Jennifer, Jane Watson, and Kevin Collis. "Volume Measurement and Intellectual Development." Paper presented at the 13th Annual Conference of the Mathematics Education Research Group of Australasia Conference, Hobart, Australia, July 1990.

Carpenter, Thomas P., Mary M. Lindquist, Catherine Brown, Vicki L. Kouba, Edward A. Silver, and Jane O. Swafford. "Results of the Fourth NAEP Assessment of Mathematics: Trends and Conclusions." *Arithmetic Teacher* 36 (December 1988): 38–41.

Doig, Brian, Jill Cheeseman, and John Lindsay. "The Medium Is the Message: Measuring Area with Different Media." In *Proceedings of the 18th Annual Conference of the Mathematics Education Research Group of Australasia*, edited by Bill Atweh and Steve Flavel, Vol. 1, pp. 229–40. Darwin, Australia: MERGA, 1995.

Hart, Kath. "Volume of a Cuboid." In *Children's Mathematical Frameworks 8–13: A Study of Classroom Teaching*, edited by Kath Hart, David Johnson, Margaret

Brown, Linda Dickson, and Rod Clarkson, pp. 126–50. London: NFER-Nelson, 1989.

Lokan, Jan, Phoebe Ford, and Lisa Greenwood. *Mathematics and Science on the Line: Australian Junior Secondary Students' Performanceæ—Third International Mathematics and Science Study (TIMSS).* Melbourne, Australia: Australian Council for Educational Research, 1996.

New South Wales Department of Education. *Mathematics K–6.* Sydney, Australia: New South Wales Department of Education, 1989.

Outhred, Lynne. "The Development in Young Children of Concepts of Rectangular Area Measurement." Ph.D. diss., Macquarie University, Australia, 1993.

Outhred, Lynne, and Diane McPhail. "A Framework for Teaching Early Measurement." In *Proceedings of the 23rd Annual Conference of the Mathematics Education Research Group of Australasia, Fremantle, Australia,* edited by Jack Bana and Anne Chapman, Vol. 2, pp. 487–94. Sydney, Australia: MERGA, 2000.

Outhred, Lynne, and Michael Mitchelmore. "Young Children's Intuitive Understanding of Area Measurement." *Journal for Research in Mathematics Education* 31 (March 2000): 144–67.

Schwartz, Sydney L. "Developing Power in Linear Measurement." *Teaching Children Mathematics* 1 (March 1995): 412–16.

Wilson, Patricia, and Ruth Rowland. "Teaching Measurement." In *Research Ideas for the Classroom: Early Childhood Mathematics,* edited by Robert J. Jensen, pp. 171–94. New York: Macmillan Publishing Co., 1993.

8

Developing an Understanding of Measurement in the Elementary Grades

Richard Lehrer

Linda Jaslow

Carmen Curtis

Measurement is a process that students in grades 3–5 use every day as they explore questions related to their school or home environment (National Council of Teachers of Mathematics [NCTM] 2000, p. 171). As this Standard suggests, developing an understanding of the mathematics of measure should originate in children's curiosity and everyday experience. Yet these pragmatic groundings must be accompanied by imaginative reach, so that children develop a *theory of measure* rather than simply collecting measures. A theory of measure develops from a grounding in contexts that highlight recognizable goals and functions but extends beyond these contexts to provide flexible adaptability to novel conditions of application (e.g., new circumstance of measure) and to serve as a foundation for future learning. This emphasis accords with our view of understanding as a path to generative and flexible learning, so that students' learning does not remain bound to the tools or situations that were instrumental in helping them develop understanding in the first place.

Our approach to designing measurement instruction for children is guided by establishing a productive tension between understanding and doing. On the one hand, we hope to help students understand "big ideas" about measurement that have repeatedly been underscored as important stepping stones in studies of children's learning. For example, units of measure may seem transparent to children because one can literally point to them on measurement devices like rulers, but understanding the nature and properties of units is the endpoint of a long process of learning (Piaget, Inhelder, and Szeminska 1960). On the other hand, insight about important qualities of the world is often generated by the practical activity of trying to measure them or by trying to design appropriate measuring tools. Moreover, practical measurement raises the prospect of error, which in turn leads to new investigations: How can the error be reduced? Does it have a structure?

To illustrate the interplay of practical activity and imaginative grasp for developing an understanding of measurement, we describe sequences of activities where children in the elementary grades explored measures of length, area, and volume. Each sequence is grounded in everyday experience but then progressively lifts away from the immediate context to push toward a theory of measure. Our focus is on students' learning about the measurement of spatial extent (length, area, and volume) because measure is a practical route to understanding spatial structure. For example, measuring an area requires visualizing a surface as a plane and structuring it as an array. Measurement is also a gateway to arithmetic. For example, rescaling a measure provides opportunities for learning about division. To illustrate, if a measure of a board is 5 m-units, and if each n-unit is half of an m-unit, the resulting measure is 10 n-units.

In the sections that follow, our primary aim is to establish grounds for children's development of theories of measure. By "development" we do not mean an expectation that particular capabilities or understandings are tightly associated with any particular age or grade. Rather, learning leads development, so that at any age, we provide children with repeated opportunities to explore an important idea in different contexts, each selected to highlight a distinctive, albeit a related, aspect of the idea. The result is a gradual emergence of an understanding of concepts, procedures, and their mutual relationships.

INVESTIGATING LENGTH

The measure of length provides an opportunity to explore conceptual foundations of measurement. Children's explorations of length can be staged within everyday experiences, like walking, and extended by involving them in the design of tools for measuring length. Children's investigations provide grounds for developing arithmetic understandings as well, especially of fractions and of operations with fractions, such as multiplication and division.

Conceptual Foundations

The NCTM Standards place appropriate emphasis on early investigations of length measure. Table 8.1 displays a set of interrelated, important ideas in measurement that should serve to orient these investigations. These ideas are roughly divided into two conceptual accomplishments: the construction of unit and the construction of scale. The research backing for each of these ideas is extensive (e.g., see the review by Lehrer [in press]).

TABLE 8.1.

Central Concepts in Linear Measurement

	Idea	Description
Conceptions of Unit	Iteration	A subdivision of a length is translated to obtain a measure.
	Identical unit	Each subdivision is identical.
	Tiling	Units fill the space.
	Partition	Units can be partitioned.
	Additivity	Measures are additive, so that a measure of 10 units can be thought of as a composition of 8 and 2, and so on.
Conceptions of Scale	Zero-point	Any point can serve as the origin or zero point on the scale.
	Precision	The choice of units in relation to the object determines the relative precision of the measure. All measurement is inherently approximate.

One of the important ideas about a unit of length measure is that the measure is obtained by first subdividing a length and then repeatedly translating or "iterating" this subdivision. Although both these characteristics of unit may seem obvious to all users of rulers, they often are not apparent to young children, even those who are proficient at using rulers (Lehrer, Jenkins, and Osana 1998). For example, figure 8.1 displays a facsimile of a "foot strip" tape measure designed by James, a student in Ms. Vollenweider's third-grade class. James indicated that the measure of the ruler's length was 4 because four footprint units fit on the tape. However, note the gaps between the units. Clearly, for James, some components of a concept of iterating a unit are salient (the units are all alike and they are sequenced). However, the process to be repeated appears to be a count rather than a measure.

In a similar vein, some children are stymied when they run out of units, so they claim they cannot measure something that is longer than the number of

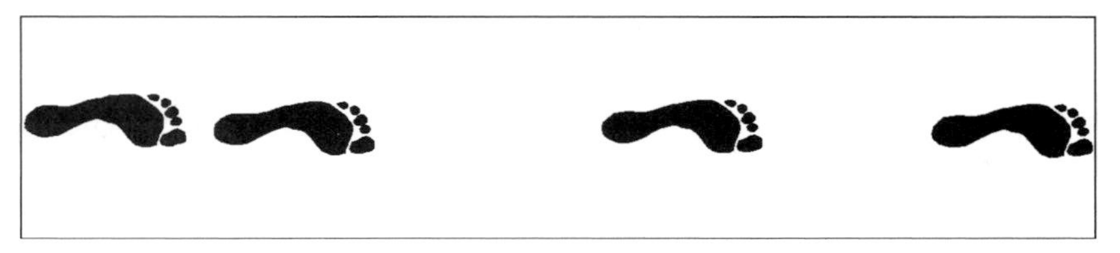

Fig. 8.1. A window to students' thinking about unit: a third grader's foot strip

units in their possession. They apparently do yet not consider that iterating a unit implies an infinitely repeatable process with an identical unit, like an inch, although they might understand that a unit is repeated in a way that fills the space. Some children solve the problem of exhausting units by supplementing the original unit (e.g., inches) with others that are handy on their rulers (e.g., centimeters). These partial understandings often constitute a resource for developing an appreciation of the roles played by iteration and identical units in a conventional theory of measure.

Length measurement also furnishes opportunities for investigating the conceptual foundations of scales of measurement. For instance, children often have difficulty conceiving of the origin of the scale as 0 (zero-point). This may be because scales blend notions of a rigid unit, like an inch, with motion (iteration) of that unit. Hence, it is often confusing for children to consider the beginning as zero when it seems obvious to them that the beginning is a movement of the first unit, or one. Moreover, any location on the scale can serve as the origin, so that, for example, a measure of 10 inches is obtained over any interval of ten (e.g., 0 to 10, or 10 to 20). Children often find the idea of an arbitrary starting point problematic, perhaps because understanding that distance is preserved by a translation requires mental coordination of the endpoints of the scale, not simply counting units.

Conceptual attainments like those just described are related, so an understanding of linear measure emerges gradually as children create a web of connections. For example, as children come to understand more about the nature of units, ideas like the zero-point of a scale make more sense. Moreover, as we have suggested, these ideas are not acquired in an all-or-none manner. A child may understand that the measure between 2 and 5 is 3, but yet may also believe that the measure between 2 and 5 1/2 is 4 1/2. In this instance, counting the number of intervals rather than the distances suggests 1 (2–3), 2(3–4), 3 (4–5), 4 (the interval between 5 and 5 1/2) and 1/2 of the fourth counted interval (see Lehrer et al. 1999).

In summary, the goal of our instruction is to help children develop a theory of measure, as well as practical knowledge of tools like rulers. Our rationale is straightforward: children who understand measure can invent or adapt their ideas to new situations.

Prototypical Investigations of Length

We describe a prototypical sequence of investigations of length measure. Our intention is to depict a potential pathway for developing an understanding of measurement that later can be profitably extended to other forms of measure.

Walking

We begin this investigation of length measure with the everyday activity of walking. Walking blends two commonplace metaphors of measure (Lakoff and Nunez 2000). The first is that of a measuring stick, with one's foot serving as the stick, a good grounding for the idea of measuring with identical units. The second metaphor is a movement in space, so that one translates the measurement stick, a good grounding for the idea of iterating these identical units. Children use their feet to measure a length, like a distance between two locations or the length of their classroom, and we begin by posing a question that focuses children on the apparent paradox that measurements conducted by different students produce results that are somewhat different: "If the wall is one length, why are the number of feet different?" (Children's feet vary in length. Different-sized feet yield different measures.) Questions like these help us find out how children think about properties of units, like identity; processes of measure, like iteration; and qualities of scale, like the origin (e.g., "What do I call the first step?").

Foot strips

After thorough investigation of measure as movement, children use paper cutouts of their feet to make foot-strip rulers. This activity serves the important cognitive function of helping children select, fix, and reflect about which aspects of walking should be embodied in their manufacture of a tool. Tools help children make the transition from movement to measurement and often provide the added bonus of revealing much about students' thinking, as illustrated earlier with James's foot strip.

Personal units

The third set of investigations employs collections of paper-strip units of different lengths, one for each child in the class (a "personal" unit). (We occasionally precede the exploration of personal units by investigating other nonstandard units, like books or paper clips, as measurement units.) Children use their strips to measure the lengths of different objects. The objects are chosen so that some are longer than the set of units and others involve fractional lengths. These investigations help us understand how children are thinking about iteration (Can they reuse units? Do they still leave spaces between units?), and parts of units (How do they think about units that are not whole numbers?).

Personal unit rulers

Children go on to design personal unit rulers. This design activity foregrounds thinking about a number of important ideas. These include "splitting" (Confrey and Smith 1995) units to obtain similar fractional units (e.g., 1/2 of 1/2), mastering the correspondence between number and measure (e.g., Which direction do the numbers go and what do they signify?), understanding scale (e.g., What happens if the measure starts at a place other than the end of the scale?), and appreciating the value of a standard (e.g., What kinds of units would best facilitate communication about lengths?).

Extensions to arithmetic

The learning sequence we have described so far can be elaborated at varying points. For example, partitioning lengths is a natural context for considering the multiplication of fractions (e.g., 1/2 of 1/2 of 1/2 of folded strips of paper) and corresponding symbolization (e.g., $1/2 \times 1 = 1/2$, $1/2 \times 1/2 = 1/4$, $1/2 \times 1/4 = 1/8$). Rescaling lengths sets the stage for division. For example, if a length is 5 a-units long, and a b-unit is half (or twice) an a-unit, what is the same length in b-units? We frequently choose to emphasize investigations like these because they provide an alternative entrée for the understanding of rational number. Multiplication can also be considered an iteration of a composite unit. For example, a 12 d-unit length can be reconsidered an iteration of composite units, where there are (perhaps) 3 d-units in every composite unit, so that the 12 d-units can be reconstituted as 4 iterations of the composite unit (4×3). Sometimes this unit-of-units approach makes more sense to children as a foundation for multiplication than other traditional approaches. The measurement of length in contexts of motion can illuminate the number line as well. For example, footsteps forward and back from the origin can correspond to positive and negative numbers, respectively. Thus, signed measures indicate distance and direction, serving as a platform to extend arithmetic operations to negative numbers (Thompson and Dreyfus 1988).

A Classroom Example

The NCTM Standards emphasize the importance of early investigation of length measure, and we have explored children's thinking extensively in the primary grades. (See Horvath and Lehrer 2000 and Lehrer et al. 1999 for examples of young children's development of ideas about linear measure in the early grades.) However, we illustrate investigations of length in an urban fifth grade to underscore an important point: If these investigations are not undertaken at an earlier age or grade, they should be undertaken later, rather than simply skipping over them. In the fifth-grade classroom where this work occurred, students' instruction in measurement in earlier grades focused tightly on using rulers. We highlight some of the thinking of these fifth-grade students to buttress our claim that development does not occur spontaneously with age.

The fifth graders first explored length measure by pacing and then by creating foot strips to represent their feet as units of measure. These beginning activities helped us understand children's thinking about qualities of units and scale. During their initial explorations, students asked: "If I am pacing, where do I start counting? Do I count each step? Do I count where my feet begin? When do I count each unit? What do I do with partial units? What do I call these partial units? How can I figure that out?" These questions revealed that, like younger students, the fifth graders found it problematic to coordinate the related ideas of movement and iterating fixed units as a measure of movement. There seemed to be a gap between counting and thinking of distance traversed, so that, for example, children had considerable difficulty reconciling the beginning of the path (0) with the first step (1).

One of the apparent paradoxes that students needed to resolve was how to reconcile continuous distance traveled with counts of discrete units (see Clements, Battista, and Sarama 1998). A popular strategy for creating a measuring tool with the foot ruler was to draw lines on the foot strips that mimicked a ruler, although the subdivisions were not identical. Only a few students could identify any part of a foot other than one-half, and one-half was used to refer to any partition into two subdivisions. One group's foot strip even featured different-sized feet.

One avenue toward bridging the gap between distance traveled and count was to involve students in constructing more finely articulated subdivisions of length. These subdivisions accentuated the continuous nature of distance. Creating subdivisions helped students develop more differentiated ideas of parts-of-units, rather than simply labeling all parts as "one-half." For example, as students were splitting a paper-strip length repeatedly by two (1/2 of 1, 1/2 of 1/2, and so on), their teacher, Ms. Griff, attempted to understand their thinking:

Mary: It is 4/16 and 1/2 of 1/16 units.

Ms. Griff: Okay, so 4… 4/16 + 1/2 of 1/16 units. So we have one, two, three, four sixteenths and half of a sixteenth… what is that piece called? What if we have sixteenths but we need to go further?

Carlos: Four sixteenths plus… [*short pause*] …1/32 units.

Ms. Griff: Where does that thirty-seconds come from?

Carlos: If you fold your units in half one more time, it will be thirty-seconds.

Ms. Griff: Why would I end up with thirty-seconds?

Carlos: Because 16 + 16 are 32….

Ms. Griff: So what am I doing when I fold it? I am doubling my sixteenths, right? So is there something else that I can call this instead of 4/16 + 1/32 units?

Ricardo: Five thirty-seconds.

Ms. Griff: Is 4/16 the same as 4/32?

Carlos: Nine thirty-seconds.

Ms. Griff: You think it is 9/32? So you think it should be 9/32 units? Where did you get the 9?

Andy: Because 2/32 = 1/16. [*Andy looks at the strip to verify.*]

Carlos: You have 4/16 and if you times it by 2, then you end up with 8/32.

Ms. Griff: So you are saying that 4/16 = 8/32, plus if you add 1/32, then you end up with 9/32?

In spite of the sophistication that the team developed about partitioning units, they continued to experience difficulty coordinating these parts of units into a scale. For example, Jessie was uncertain about what to call the halfway point of the scale: Should it be 16/32 or 17/32? (See fig. 8.2.)

Ms. Griff: What did you get?

Jessie: Seventeen thirty-seconds ... it's ...

Ms. Griff: How many thirty-seconds is that? [*Jessie tries to find how many thirty-seconds are equal to a half by folding her tape in half repeatedly. Then she counts the boxes.*] So you are going to fold it in half to figure out how many thirty-seconds are equal to 1/2?

Jessie: 16 1/2 thirty-seconds?

Jessie's responses suggest that she perceived the iteration of partial units as a count, not as a measure of distance, so her count began at one. She

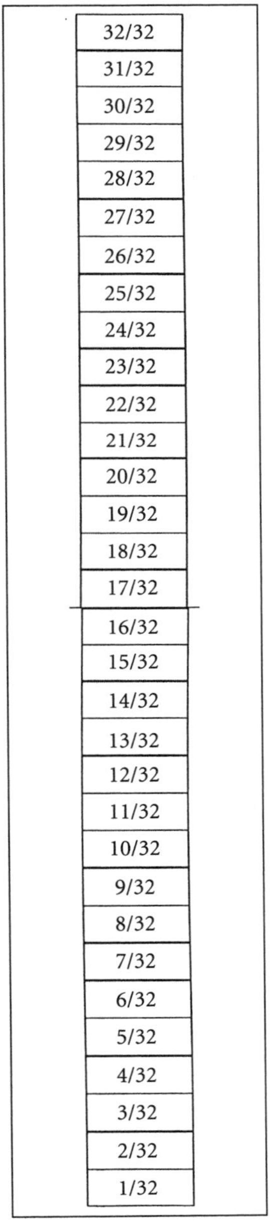

32/32
31/32
30/32
29/32
28/32
27/32
26/32
25/32
24/32
23/32
22/32
21/32
20/32
19/32
18/32
17/32
16/32
15/32
14/32
13/32
12/32
11/32
10/32
9/32
8/32
7/32
6/32
5/32
4/32
3/32
2/32
1/32

Fig. 8.2. Coordinating partitions of units with counts of units in a fifth-grade length activity

proposed splitting the difference between the sixteenth and seventeenth count to label the halfway point of her ruler as a reasonable compromise between adjacent counts. At this point, other members of her group chimed in:

Rachel: It should be 16, because that is half of 32.

Ms. Griff: Jessie, what are you counting, the lines or the boxes?

Jessie: The boxes…

Ms. Griff: So what is the answer?

Chris: It's 16.

Ms. Griff: It's 16? Why?

Chris: Because there are 16/32 on one side and 16/32 on the other side.

Ms. Griff: So this brings up that never-ending question: Where do you number your measuring tape? Do I label the boxes or the lines? Or does it not matter? Why?

As these examples illustrate, important ideas about units of measure need to be revisited repeatedly if children are to grasp their implications. Jessie's team had already successfully constructed a personal tape measure with distances represented by integer counts. Yet, when the context shifted to include parts of units, children reverted to counting discrete objects rather than iterating units of length measure.

Later, we explored the potential of walking a measure as a context for expanding conceptions of the number line. We asked children how they would think of "showing" or symbolizing walks both forward and backward from the origin. Students proposed that moving one step from the origin followed by four more could be symbolized with a diagram of the corresponding foot-steps. When Rich asked if the action could be shown without drawing the feet, Rosalie quickly suggested: 1 + 4 = 5. The class also agreed that moving backward three steps could be represented as 5 – 3 = 2. Rich then asked the class to consider taking one step back from the origin. Jamie's drawing is displayed in figure 8.3. Note that zero is represented twice. Rich asked if there were another way to show Jamie's thinking.

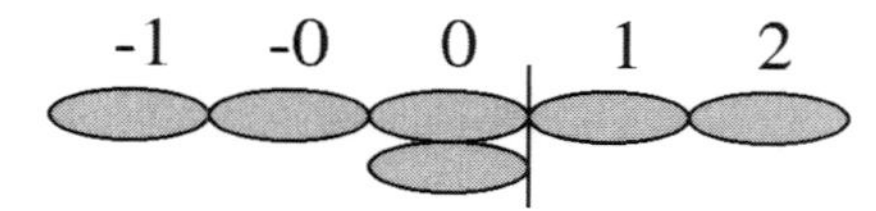

Fig. 8.3. Jamie's model of walking backward

Mark: –1, –0, 0, 1, 2.

Rich: What is the difference between 0 and –0?

Mark: Zero is me standing in the middle, and –0 is stepping back one.

Rich: I'm confused. How come the picture is not showing that?

Jamie: One side is going forward, and one side is going backward.

Ricardo: –2, –1, 1, 2. I took a step backward, and that was something, not nothing!

At this point, Jose reiterated Mark's position, and Ricardo responded by modifying his to incorporate zero: "–3, –2, –1, 0, 1, 2, 3." There was some further discussion of the relative merits of each point of view, and then Jamie exclaimed, pointing to a number line displayed on the far wall of the class: "Look at how they do it!" This discovery set the stage for coming to regard zero as the origin of the number line and introduced the concept of a directed quantity. As Jamie explained, "You step backwards, and that's –1!"

In summary, these illustrations from an urban, fifth-grade classroom suggest the importance of helping even older students "unpack" central ideas in linear measure. Although their previous instruction had emphasized the use of rulers, these students retained an image of a ruler without much understanding of its mathematical design. Our examples suggest further that linear measurement is a central route to better understanding of fractions and operations on them. Moreover, contexts like travel and its measure can embed related ideas, for example, symbolizing actions mathematically. Our examples also illustrate that developing an understanding of these ideas requires a history of learning and cannot be assumed to occur spontaneously as a result of participation in everyday activity, even though everyday activities serve as a good departure point.

INVESTIGATING AREA

Students' investigations of area extend conceptions of unit to the plane. This is an important step both for elaborating measurement concepts and for establishing concepts of arrays. Arrays figure prominently in most models of multiplication and division.

Conceptual Foundations

Central conceptions for the measurement of area parallel those shown in table 8.1, although students must constitute each of these ideas in two dimensions. One of the additional challenges of area measure is the conception of a two-dimensional region as an array composed of units of area measure, typically squares. This spatial restructuring of the plane is at the heart of its measure, since area measure associates with every plane figure a number indicating the ratio between its area and the area of a given partition of it (e.g., a square). Constructing the array is also the cornerstone of thinking about area as a product of lengths. Unfortunately, many children simply learn to multiply lengths to generate areas without understanding that these products are generating arrays (Battista et al. 1998).

Another challenge in measuring area is a tendency for children to prefer units of measure that resemble the surface being measured (Lehrer et al. 1998). For example, an assortment of beans is considered a better match to the contour of a hand than a collection of squares, so beans are preferred as units in spite of their limitations with respect to properties like identity and space-filling. Many children prefer to treat object boundaries as absolute, so that in their view, units of measure cannot in principle overlap the boundaries of the object being measured. Returning to the example of the contour of the hand, unit squares overlap the boundaries of the hand and thus are often rejected as appropriate tools for measure.

Prototypical Investigations of Area

Investigations of area again begin with everyday experience, using area measure as a tool for resolving the difference between what a shape looks like and how much space it covers. This investigation introduces children to partitioning an area, a conceptual precursor to the idea of an array. We proceed then to involve children in reconsidering concepts of unit in situations that they might not ordinarily consider, like rank ordering the areas of their handprints. We conclude with revisiting area measure by structuring an array.

Reallotment

We initiate an investigation of area by formulating a comparison among three or more different-looking rectangles, asking students which covers the most [surface]. In spite of their differences in dimension and appearance, the rectangles have the same area measure. Some students invariably focus their preference on one dimension (e.g., the "skinny one" or the "fat one"), so the stage is set for contested claims. Measurement becomes a way to resolve these claims. What we hope to accomplish in this problem context is to help students take the first steps toward developing a unit of area measure. Consequently, we restrict students' activity to folding or otherwise partitioning the rectangles. Folding and partitioning create literal arrays, and students often resort to reallotment to compare areas. If the partitions of rectangle A can be rearranged so that they are congruent with those of rectangle B, then the rectangles obviously cover the same space. One such reallotment is displayed in figure 8.4. Students typically invent many potential partitions and ways of reallotting the pieces. They take the first steps toward constructing units of area measure when they use one kind of partition (e.g., a square or a rectangle) as the basis for comparing all three rectangles. We avoid strategies like covering each rectangle with square units and then counting them, since

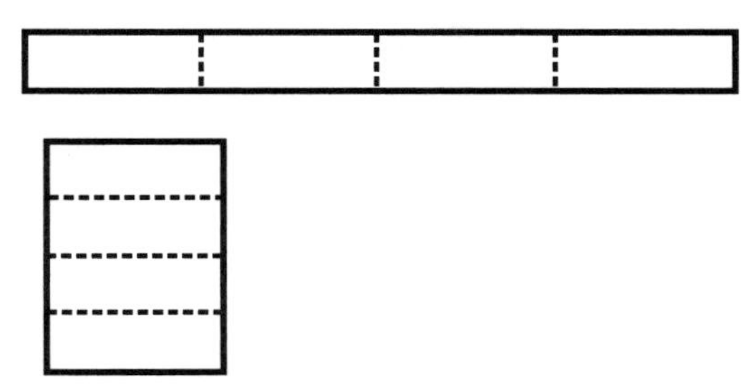

Fig. 8.4. Realloting pieces of rectangles to establish equivalent area

covering and counting forestalls students' own efforts to restructure the plane, which we consider foundational to understanding the qualities of those partitions as units of measure.

Considering qualities of units

After these initial steps at constructing a unit of area measure, we pose tasks to make other qualities of units of area measure more apparent. For example, students are asked to compare the space covered (area) by the hands of their classmates and to rank these areas from smallest to largest. Students typically invent strategies involving direct comparison of the congruence of the hands of any two of their classmates, but this strategy is ultimately defeated by individual differences in the shapes of their hands. Some students have thicker thumbs or tinier pinkies.

Because the reallotment of hands is not possible, students try to find area by covering and counting handprints with units that we provide. These units include pots of dried beans, pasta, and other materials with curvature. One might anticipate that students would readily adapt square units of measure to this task and then proceed to estimate areas. Yet this seldom occurs immediately. Instead, younger students are usually drawn to the units that resemble the contour of the hand, often choosing beans or beads as units of area measure. These units have the additional virtue that they "don't overlap" or are bounded by the outline of the figure.

When students make these choices, we ask different groups of students to use beans or beads to find the measure of the same handprint. They are invariably disconcerted to find that the values are not the same, and the resulting discrepancy leads to a search for an explanation. Properties like identical units ("all the same") and space-filling (tiling the plane—"no cracks") are generated by students, or in rare instances, prompted by teachers. These properties were implied by the units developed in the previous task (the partitioned squares or rectangles), but their virtues were not made explicit at that point. Children usually suggest eventually that a solution to the "cracks" problem is to use square units.

Parts of units

Square units, however, raise new difficulties for prospective measurers, since the outline of the hand partitions the units unequally, leading to new problems in how to describe, combine, and iterate these partial units. Children's invented strategies are diverse and usually very productive. Some propose using color or shadings to combine parts of units estimated to constitute whole units, which can then be added to the number of whole units already counted to arrive at a total. Others propose that parts of units can be described as fractional quantities, like 1/2, 1/4, and 1/3, and that these quantities can then be summed (e.g., 1/2 + 1/4 + 1/4) and added to the number of whole units already counted. Opportunities abound to consider different ways of partitioning a unit so that it is represented by 1/4 or 1/3, and these different ways have practical value because the handprint often cuts units at unique angles. Hence, there is immediate connection to fractions and sums of fractions. Students rarely make errors like 1/4 + 1/4 = 2/8 because there is a visual counterpart to, as well as a practical goal being served, by addition. (In this context, students readily note that it does not make sense that an addition would result in 8 partitions.)

Unit and error

We repeat this experience with other nonpolygonal forms (like "islands" or "cactus plants") to give students additional experience with units of area measure and also to raise additional issues about error of measure. To address error, we ask several students to make measures of the same form and ask why their answers are still slightly different. This kind of conversation leads to attributions about potential sources of error (e.g., potential miscounts of fractional pieces, different ways of estimating parts of units). We also engage students in discussion about the "real" area measure of the form. In this context, students often propose indicators of the center of the set of measures, like the median or mode, because the center is seen as an unvarying attribute (the area) and the spread of scores regarded as due to different sources of error in measure. Repeated measures help students begin to explore the approximations at the heart of the practical activity of measurement.

Reconstructing area as products

We engage students in activities designed to foster further reallotment of areas. For example, the area of a parallelogram can be compared to that of a corresponding rectangle by rearranging its parts. Once we are convinced that students are comfortable with partitions and reallotment of areas, we go on to pose problems comparing areas that do not involve literal folding and partitioning. We begin with simple forms like rectangles and again ask students which of several rectangles has the greatest area. But now students are provided only rulers. This goal fosters a mental structuring of the plane by partitioning it according to the units of length of the sides. For example, a 5″ × 8″ rectangle is subdivided into 5 inches and 8 inches, and the intersections of these subdivisions form square inches. Students often invent this procedure spontaneously or with a reminder from a classmate or teacher about how earlier efforts to find solutions to this problem were assisted by partitioning and rearranging parts. The additional challenge provided by this problem is the need to coordinate partitioning in two dimensions, a process familiar to students from their previous explorations (e.g., rectangles or square units of measure are often generated in the first task by folding along vertical lines, followed by folding along horizontal lines). However, in this example length units guide the partitions, and students invent ways to efficiently count the resulting unit squares (e.g., skip counting by rows or columns). The important accomplishment is constituting an array by units of length measure.

Although counting squares does not produce a product, it invariably occurs to some students that the result is the same as a multiplication of lengths. We find it useful to orient students toward finding an explanation of this equivalence, which is readily visible as an array. Having established multiplication as an efficient form of counting, we go on to emphasize the continuous nature of arrays. Students consider arrays with mixed number dimensions (e.g., 4 1/2 × 8 1/2) and also arrays where original measures are rescaled (e.g., measure in square units of 1/2 inch instead of a single inch). This progressive play leads students toward ideas about area as the product of lengths, which is a reinterpretation of multiplication. Students investigate properties of multiplication, like commutativity and distributivity, by interpreting areas. For example, figure 8.5 depicts a student's solution of the commutativity of multiplication. This third-grade student suggested that this instance must be true because the area of the rectangles is invariant under rotation. (Our students usually have experience with transformational geometry.) It was a short step for the class to conclude that this property of transformation implied that this would be true for any product.

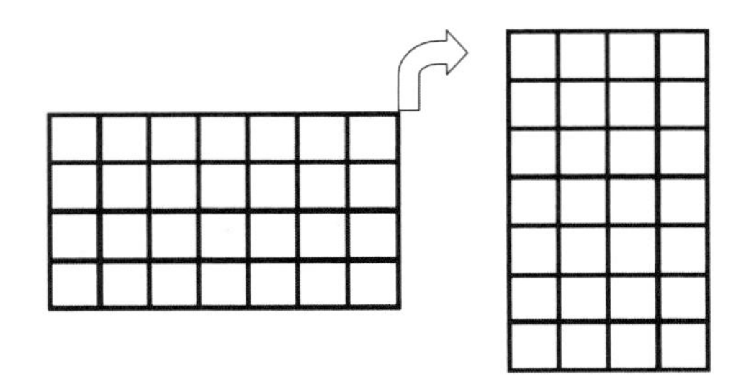

Fig. 8.5. An illustration of the commutative property of multiplication by rotation

A Classroom Example

Investigation of area measure occurs throughout the school year. Because the construction of an array is crucial to conceiving of area measure as a product of lengths, we illustrate how children in a second-grade classroom developed this idea when they were considering whether or not some figures representing zoo cages covered the same amount of space. Children at this point had already investigated all the forms of activity outlined previously (e.g., folding and partitioning three rectangles, rank ordering the area of their handprints). Our aim is to illustrate in this extended example how Ms. Curtis orchestrated children's exploration of the properties of shape as children developed understanding of area measure, so that practical activity of measure was coordinated with imagining qualities of space represented by chalk and chalkboard.

Considering zoo cages drawn on chart paper in front of the class, two of which are displayed in figure 8.6 as E and F, children first considered whether or not the figures were "really rectangles" or "just looked like rectangles." Ms. Curtis orchestrated discussion to consider the implications of properties of rectangles for the measure of their sides. At this point in the conversation, Suzanne had just proposed that if the two rectangles both have "short sides" of the same length, then they will be the same, and thus cover the same space.

Ms. Curtis: OK, if the short side on E was the same size—length—as the short side on F, would that mean that the two rectangles would be exactly the same?

Suzanne: No, not really.

Ms. Curtis: What else would I have to know? What else would I have to look at?

Thomas: The wide [*gestures*].

Ms. Curtis: What if this and this were exactly the same [*gestures to the corresponding short sides*], and if this and this were exactly the same [*gestures to the corresponding long sides*] were exactly the same? Would that make the two shapes exactly the same?

Class: No. Yes.

Ms. Curtis: Some people say no; some people were saying that they would have to be the same.

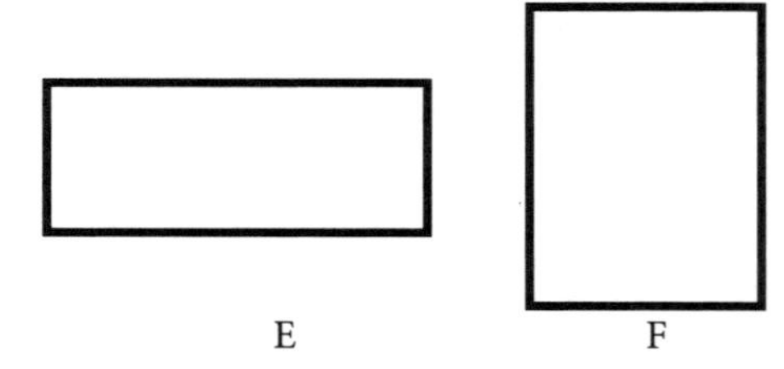

Fig.8.6. Figures E and F, representing zoo cages

Becky: Well, this long part and this long part could be the same [*gestures E, F*], and this short part and this short part could be exactly the same [*gestures, E, F*] but the middle would be different.

Ms. Curtis: So if this and this were exactly the same [*points to short sides*] and this and this side were exactly the same [*repeats Becky's motions*], could this one [F] have more space in the middle?

Becky's suggestion indicated that not all students shared the assumptions implicit in a diagram like those displayed in figure 8.6. To resolve this issue, the class first measured the lengths of the sides of the rectangles, finding that E was 4″ × 10″ and F was 5″ × 8″. Carl proposed measuring the middle to resolve the question, but Thomas emphatically disagreed, noting that the measure "couldn't be" different because the figure was a rectangle. He went on to explain to the class how the idea of a changing distance contradicted the necessary properties of a rectangle, and Ms. Curtis drew the contradiction that he proposed, displayed in figure 8.7 (if it were true that the measure of the side varied). This appeal to logic satisfied most of the class, although a few continued to insist on measuring to be certain that there was no sleight of hand.

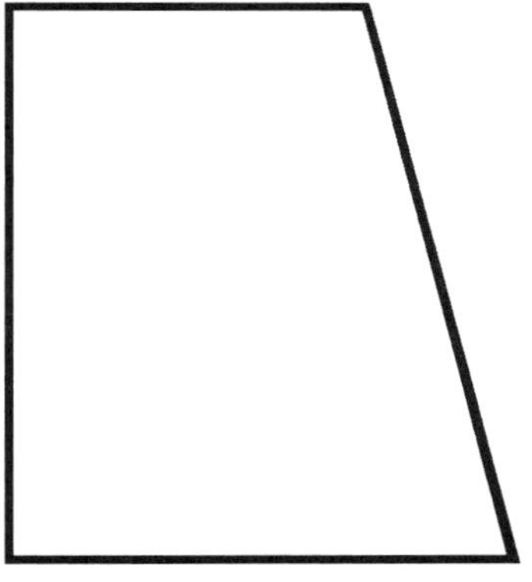

Fig. 8.7. Exploring the relationship between necessary properties of rectangles and their diagrams

After the meaning of a diagram was resolved, the class again returned to considering the space covered.

Craig: 4 times 10 is 40. So that means it covers up 40 inches. And then, 8 times 5 is 40. So it covers up 40 inches.

Ms. Curtis: Um, how do you know that? Tell me more about that covering up 40 inches.

Craig: Well, wait, like 40 one-inch cubes. [*Craig gestures the outline of a square with his hands.*]

Ms. Curtis: Let's see, how big would that be? [*She draws a one-inch square with Craig's tutelage.*] Craig is asking you to think about a square or a cube, this, it's actually a square, right?

Craig: Yeah.

Ms. Curtis: Not a cube? A square that big, one inch on each side. So think about that, and then, Craig, can you tell us how thinking about one of those helped you think about shapes E and F?

Craig: If it, well, if like, use a, um, like, make one-inch squares [*gestures filling the starting at the lower right corner of E*]. I can see it in my head.

Ms. Curtis asked Craig to illustrate what he saw in his head, and he suggested making one-inch lines "in a like a straight row going down." Ms. Curtis offered the use of a ruler, and together they partitioned the lower side of E into ten one-inch parts.

Craig: 1, 2, 3, 4, 5, 6, 7, 8, 9, 10 [*counting the partitions of the bottom of E*].

Ms. Curtis: Huh? I only see 10, not 40.

Craig: Well then, now we have to do, um, inches like this way [*gestures toward the left short side of E*].

Ms. Curtis: Oh, okay. To make them one-inch squares. Okay, all right. And how many rows will there be there? Can I just make any number as long as I use a ruler?

Class: No. You just put 4. [*Ms. Curtis draws these, resulting in a 4 × 10 array of one-inch squares.*]

Conversation then turned to different ways of counting the total number of squares. Different students generated a variety of strategies, including count-all, skip counts of ten columns of four, and skip counts of four rows of ten. After thorough exploration of this array, Ms. Curtis reminded the children of the purpose of the count, and asked, "Well, I see 40 squares in there (E). How does it help me think about those other shapes? How does that help me think about F to prove they're the same?" The class replied, "Forty," so Ms. Curtis probed, "Any-sized squares?" Students thought this was amusing and noted that the squares would need to have dimensions of one inch if the counts were to mean the same thing.

Ms. Curtis's question sparked the imposition of an array on F by another student, who proposed thinking about F as 8 rows of five-inch squares. He noted that he had considered changing the dimension of the unit itself to five-inch squares, but "five-inch squares would be too big." (He gestured to indicate that they don't fit the 4 × 10 evenly.) The class again considered different strategies for finding the total area, linking the equivalence of the strategies to the commutative property of multiplication (8 × 5 = 5 × 8). In related lessons, the class revisited the commutative property as a necessary feature of the invariance of the space covered in a rectangle under rotation.

Later in this lesson, the class began to blend strategies of structuring rectangles as arrays with reallotment of their areas. For example, Jordan proposed that the shape depicted by figure 8.8a, which the class called C, could

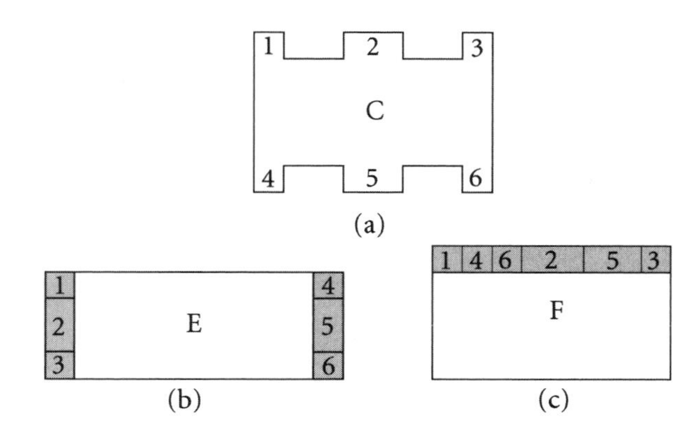

Fig. 8.8. Drawing parallels between arrays and reallotment of areas.

be realloted to create shape E (of fig. 8.6) by rearranging the pieces to create a 4 × 10 rectangle, as illustrated in figure 8.8b.

Ms. Curtis asked, "How can we test that and be really sure? How long is it from here to here [upper left piece of C]?" Jordan first established that the pieces accumulated to dimension 4 (1, 2, and 1), so that they would indeed fit on the side 4 of E. Ms. Curtis and another student asked about how far the side could be extended. Jordan responded by showing that the total area of the top pieces was four squares, so she extended the left (short side) of C by one, and by symmetry of the top and bottom pieces, proposed that the right side could be similarly extended. The result was rectangle E. By this time, the array of 40 square inches was taken for granted, so all accepted this argument. Tom proposed an alternative reallotment of moving the protruding rectangular pieces on the bottom of C to the top. He then showed how the reallotment of those squares would result in a 5 × 8 rectangle, which by rotation can be made congruent with F (of fig. 8.6), "turning C into F," as he stated it, displayed in figure 8.8c. The work here extended the partitioning of the rectangular arrays by showing how partitioning was related to the reallotment of the space of C to produce rectangles E and F.

In summary, this extended tour of one lesson illustrates several important developments in a child's theory of measure. First, by considering the measure of a shape, children reconsidered its properties and the implications of these properties for corresponding measures (i.e., What does rectangle imply about the distance between sides?). Second, rather than using ready-made units and procedures, students were guided to reinvent some of the foundations of area measure. To be sure, this example relied heavily on one child's insight, but the foundations for this insight had been set in a previous history of partitioning and rearranging parts of shapes. Third, the teacher, Ms. Curtis, continually helped students explore the implications of their actions and their beliefs (e.g., "If this is a rectangle, what must be true?" Later she asked, "Any-sized squares?"). Fourth, students invented and explored a variety of strategies that provided alternative means of restructuring the plane as an array. These restructurings put them on the road toward conceiving of areas as products of lengths and toward developing operational sense of multiplication as commutative.

INVESTIGATING VOLUME

Investigations of volume measure can proceed in parallel with investigations of area measure, and students' emerging ideas about area measure can bootstrap their understanding of volume. The conceptual attainments about unit and scale described in the measure of length and area are recapitulated in volume measure. Our goal is again to help students structure space as units of measure, albeit in three, rather than two or one, dimensions. Of course, mental coordination of units in three dimensions is more challenging than in its two-dimensional counterpart (Battista 1999). Volume measure is yet another fertile context for related explorations of arithmetic, especially because children can literally see the consequences of principles like commutativity. We next briefly outline the motivation for the series of problems in volume measure.

Prototypical Investigations

Our sequence of investigations again begins with volume measure as a solution to a problem of finding the space taken up by different "apartments" made of cubes, some of which are hidden. As with the three-rectangle problem in area, we encourage students to find different rearrangements of the space so that they can resolve contested claims about which takes up the most space. Students' work with these cube towers usually focuses on how to account for hidden cubes and how to avoid double-counting particular cubes, especially those at the corners of the towers. These are important first steps in developing ideas about these forms as structured arrays rather than as simple collections of cubes. Sharing strategies is an important resource for students who struggle with these ideas.

Students' initial ideas about rearrangements of cubes to create volumes are extended by investigations with open boxes. We supply students with multiple potential units, ranging from marbles to cubes, to highlight again the comparative advantages of different choices of unit. Students' strategies are typically diverse. Some count horizontal layers; others, vertical layers of cubes. Some count all cubes, and for these students, we work explicitly toward helping them think about the total count as being decomposable into counts of layers.

When students are comfortable with different ways of partitioning and layering cube arrays to find the volume of boxes, we raise the ante by including at least one box with a mixed-number dimension (e.g., a height of 4 1/2 edges of the unit cubes). This promotes a revision of the layering strategy in a direction that leads to the consideration of what happens as the layer becomes increasingly like a "slice" rather than a layer. Slices of ever decreasing thickness lead to the idea of volume as an iteration of area, making sensible product formulas for the volume of rectangular prisms.

We typically conclude these initial investigations of volume with problems involving cylinders. Posing the problem of volume measure with potential units of measure like marbles and cubes often triggers fruitful discussion. For example, some students are again drawn to resemblance between units of measure and the object being measured, so that marbles are preferred. Others propose using cubes but are stymied by the lack of ready fit, a problem that they have encountered with boxes of mixed-number dimension, but here magnified on a scale that makes physical fitting impossible. We ask students if it might be possible to use cubes as units of measure in spite of these physical obstacles. Reasoning by analogy to slices of boxes, students often generate the idea of finding the volume of a cylinder by iterating slices of the base. Because students have had experience at this point of finding the areas of hands, islands, and so on, they extend this strategy to estimate the area of the base of the cylinder. The pivotal recognition is that area of the base is the culmination of the process of slicing the cylinder.

POINTS OF DEPARTURE

With the understandings of measurement, space, and arithmetic developed in contexts of measuring length, area, and volume, children are poised to extend and elaborate the mathematics of measure in multiple ways. We have pursued several related agendas that we can merely mention here. First, children's theories of measurement can be extended as tools for considering the natural world. For example, in a third-grade class taught by Ms. Curtis, students were presented with twenty-four objects composed of different materials (e.g., brass, Teflon, wood, aluminum) and asked to rank them by weight. Available tools included balances and washers of different sizes, but no scales. Students initially objected that they could not measure weight without scales, but several students broke this impasse by suggesting a series of pairwise comparisons using the balances. This procedure resulted in a ranking and also posed considerable problems of organizing systematic comparison! Their teacher, Ms. Curtis, next raised the ante by asking, "How much heavier is each object than the immediately preceding one?" With the introduction of this question, ranking was no longer sufficient.

Students again were initially confounded by this request. Several proposed using washers as units of measure, finding the measure of each object as a number of washers. Several objected, "We have to know how much the washers weigh!" Nadia asked, "What do the washers stand for?" Sara replied, "Washers!" Having adopted the proposal that washers might serve as units of measure, the children next considered whether or not big and little washers could be mixed. Many children made analogy to other instances of measure in which the units needed to be the same. As Anna summarized, "They have to be the same size, or else!" So, a count of identical washers resulted in a measure of weight, but for some of the objects, the number of washers available was not sufficient to balance the pan. The third graders resolved this issue by adopting a unit-of-units perspective, combining objects with known weights (in washers) with additional washers to obtain the measure of the weight of more massive objects. Hence, within a single lesson, students revisited many of the issues that they first encountered developing the measure of spatial extent. Drawing on their theories of measure, they developed appropriate means for measuring weight, although their application of this knowledge resulted in a comparative "speed up" in encountering and resolving challenges of measure, not the complete elimination of these challenges. In other words, children did not simply transfer their knowledge, but instead used it to bootstrap reasonable solutions to the measurement of weight. They were initially drawn to the idea that weight meant pounds, and as they invented alternative means of measurement, they again reconsidered the very nature of unit and scale.

Second, practical grasp of the measurement of the natural world, like the measure of weight just described, provides access to the mathematics of data and statistics. For example, repeated measurements of an attribute, like the weight of an object or the height of a plant, often result in a distribution of measures. The center of the distribution provides a best guess of the value of the attribute being measured, and its spread provides an estimate of the relative precision of the measurement process (Petrosino, Lehrer, and Schauble in press). Measurement thus serves as a gateway to new realms of mathematical inquiry, effectively suggesting the practical wisdom of promoting a mathematics of measure throughout schooling.

REFERENCES

Battista, Michael T. "Fifth Graders' Enumeration of Cubes in 3D Arrays: Conceptual Progress in an Inquiry Classroom." *Journal for Research in Mathematics Education* 30 (July 1999): 417–48.

Battista, Michael T., Douglas H. Clements, Judy Arnoff, Kathryn Battista, and Caroline Van Auken Borrow. "Students' Spatial Structuring of 2D Arrays of Squares." *Journal for Research in Mathematics Education* 29 (November 1998): 503–32.

Clements, Douglas, Michael T. Battista, and Julie Sarama. "Development of Geometric and Measurement Ideas." In *Designing Learning Environments for Developing Understanding of Geometry and Space*, edited by Richard Lehrer and Daniel Chazan, pp. 201–25. Mahwah, N.J.: Lawrence Erlbaum Associates, 1998.

Confrey, Jere, and Eric Smith. "Splitting, Covariation, and Their Role in the Development of Exponential Functions." *Journal for Research in Mathematics Education* 26 (January 1995): 66–86.

Horvath, Jeff, and Richard Lehrer. "The Design of a Case-Based Hypermedia Teaching Tool." *International Journal of Computers for Mathematical Learning* 5 (2000): 115–41.

Lakoff, George, and Rafael E. Núñez. *Where Mathematics Comes From: How the Embodied Mind Brings Mathematics into Being.* New York: Basic Books, 2000.

Lehrer, Richard. "Developing Understanding of Measurement." In *A Research Companion to "Principles and Standards for School Mathematics,"* edited by Jeremy Kilpatrick, W. Gary Martin, and Deborah E. Schifter. Reston, Va.: National Council of Teachers of Mathematics, in press.

Lehrer, Richard, Cathy Jacobson, Greg Thoyre, Vera Kemeny, Dolores Strom, Jeff Horvath, Steve Gance, and Matt Koehler. "Developing Understanding of Geometry and Space in the Primary Grades." In *Designing Learning Environments for Developing Understanding of Geometry and Space*, edited by Richard Lehrer and Daniel Chazan, pp. 169–200. Mahwah, N.J.: Lawrence Erlbaum Associates, 1998.

Lehrer, Richard, Cathy Jacobson, Vera Kemeny, and Dolores Strom. "Building on Children's Intuitions to Develop Mathematical Understanding of Space." In *Mathematics Classrooms That Promote Understanding*, edited by Elizabeth Fennema and Thomas Romberg, pp. 63–87. Mahwah, N.J.: Lawrence Erlbaum Associates, 1999.

Lehrer, Richard, Michael Jenkins, and Helen Osana. "Longitudinal Study of Children's Reasoning about Space and Geometry." In *Designing Learning Environments for Developing Understanding of Geometry and Space*, edited by Richard Lehrer and Daniel Chazan, pp. 137–67. Mahwah, N.J.: Lawrence Erlbaum Associates, 1998.

National Council of Teachers of Mathematics (NCTM). *Principles and Standards for School Mathematics.* Reston, Va.: NCTM, 2000.

Petrosino, Anthony, Richard Lehrer, and Leona Schauble. "Structuring Error and Experimental Variation as Distribution in the Fourth Grade." *Mathematical Thinking and Learning*, in press.

Piaget, Jean, Bärbel Inhelder, and Alina Szeminska. *The Child's Conception of Geometry.* 1948. Reprint, New York: Basic Books, 1960.

Thompson, Patrick W., and Tommy Dreyfus. "Integers as Transformations." *Journal for Research in Mathematics Education* 19 (March 1988): 115–33.

9

Understanding Students' Thinking about Area and Volume Measurement

Michael T. Battista

TEACHING area and volume measurement so that students learn the underlying ideas in personally meaningfully ways requires a firm understanding of students' thinking about these ideas. This understanding is essential for choosing appropriate instructional tasks, guiding students' discussions, understanding students' learning difficulties, and monitoring as well as assessing students' learning progress. This article first examines the fascinating realm of students' thinking about area and volume, then it suggests instructional activities to help students develop genuine understanding of these important mathematical ideas.

UNDERLYING MENTAL PROCESSES

To measure area and volume in standard measurement systems, we determine the number of unit squares or cubes in the region we are measuring. Thus, the foundation for developing competence with measuring area and volume in standard measurement systems is understanding how to enumerate meaningfully arrays of squares and cubes such as those shown in figures 9.1a and 9.1b. (Other shapes can be used as units, but squares and cubes are the standard units because usually they are easiest to reason with.)

Four mental processes are essential for meaningful enumeration of arrays of squares and cubes and will be used throughout this article to explain students' thinking: forming and using mental models, spatial structuring, units-locating, and organizing-by-composites. In the *forming and using men-*

Support for this work was providded by grants RED 8954664 and ESI 0099047 from the National Science Foundation. The opinions expressed, however, are the author's and do not necessarily reflect the views of that foundation.

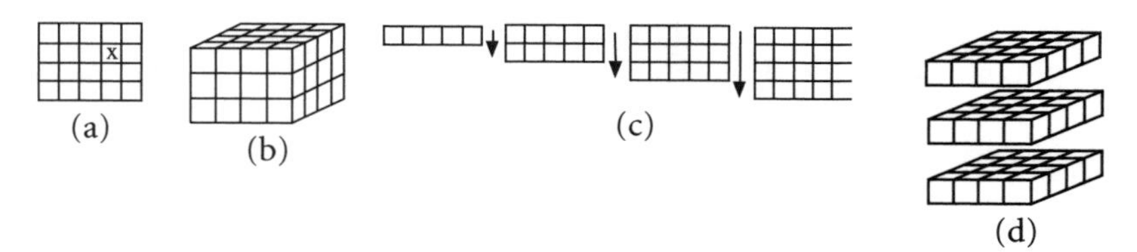

Fig. 9.1. Arrays of squares and cubes, (a) and (b); composite units, (c) and (d)

tal models process, individuals create and use imagistic or recall-of-experience-like mental representations to visualize, comprehend, and reason about situations. For instance, to give someone travel directions without the use of a map, you reflect on your mental model of the locality to visualize and describe a route to follow. In the *spatial structuring* process, individuals abstract an object's composition and form by identifying, interrelating, and organizing its components. For instance, to structure a rectangular array of squares spatially, you might see it as rows and columns. Fundamentally, students can meaningfully enumerate arrays of squares and cubes only if they have developed properly structured mental models that enable them to locate and organize the squares and cubes correctly.

To develop such properly structured mental models, two additional processes are required. The *units-locating* process locates squares and cubes, and composites of squares and cubes, by coordinating their locations along the dimensions of an array. For instance, to understand the location of square X in the array shown in figure 9.1a, an individual must see the square in a two-dimensional, coordinate-like system—for example, it is in the fourth column and the second row, or it is the fourth unit to the right and the second unit down.

The *organizing-by-composites* process combines an array's basic units (squares or cubes) into more complicated, composite units that can be repeated or iterated to generate the whole array. (For brevity, the term *composite unit*—which is a unit consisting of more basic units—will be shortened to *composite.*) For instance, in a 2-D array, a student might mentally unite the squares in a row to form a composite unit that can be iterated in the direction of a column to generate the array (see fig. 9.1c). In a 3-D array, the cubes in a horizontal layer can be grouped into a layer-composite that can be iterated vertically to generate the array (see fig. 9.1d).

LEVELS OF SOPHISTICATION IN STUDENTS' STRUCTURING AND ENUMERATION OF ARRAYS

The four mental processes described above will now be used to describe levels of sophistication in students' understanding of area and volume measurement. Students' thinking about area is illustrated with second graders' work; students' thinking about volume is illustrated with fifth graders' work.

Level 1: The Absence of Units-Locating and Organizing-by-Composites Processes

Students do not organize units into composites, and, because they do not properly coordinate spatial information, they are unable to locate all the units in arrays.

Area (Battista et al. 1998)

Katy was shown that a plastic inch square was the same size as one of the indicated squares on the 7-by-3-inch rectangle displayed in figure 9.2a. She was then asked to predict how many plastic squares it would take to completely cover the rectangle. Katy drew squares and counted 30 as shown in figure 9.2b.

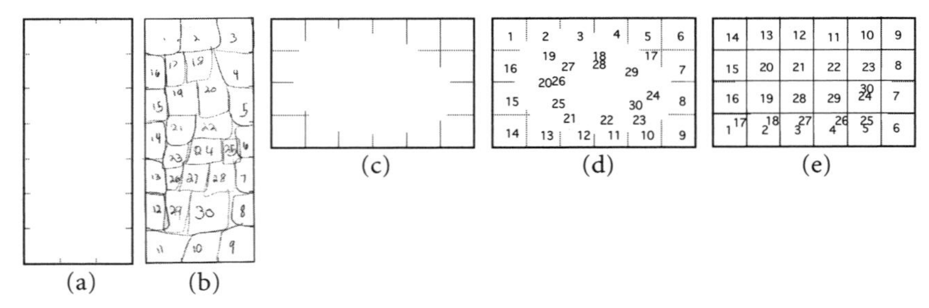

Fig. 9.2. Katy's work

On a similar problem, Katy was asked to predict how many squares would cover the rectangle shown in figure 9.2c (making her prediction without drawing). Katy pointed and counted as in figure 9.2d, predicting 30. When checking her answer, she pointed to and counted plastic squares as shown in figure 9.2e, getting 30. When she counted the squares again, first she got 24, then 27.

Although, as educated adults, we instantly "see" the squares covering these rectangles arranged by rows and columns, Katy had not yet mentally constructed a row-by-column structuring to organize and locate the squares properly. Instead, because Katy's mental model located squares along an almost random path, she got lost in her counting.

Volume (Battista and Clements 1996)

When asked how many cubes were needed to completely fill the box shown in figure 9.3a, Bob counted the 8 cubes shown in the box, then pointed to and counted 6 imagined cubes on the box's left side, 4 on the back, 4 on the bottom, and 5 on the top. Thus, Bob's units-locating process was insufficient to create an accurate mental model of the cube array.

Randa was shown a picture of a box with the length, width, and height

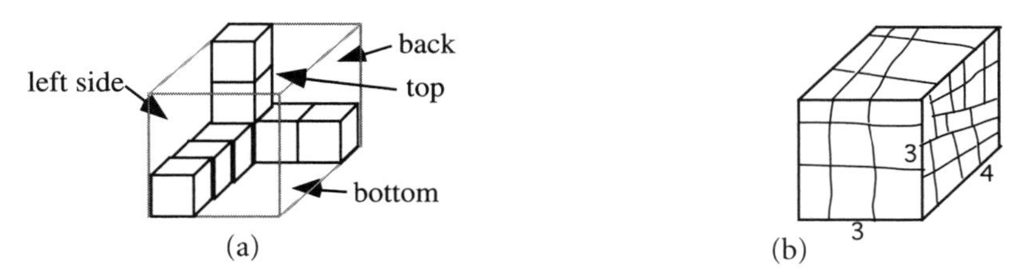

Fig. 9.3. Level 1, work of Bob (a) and Randa (b)

labeled and was told, "This box contains three cubes along the bottom, three up from here to here, and four from here to here [*pointing appropriately at the box picture*]." When Randa said she could not find the number of cubes needed to fill the box, she was asked to "draw what the cubes look like on the outside." After about ten to fifteen minutes and many erasures, Randa's drawing looked like figure 9.3b, showing a clear lack of coordination of spatial information. (For instance, the right side shows four cubes in horizontal rows; the top shows only three.)

Further evidence that insufficient coordination causes major difficulty in the units-locating process is the ubiquitous "double counting" error. For instance, when Jeff tried to enumerate the cubes needed to make the building shown in figure 9.1b, he counted all cube faces that appeared on the six sides of the building, double-counting edge cubes and triple-counting corner cubes; he said there were two additional cubes in the interior. Because he did not properly coordinate what he saw on the different sides of the building, Jeff failed to see when adjacent cube faces were part of the same cube.

Level 2: Beginning Use of the Units-Locating and the Organizing-by-Composites Processes

Students not only start to form composite units, they use the units-locating process to see equivalent composites.

Area

Bill: First I count the bottom and there's 6 [*moving his hands inward as shown in fig.9.4.*] So the top and bottom would equal 12. And these 2 [*pointing to the middle squares on the right and left sides*] would be 14. [*Using fingers to estimate where individual squares were located*] I'd say maybe 12 in the middle; 12 + 12 = 24. So I'd say 24.

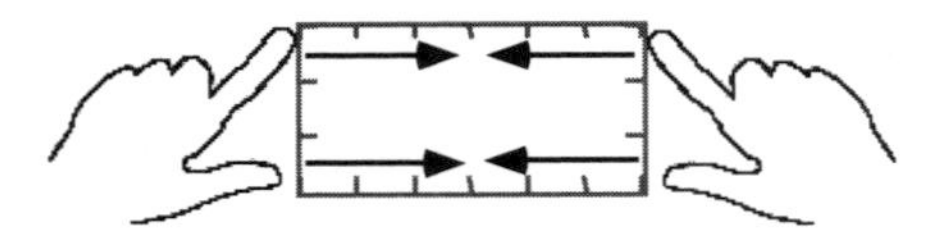

Fig. 9.4. Bill's work

Bill was beginning to structure the array into composites (the top and bottom rows). Although he was unable to use the units-locating process to locate interior squares correctly, he did use it to see the numerical equivalence of his composites.

Volume

For the building shown in figure 9.1b, Fred counted 12 cubes on the front, then immediately said there must be 12 on the back; he counted 16 on the top and immediately said there must be 16 on the bottom; finally, he counted 12 cubes on the right side, then immediately said there must be 12 on the left side. In each instance, after counting the cubes visible on one side of the building, he inferred the number of cubes on the opposite side, clear evidence that he was organizing cubes into composites and that he was using the units-locating process to relate these composites spatially and numerically.

Level 3: The Units-Locating Process Becomes Sufficiently Coordinated to Recognize and Eliminate Double-Counting Errors

A major breakthrough in thinking occurs when a student's units-locating process coordinates single-dimension views (e.g., top, side, front) into a mental model that is sufficient to recognize the same unit from different views. This refined mental model enables students to eliminate double-counting errors caused by insufficient coordination.

Area

Bill was enumerating the squares in a 6-by-4 rectangle in which he had correctly drawn the array of squares. He counted the 6 squares in the left column, then immediately said there must be 6 in the right column. He counted 4 squares on the top, then said there must be 4 on the bottom. He counted 8 more squares in the middle, getting a total of 28. But when he explained his strategy to the teacher, Bill changed his mind. He then said there were 6 squares in each of the left and right columns and counted 2 on the top, 2 on the bottom, and 8 in the middle. Through an increase in coordination, Bill could simultaneously see a corner square as part of a row and as part of a column, enabling him to realize that he had initially double-counted such squares.

Volume

As shown in figure 9.5, Juan coordinated spatial information sufficiently to avoid double-counting edge cubes. However, his coordination was still insufficient to build a mental model that properly located interior cubes.

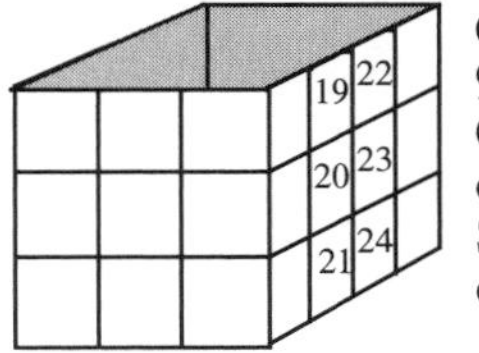

Counts 9 on the front, infers 9 on the back, making 18. Counts the 6 uncounted cubes on the right side, 19–24. Says there are 6 (uncounted) cubes on the left side.

Fig. 9.5. Juan's work

Level 4: The Use of the Organizing-by-Composites Process to See a Whole Array Composed from Maximal Composites, but with Improper Iteration

Students see arrays as "maximal" composites (rows and columns for area, layers for volume) that can be iterated in a single direction to generate an array. But because of insufficient coordination, students cannot precisely locate these maximal composites, and they consequently make iteration errors.

Area

Joe was shown that 5 plastic squares fit across the top of a rectangle and that 7 fit down the middle (then the squares were removed). See figure 9.12k.

> *Joe:* 5, 10, 15, ..., 45 [*motioning along estimated row positions inside the rectangle*].
>
> *Teacher:* How did you get that?
>
> *Joe:* I was trying to guess where the bottoms of the squares were.

Joe structured the array into row composites of five. However, his coordination of rows and columns was insufficient to enable him to properly imagine the locations of the rows.

Volume

Randa was shown a picture of a $5 \times 3 \times 4$ cube array and was asked to build the bottom layer for the array. She built a 5×3 array of cubes and said that there were 15 cubes in the layer. When asked how many cubes were in the entire building, Randa counted from the bottom up on the picture, but continued to count on the top, getting seven layers (see fig. 9.6). She gave an answer of 105 cubes. When the interviewer asked Randa to make the building with cubes, she built and stacked four layers, and, was going to continue to build three more until the interviewer asked her to compare what she already built to the picture. Surprised, Randa concluded that the building was complete, then pointed to each layer saying, "15 here, 15 here, 15 here, 15 here; 15 × 4 = 60." Initially, Randa was unable to determine where the layers occurred. She could not coordinate the horizontal layers with the third dimension that was the prism's height.

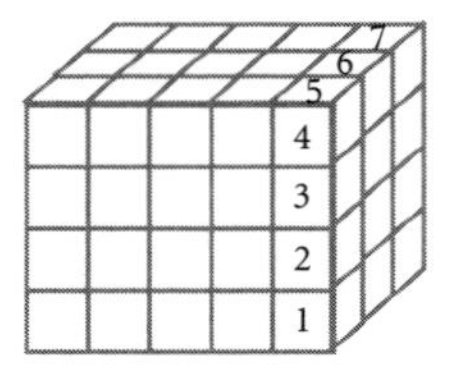

Fig. 9.6. Randa's work

Although both Joe and Randa took major steps by structuring arrays in maximal composite units, neither student had yet developed an accurate enough mental model in which these composites could, in imagination, be properly iterated to form the whole array.

Level 5: The Use of the Units-Locating Process Sufficient to Locate All Units Correctly, but with Less-than-Maximal Composites Employed

This is the first level in which the units-locating process is sufficient to create a mental model that correctly locates all squares or cubes in an array. However, although students sometimes obtain correct answers (as in the next two examples), because students inefficiently or inconsistently organize arrays into composites, they quite frequently lose their place in counting or adding and make enumeration errors. Furthermore, students' structuring and enumeration strategies are not generalizable and are inadequate for large arrays.

Area

Given rectangle (i) in figure 9.12 and asked to predict how many unit squares would cover it, Billie correctly employed the units-locating process (see fig. 9.7a). However, although she organized the array as composites of two (which was sufficient for correct enumeration), these composites were not the maximal ones needed to give the array its row-by-column structure.

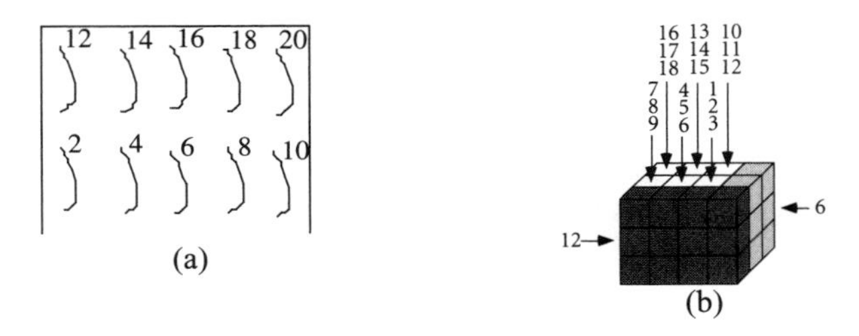

Fig. 9.7. Level 5, work of Billie (a) and Mary (b)

Volume

Mary counted the cubes visible on the front face (12), then counted those on the right side that had not already been counted (6). (See fig. 9.7b.) She then pointed to the remaining cubes on the top, and for each, counted cubes in columns of three: 1, 2, 3; 4, 5, 6; ...; 16, 17, 18. She then added 18, 12, and 6.

Level 6: Complete Development and Coordination of both the Units-Locating and the Organizing-by-Composites Processes

Students' mental models fully incorporate row-by-column or layer structuring so that students can accurately reflect on and enumerate arrays without physical or perceptual material. An important point from a mathematical perspective is that such structuring is more general and powerful than using standard area and volume formulas. For example, layer structuring is extremely useful for thinking about the volumes of cylinders and many problems in calculus.

Area

Paul was shown that 5 plastic squares fit across the top of a rectangle and that 7 fit down the middle (then the squares were removed). See figure 9.12k.

Paul: [*Counting and pointing across the top row by ones*] 5 across; 7 down. [*Motioning across the top three rows*] 5, 10, 15. [*Counting on seven fingers*] 5, 10, 15, 20, 25, 30, 35; 35.

Teacher: How did you know to stop at 35?

Paul: There are only 7 down that way [*motioning vertically down the middle of the rectangle*].

Volume

For the cube array given in figure 9.1b, Julie counted cubes in the top layer, 1–12, pointed to the middle layer and said 24, then pointed to the bottom layer and said 36. Her partner, Juanita, pointed to the right side of the array and said 9. She then counted 4 columns of 3 on the top and said, "So it's 9 times 4 equals 36."

FURTHER DISCUSSION OF STUDENTS' STRUCTURING AND ENUMERATION DIFFICULTIES

The Root of the Problem: Lack of Coordination

Properly coordinating spatial information is extremely difficult for many students. The first episode below illustrates how deep-seated this difficulty can be. The second episode shows the reasoning that students must use to resolve coordination difficulties.

A deep-seated problem

After Randa, a fifth-grader, built several rectangular buildings with interconnecting cubes, she was asked how many cubes it would take to make the building shown in figure 9.8a.

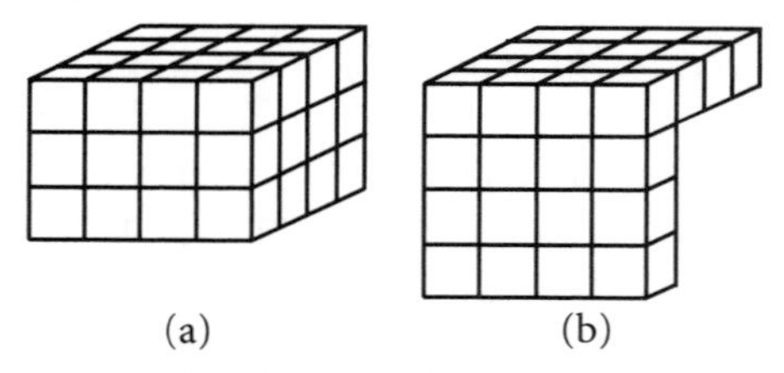

Fig. 9.8. Randa's cube configuration

Randa: It has 4×3 and 4×4; $12 \times 4 = 48$, $+ 16 = 64$. There's some you can't see in the picture. [*She multiplied 12, the number of cubes in each lateral face, by 4, the number of lateral faces. She then added 16 for the top (ignoring the bottom). Note that Randa's strategy double-counted cubes along most of the building's edges.*]

Randa then spent about twenty minutes trying to construct the building with cubes. She built the configuration shown in figure 9.8b several times, making the 4-by-4 top and the 3-by-4 front, then joining them. Each time, however, she stopped and started over when she noted that the front or top of her configuration did not match the building pictured in figure 9.8a. Randa was then asked to find the number of cubes in the configuration shown in figure 9.8b.

Randa: Four rows of 4 on this side [*the top*] and 4 on this side [*the front*]. So it's 16 + 16 = 32. [*Note that she still double-counted cubes along the top front edge.*]

Both in counting and in building, Randa showed a striking inability to coordinate different views of cube configurations. For the configuration shown in figure 9.8b, she constructed the top with cubes, then the front, but she could not figure out how these two parts fit together to make the configuration shown in figure 9.8a. To coordinate these views, Randa would have to decompose the top into individual cubes, do the same with the front, then establish a spatial relationship between the views that recognized that the four edge cubes were part of the front and top faces. Randa's difficulties are common among elementary and middle school students (Battista and Clements 1996).

Resolving coordination difficulties

Amanda, a second-grader, is predicting how many squares will cover a 4-by-3 rectangle, having been shown only that 4 plastic squares fit across the top and 3 fit down the left side (fig. 9.12j).

Amanda: [*Making her original prediction for an unmarked 4-by-3 rectangle; (see fig. 9.9a)*] There's 4 here [*top row*] and 4 here [*bottom row*] plus 2 here [*one on the left side, one on the right*], and that equals 10. [*Pointing to what would be the two interior squares*] But I'm not sure if there's 2 in the middle or 1 in the middle. [*Amanda draws the picture shown in figure 9.9b.*]

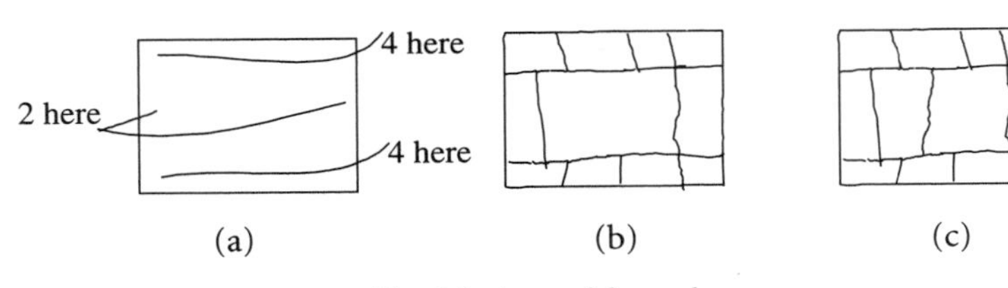

Fig. 9.9. Amanda's work

Interviewer: Where are the 2 in the middle? [*Amanda draws the middle vertical segment in the second row and points to the 2 squares formed (see fig. 9.9c).*] So how many do you have in there now?

Amanda: So 12 or 11. I think there's 2 in the middle or 1 in the middle. [*Motioning across the second row*] I just changed my mind. ... There's going to be 2 in the middle because if there's only 3 here [*motioning to the right column*] and 3 here [*motioning to the left column*], then going across would be 4 [*motioning across the second row*]; 12.

Interviewer: Why do you now think 12?

Amanda: Because there's 4 on the top [*motioning to the top row*], so going across down here [*motioning to the second row*] would be the same as going across up here [*motioning to the top row*], so there's probably 2 in the middle.

Initially Amanda structured the array as a set of disjoint components—top, bottom, sides, and an amorphous interior. She was able to structure the rectangle's interior as two squares, and thus its middle row as four, only when she saw the two middle squares on the lateral sides, not as separate, but as part of the right and left columns. This coordinating action enabled her to infer the equivalence of the first and second rows because it vertically aligned their squares and set up a one-to-one correspondence between them. Such reasoning was crucial to Amanda's proper structuring of this array.

Difficulties with Numerical Procedures

Most students do not fully understand the spatial structuring that underlies traditionally learned numerical procedures for determining area and volume, and consequently they improperly apply these procedures to new problems.

Insufficient understanding

Bethany, a fifth grader, regularly determined the number of cubes in 3-D arrays using layers. She also had discovered that the number of cubes could be found by multiplying the length, width, and height. But as her class discussed this procedure, Bethany questioned its validity. She told her classmates that she was puzzled because "the corner cube gets counted once when you find the length, once for the width, and once for the height." Although almost every student in the class had discovered, and was routinely employing, a layer approach, not one of the students had an answer for Bethany's question. Even when the teacher posed Bethany's question in the context of area, the students had no answer.

The teacher took advantage of this "teachable moment" by having students work on the problem in pairs. When I asked Bethany and her partner how they were thinking about the problem, they said that they were "stuck." So I posed questions that I thought might help them clarify their thinking.

Interviewer: [*Arranges the 3-by-3 set of cubes that the students had been working with into 3 rows of 3 and points successively to the cubes in one row*] 1, 2, 3. What am I counting here?

Partner: Cubes.

Bethany: Yeah.

Interviewer: [*Pointing to the three rows*] 1, 2, 3. What am I counting here?

Bethany: [*Excitedly*] Rows of cubes. You're not counting cubes this time. So, first, you count cubes, then you count rows.

Partner: So you're not really counting the cube twice. We got it!

Bethany's question posed a real conundrum for the students. They knew that multiplying the length times the width gave the number of cubes in a rectangular array. Almost all the students justified this procedure by saying that they were multiplying the number of cubes in a row times the number of rows, thus satisfying the traditional criterion that they had learned the procedure "meaningfully." But initially, their understanding of this enumeration strategy did not clearly identify exactly what was being counted. To overcome their difficulty, Bethany and her partner not only had to structure the array properly but also had to conceptualize that structuring properly. Whereas other students made, without aid, the same discovery as Bethany and her partner, still others gave explanations that lacked genuine appreciation of the difficulty. For instance, one pair of students argued that, since "a cube is in both the width and the length, it's okay to count it twice."

Misapplication

Pam, a bright eighth grader who was three weeks from completing a standard course in high school geometry, responded as follows on the problem shown in figure 9.10.

Pam: It's 45 packages. And the way I found it is I multiplied how many packages could fit in the height by the number in the width, which is 3 times 3 equals 9. Then I took that and multiplied it by the length, which is 5, and came up with 9 times 5, which is 45.

Observer: How do you know that is the right answer?

Pam: Because the equation for the volume of a box is length times width times height.

Observer: Do you know why that equation works?

Pam: Because you are covering all three dimensions, I think. I'm

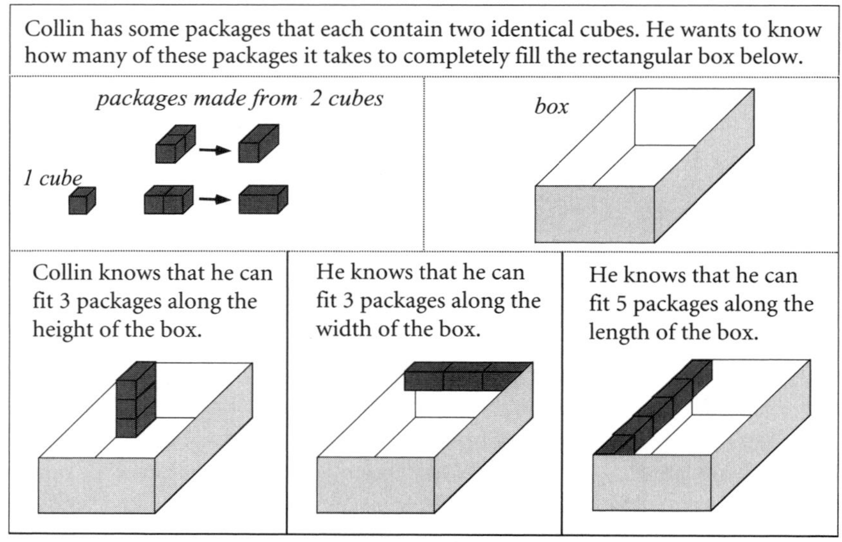

Fig. 9.10. Two-cube problem

not really sure. I just know the equation.

Students in grades 3–5 made an identical structuring error on a simpler problem in which the box was only one layer high.

Student 1: We know 3 will fit across; we know 5 will fit downward; just multiply.

Student 2: I kept counting 5 along the side of the package until I counted 3 times because you can only fit 3 packages along the top of the box.

On both the one-layer and three-layer problems, students used numerical procedures without first mentally constructing appropriate spatial structurings of the situations. Apparently, the salience of the dimension-like information given in the problem automatically activated students' use of a familiar numerical procedure, causing them to bypass careful spatial analyses of how the packages fill the box. A lack of depth in students' understanding of procedures caused them to misapply the procedures.

Even when using packages made from interlocking cubes and paper boxes, many students still have substantial difficulties with the problem in figure 9.10. For instance, Anita, a fifth-grader, asserted that 5 packages fit in the box across the length and 3 across the width. To test her assertion, she placed 3 packages along the width. But after placing 4 packages along the length, she was perplexed that the fifth package would not fit in the bottom of the box while the 3 packages along the width remained (see fig. 9.11). Eventual-

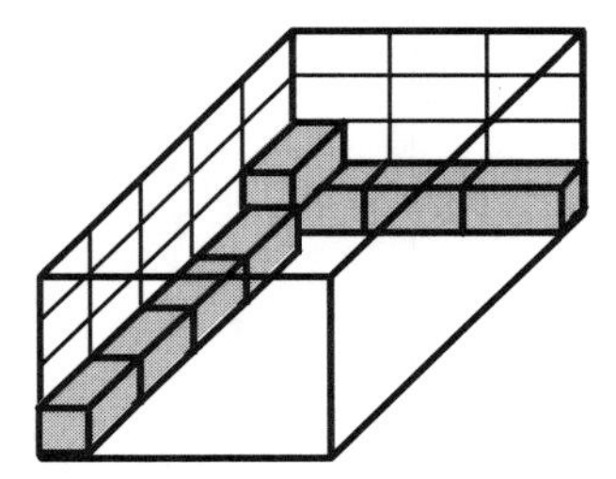

Fig. 9.11. Anita's work on the two-cube problem in figure 9.10

ly, Anita decided that she could place the packages into the box so that their longest sides were parallel to the longest side of the box. But she was unable to determine the number of packages that covered the bottom of the box until she completely covered it with packages.

INSTRUCTIONAL ACTIVITIES

The ultimate goal of instruction on area and volume should be for students to develop properly structured mental models that enable them to reason powerfully about these concepts in a wide variety of situations. To cultivate students' creation of these mental models, instruction must nurture students' development of the four mental processes described at the beginning of this article. Instructional tasks must also encourage and support students' construction of personally meaningful enumeration strategies (i.e., those that are based on properly structured mental models). Students' construction of such strategies is facilitated, not by "giving" them formulas, but by encouraging students to invent, reflect on, test, and discuss enumeration strategies in a spirit of inquiry and problem solving. (Additional information on instruction can be found in Akers et. al 1997, Battista and Clements 1995, and Battista and Berle-Carman 1996.)

Teaching Area Measurement: Enumerating Squares in Rectangular Arrays

To construct proper spatial structurings of two-dimensional arrays of squares, students need numerous opportunities to structure such arrays and to reflect on the appropriateness of their structurings. One good way of presenting such opportunities is to use, in inquiry-based instruction, problems similar to those already described. As you give students the rectangles shown in figure 9.12 (which should have dimensions in inches), show students how a plastic inch square fits on one of the indicated squares. Students first pre-

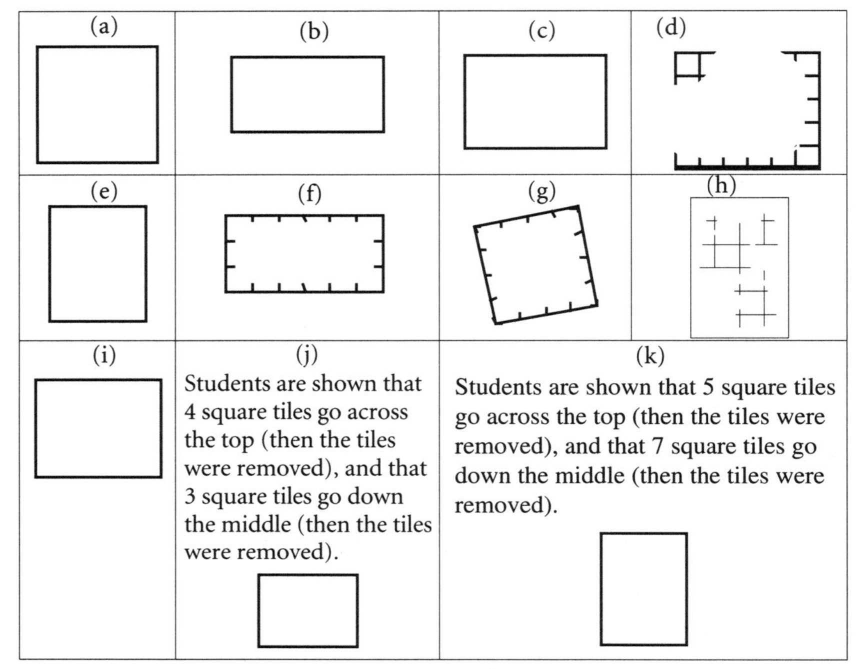

Fig. 9.12. Rectangle tasks

dict how many squares it takes to cover the rectangle, then check their predictions with plastic squares.

Start with rectangles that give the most graphic information about the location of squares, then gradually move to rectangles that give less information (the rectangles in fig. 9.12 are listed roughly in this order). Give students several problems of each type so that they have an opportunity to develop adequate structuring for that type before moving on to more difficult problems. As a variation, after students have made their first prediction, have them draw how they think squares will cover the rectangle, make another prediction, then check their predictions with plastic squares. Many students will be able to make a correct prediction after drawing squares on a rectangle, but their structuring will be inadequate for them to make a prediction without drawing. Students' explanations and drawings provide insights into the levels of sophistication of their thinking, information that can be invaluable in choosing appropriate instructional tasks, guiding discussions, and assessing students' progress.

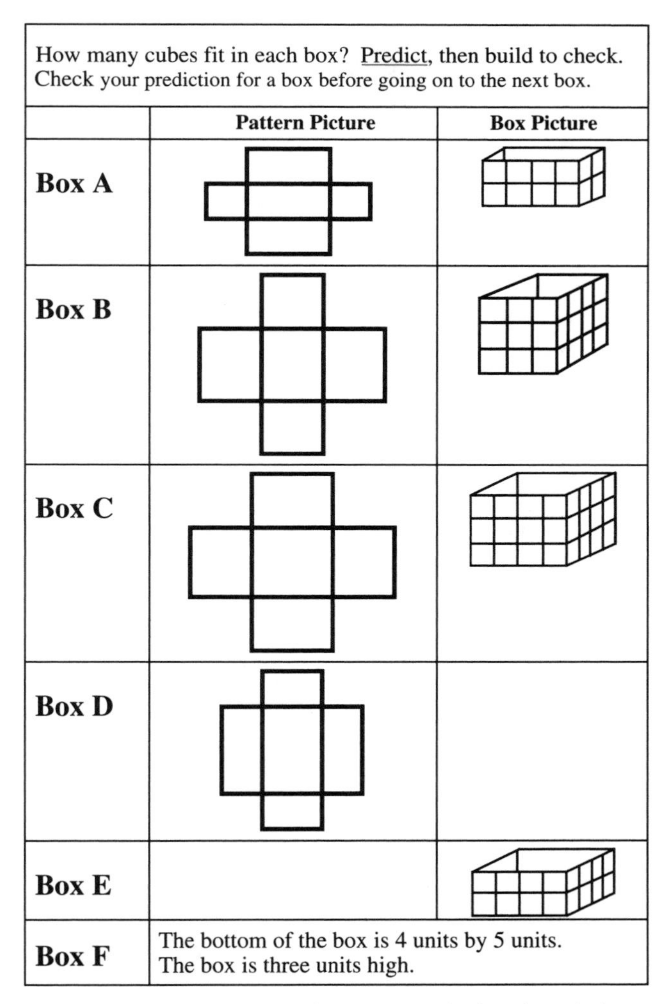

How many cubes fit in each box? Predict, then build to check.
Check your prediction for a box before going on to the next box.

	Pattern Picture	Box Picture
Box A		
Box B		
Box C		
Box D		
Box E		
Box F	The bottom of the box is 4 units by 5 units. The box is three units high.	

Fig. 9.13. How Many Cubes? activity (reduced in size)

Teaching Volume Measurement: Enumerating Cubes in Boxes

Once students can accurately determine the number of cubes in buildings using actual cubes (being allowed to take the buildings apart), they can move on to activities like that shown in figure 9.13 (Battista and Berle-Carman 1996). The goal for this activity is for students to use the four basic mental processes previously described to develop correct strategies for enumerating the cubes. To illustrate how students' thinking progresses with appropriate instruction, one student-pair's work on this activity is summarized (Battista 1999). Especially important in this example is how students gradually overcome their spatial coordination difficulties as they refine their units-locating and organizing-by-composites processes.

The teacher distributed the How Many Cubes? activity sheet (figure 9.13) to each student and explained that the students' goal was to find a way to predict correctly the number of cubes that would fill boxes described by pictures, patterns, or words. Students worked collaboratively in pairs, predicting how many cubes would fit in a box, then checking their answers by making the box out of grid paper and filling it with cubes. Students predicted and then checked their results for one problem before going on to the next.

Episode 1, Day 1

For Box A, Nate counted the 12 outermost squares on the 4 side flaps of the pattern picture (see fig. 9.14), then multiplies by 2. Pablo counted the 12 visible cube faces on Box Picture A, then doubled that for the hidden lateral faces of the box. The boys agreed on 24 as their prediction.

Pablo: [*After putting 4 rows of 4 cubes into Box A*] We're wrong. It's 4 sets of 4 equals 16.

Nate: What are we doing wrong? [*Neither student has an answer, so they move on to Box B.*]

Pablo: [*Pointing at the 2 visible faces of the cube at the bottom right front corner of Box Picture B*] This is 1 box [cube], those 2.

Nate: Oh, I know what we did wrong! We counted this [*pointing to the front face of the bottom right front cube*] and then the side over there [*pointing to the right face of that cube*].

Pablo: So we'll have to take away 4 [*pointing to the 4 vertical edges of Box Picture A*], no wait, we have to take away 8.

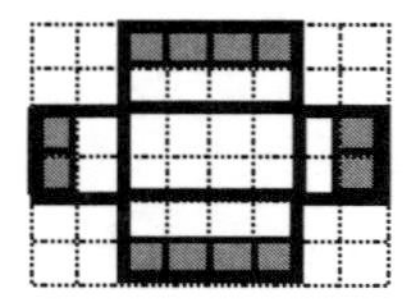

Fig. 9.14. Work of Nate and Pablo

Although Nate and Pablo agreed on a prediction of 24 for Box A, the discrepancy between their predicted and actual answers caused them to reevaluate their reasoning. As the boys reflected on Pablo's strategy, they realized (through the type of mental coordination Randa had so much difficulty with) that he had mistakenly counted the front and right faces of the same cube. When they properly coordinated the positions of these faces in 3-D space, they realized that they were the front and right faces of the same cube. They had made progress in reconceptualizing the situation.

Episode 2

In their prediction for Box B, Pablo counted 21 visible cube faces on the box picture, doubled it for the box's hidden lateral faces, then subtracted 8 for double-counting (not taking into account that this box was 3 cubes high, not 2, like Box A), predicting 42 – 8 = 34. Nate added 12 and 12 for the right and left lateral sides of Box Picture B, then 3 and 3 for the middle column of both the front and back, explaining that the outer columns of 3 on the front and back were counted when he enumerated the right and left faces. He predicted 30. After Box B was constructed, the boys used cubes to determine that 36 cubes fill it. This puzzled the boys, leading them to reflect further on the situation. Nate thought that the error arose from missing interior cubes and tried to imagine the spatial organization of those cubes. Pablo thought the error arose from failing to account properly for the building height in his subtract-to-compensate-for-double-counted-cubes strategy; he attempted to adjust his strategy numerically.

In their predictions for Box B, Nate and Pablo dealt with the double-counting error in different ways. Pablo compensated for the error by subtracting the number of cubes he thought he had double-counted. He adjusted his original enumeration method by focusing on its numeric, rather than spatial, components. Nate attempted to imagine the cubes so he would not double-count them. He focused on obtaining a proper spatial structuring of the array. But neither boy had yet properly structured the array.

Episode 3

Nate and Pablo jointly counted 21 outside cube faces for Box Picture C, not double-counting cubes on the right front vertical edge. They then multiplied by 2 for the hidden lateral sides and added 2 for the interior cubes (which was how many cubes they concluded they had missed in the interior of Box B). Their prediction was 44. The boys made and filled the box and found that it contained 48 cubes. They were puzzled. As they reflected on their error, Pablo concluded that they failed to count some of the cubes in the four vertical edges. However, as in his previous adjustments, Pablo derived his correction by comparing the predicted and actual answers, not by finding an error in his spatial structuring. Nate dealt with the error by continuing his focus on spatial structuring; as a result, he made a conceptual breakthrough on the next prediction.

Nate: I think I know Box D; I think it's going to be 30; 5 plus 5 plus 5 [*pointing to the columns in the pattern's middle*], 15.

> And it's 2 high. Then you need to do 3 more rows of that because you need to do the top; 20, 25, 30 [*pointing to middle columns again*].

Episode 4

On the next problem (Box Picture E), because neither boy was able to employ Nate's layering strategy in this different graphic context, both returned to variants of their old strategies, taking a step backward in their conceptualizing. However, after the boys completed the pattern for Box E and filled it with cubes, Nate commented that his layering strategy would have worked.

Episode 5

For Box F, once the boys drew the pattern, Nate silently pointed to and counted the squares in its 4-by-5 middle section, 1–20 for the first layer, 21–40 for the second, and 41–60 for the third. The boys built the box and filled it with cubes to verify their answer. But they were not at all surprised that they were correct; they were already sure of their answer.

After two-and-a-half one-hour sessions of small-group work, Nate and Pablo—along with the other students in their class—arrived at a layer-based enumeration strategy that they could apply in various situations (Battista 1999). They had moved from the second lowest level of reasoning about cube arrays to the highest level. Of course, this learning was not easy. Nate and Pablo struggled with these ideas. But because they were accustomed to inquiry-based instruction, they intensely and productively, and without getting frustrated, maintained their inquiry spirit to attain the instructional goals.

The predict-and-check approach was essential in students' development of appropriate mental models and enumeration strategies. Because students' predictions were based on their mental models, the cycle of making predictions then checking them with cubes and boxes encouraged them to reflect on and refine those mental models. Having students merely make boxes and determine how many cubes fill them would have been unlikely to have promoted nearly as much student reflection because (*a*) opportunities for reflection arising from discrepancies between predicted and actual answers would have been greatly reduced and (*b*) students' attention would have been focused on physical activity rather than thinking.

Ensuring proper spatial structuring

To ensure that students complete their attainment of Level 6, that they properly structure sets of volume units rather than use a numerical procedure that they do not fully understand, students should next move to the How Many Packages? activity shown in figure 9.15. Rotely applying a proce-

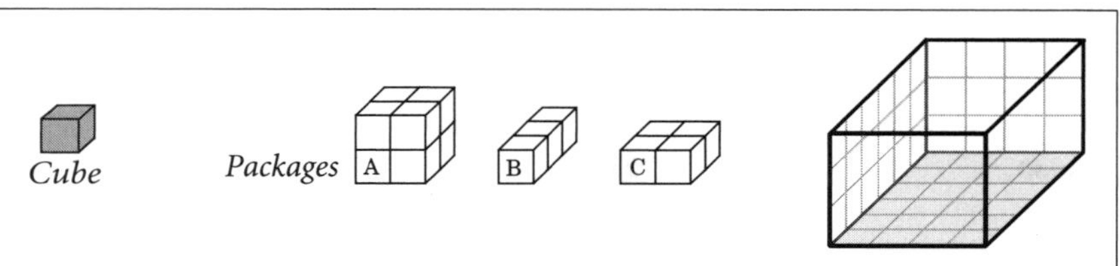

- How many of each package do you predict will fit in the box at the right? The bottom of the box is 6 cubes long and 4 cubes wide. The box is 3 cubes high.
- Use only one type of package at a time. You may not break packages apart.
- Predict, then make the box and fill it with cube packages to check your predictions.

Fig. 9.15. How Many Packages?

dure such as $L \times W \times H$ will not work for these problems. Neither will applying division—reasoning, for example, that because 72 cubes fill the box and Package A is made up of 8 cubes, there are 72 divided by 8 Package A's. Although this strategy works for Packages B and C, it does not work for A because copies of this package do not completely fill the box (since we cannot break packages apart). This activity forces students to attend explicitly to how packages fit in a box, to structure the set of packages properly. Such structuring can also be encouraged by problems such as the two-cube problem shown in figure 9.10.

Extending Students' Conceptualizations

Meaningful use of linear measurements

Once students can properly structure and enumerate squares in 2-D arrays or cubes in 3-D arrays, they should explore what linear measurements reveal about these arrays. For instance, show students an unmarked rectangle, tell them it is 5 cm wide and 8 cm long, and ask, "What is the area of the rectangle in square centimeters? Exactly how and why can length measurements be used to find the number of square centimeters that cover a rectangle?" Be sure that students clearly distinguish length and area units. Similarly, for volume, students can be asked to find the number of cubic centimeters that fit in a closed, unmarked rectangular box.

Fractional and "deformable" area and volume units

When students are fluent in enumerating whole area and volume units, they need to expand their conceptions so that they can reason about fractional units and units that can be deformed to fit in or cover regions. For fractional units, students can be given problems in which the dimensions of boxes are not whole numbers. For instance, they can be asked, "How many one-inch clay cubes will completely fill a box that measures 6 inches by 4 inches by 3 1/2 inches? The cubes can be cut apart."

As students move to more abstract situations, they must expand their conceptualizations to "deformable" area and volume units, that is, units that maintain their areas or volumes but not their shapes. Students must progress from seeing area and volume as the number of rigid squares or cubes that cover or fit in a region to thinking about the actual amount of area or volume in these units. That is, they must move from covering and filling to quantitative comparison. For instance, students can be asked to determine the volume of a cone by filling it with rice, then dumping the rice into a box made from an overhead transparency in which individual cubic centimeters are indicated.

Conclusion

Designing and implementing instruction that supports students' meaningful learning of area and volume measurement must be based on firm understanding of the development of students' thinking about these concepts. Such understanding is essential to teaching in a way that is consistent with professional recommendations and modern research on students' mathematics learning. To enable students to achieve more than superficial understanding of area and volume concepts, instruction should focus on, guide, and support students' movement through the increasingly sophisticated levels of thinking described in this article.

References

Akers, Joan, Michael T. Battista, Anne Goodrow, Douglas H. Clements, and Julie Sarama. *Shapes, Halves, and Symmetry: Geometry and Fractions.* Palo Alto, Calif.: Dale Seymour Publications, 1997.

Battista, Michael T. "Fifth Graders' Enumeration of Cubes in 3D Arrays: Conceptual Progress in an Inquiry-Based Classroom." *Journal for Research in Mathematics Education* 30 (July 1999): 417–48.

Battista, Michael. T. and Douglas H. Clements. *Exploring Solids and Boxes.* Palo Alto, Calif.: Dale Seymour Publications, 1995.

———. "Students' Understanding of Three-Dimensional Rectangular Arrays of Cubes." *Journal for Research in Mathematics Education* 27 (May 1996): 258–92.

Battista, Michael T., and Mary Berle-Carman. *Containers and Cubes.* Palo Alto, Calif.: Dale Seymour Publications, 1996.

Battista, Michael T., Douglas H. Clements, Judy Arnoff, Kathryn Battista, and Caroline Van Auken Borrow. "Students' Spatial Structuring and Enumeration of 2D Arrays of Squares." *Journal for Research in Mathematics Education* 29 (November 1998): 503–32.

10

Structuring a Rectangle: Teachers Write to Learn about Their Students' Thinking

Deborah Schifter

Jan Szymaszek

It's a safe bet that most people who go through school can still, years later, retrieve from memory that area equals length times width, or $A=lw$. Certain, too, is that, as simple as it sounds, few fully understand to what it refers: for example, that the formula applies only to the area of a rectangle; or that length and width are expressed in linear units, but the resulting product, in related square units. Thus, although it may seem simple, this compact equation has complexities embedded within it.

The conceptual issues posed for children working to understand the area of the rectangle are considered here, the medium used being the writings of teachers engaged in inquiry into their students' thinking. The article begins with some background: first about extant research into the conceptual issue at hand, then about the context of the teacher's writings we will use. The next section presents a case written by Jan Szymaszek, a third-grade teacher, in which the results of an assessment exercise she devised for her class are summarized. This is followed by an elaboration of the conceptual issues revealed in Jan's case. The article concludes with a discussion of how teachers' inquiry of their students' mathematical thinking is interwoven with their own learning of mathematics.

This work was supported by the National Science Foundation under Grant No. ESI-9731064. Any opinions, findings, conclusions, or recommendations expressed in this article are those of the authors and do not necessarily reflect the views of the National Science Foundation.

STRUCTURING ARRAYS

In recent years, researchers have been investigating the cognitive issues of understanding rectangles (Outhred and Mitchelmore 1992, 2000; Simon and Blume 1994). Of particular consequence is the recognition that the structure of a rectangular array—equal rows of congruent squares and equal columns of those same squares—is not automatically apprehended by all (Battista et al. 1998). If students do not see the row-by-column structure of an array, how can the area formula of a rectangle make sense to them?

To some adults looking at an array of squares, it may seem as though its structure is in the array, ready to be recognized just by looking. However, visualizing the structure of an array is not merely a matter of reading off the properties of a rectangle. Instead, it is one of many ways of looking at a rectangle and entails the coordination of several insights and skills: identifying a shape's spatial components (squares), combining components into spatial composites (rows and columns), and establishing interrelationships between and among components and composites.

Evidence of the work of structuring may be found in children who, when asked to count the number of squares in the array shown in figure 10.1, first count the squares around the boundary and then those in the middle row, in the process counting some squares more than once, or some not at all. In the words of researchers Battista and colleagues (1998), such children have structured the array as a one-dimensional path, "as if they were traveling along a road with no awareness of their surroundings, as if in a tunnel" (p. 528). Or another child, when asked to draw a copy of the array, might draw in squarelike objects, but not in equal rows or columns, nor covering the entire space.

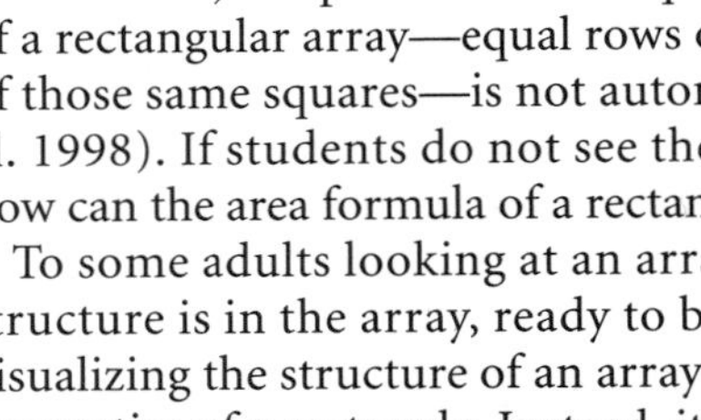

1	2	3	4	5
12	13	14	15	16 6
11	10	9	8	7

Fig. 10.1. Some children count squares in an array by following a path around the rectangle, often counting some squares more than once (Battista et al. 1998).

Through a series of interviews with second graders, Battista and his colleagues identified different levels of sophistication in students' structuring of rectangular arrays, from a complete lack of row or column structuring, to partial row or column structuring, and finally to conceptualizing an array as a set of row- or column-composites that fill the space. Furthermore, the researchers also found that as the children worked on the questions posed, they often progressed to more sophisticated structurings of arrays. As the children acted on the objects, they often recognized the inadequacies of their structurings (for example, if they got different answers on two attempts to count squares), causing them to reflect and revise.

Outhred and Mitchelmore (2000) presented structuring tasks to 115 children, grades 1 through 4, and found similar results to those of Battista and his colleagues. These researchers went on to posit four operational principles central to the development of children's abilities to represent a rectangular covering either in a drawing or in an inferred mental image (p. 161):

1. The rectangle must be completely covered by the units, without overlaps or gaps.
2. The units must be aligned in an array with the same number of units in each row.
3. Both the number of units in each row and the number of rows can be determined from the lengths of the sides of the rectangle.
4. The number of units in a rectangular array can be calculated from the number of units in each row and in each column.

Furthermore, they add, the third principle requires that "children must realize that the length of a side (in centimeters) specifies the number of 1-cm unit lengths that will fit along that side and that this number determines the number of 1-cm unit squares that will fit along that side" (p. 162).

These findings by researchers can be seen in children's classroom work, as is illustrated by the inquiries of teacher-participants in two related professional development projects—Teaching to the Big Ideas and Developing Mathematical Ideas.

TEACHERS WRITING ABOUT STUDENTS' THINKING

For the last ten years, under the auspices of the Teaching to the Big Ideas project and related seminars called Developing Mathematical Ideas (Schifter, Bastable, and Russell 1999a, 1999b), groups of elementary- and middle-grade teachers have come together to inquire into the mathematical reasoning of their students. As an inquiry mechanism, teachers write cases from their classrooms capturing the mathematical thinking of one or more students. These are read and discussed in small groups and responded to and analyzed by project staff.

One of the themes explored by these groups has been how students organize their thinking about two-dimensional arrays. For example, one first-grade teacher brought in the work of her student, Heather, shown in figure 10.2 (Schifter, Bastable, and Russell 2002b, p. 132), who represented a 3-by-5 array of cubes by first tracing its outline and then drawing in squares. Like those children described by other researchers, Heather did not recognize the row-and-column structure of the rectangle. Indeed, the number of squares across the top is not equal to the number across the bottom, nor do the numbers of squares along the sides match. Furthermore, she did not draw her squares to cover the rectangle without gaps. All these facets of spatial structuring are still ahead for Heather. Instead, like other children at the Battista group's first level of structuring, Heather drew squares along a one-dimensional path inside the boundary of her shape.

Fig. 10.2. Heather represented a 3-by-5 array as a rectangle with squares along its boundary.

Another teacher wrote about her third grader, Kalil, who, asked to represent a 4-by-5 array, drew four units across, five below the rightmost rectangle, and then tried to fill in the rest (see fig. 10.3 [Schifter, Bastable, and Russell 2002a, p. 96]). Realizing that his multiplication fact, $4 \times 5 = 20$, was supposed to tell him the number of cubes in the array, he stopped counting at 24. "I couldn't really count it," he said. "It's so messy." The teacher recorded two interviews with Kalil (pp. 95–103) in which, in response to her questions and using a handful of tiles in addition to paper and pencil, he worked his way to an understanding of the rectangle's array structure. By the second day, he could build a 3-by-4 array and demonstrate how it can be seen as 3 groups of 4 and also as 4 groups of 3 (see fig. 10.4).

After considering research findings and studying classroom cases written by others, teachers turn back to their own students' work with a more discerning eye, now aware of the complexity of the structure of an array and its relationship to the dimensions of a rectangle, multiplication, and area. Such was the experience of Jan Szymaszek whose case, brought to her peers and instructor and now presented below, inquires more deeply into what her students reveal through their work.

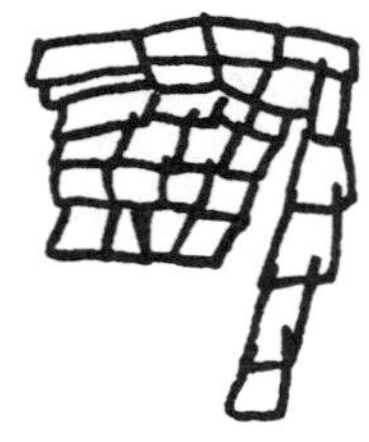

Fig. 10.3. Kalil drew a "frame" and filled in a 4-by-5 array. (From *Developing Mathematical Ideas: Measuring Space in One, Two, and Three Dimensions Facilitators' Guide* by Deborah Schifter, Virginia Bastable, and Susan Jo Russell. © 2002 by the Education Development Center, Inc., published by Dale Seymour. Used by permission of Pearson Education, Inc.)

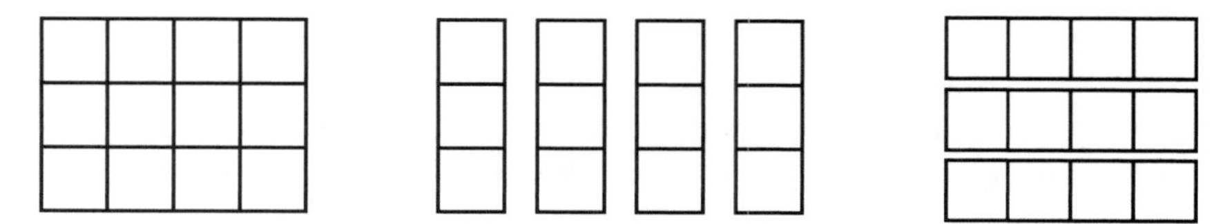

Fig. 10.4. By the end of the second interview, Kalil demonstrated that a 4-by-3 array can be seen as 4 groups of 3 and as 3 groups of 4. (From *Developing Mathematical Ideas: Measuring Space in One, Two, and Three Dimensions Casebook* by Deborah Schifter, Virginia Bastable, and Susan Jo Russell. © 2002 by the Education Development Center, Inc., published by Dale Seymour. Used by permission of Pearson Education, Inc.)

Cover Up or Use a Ruler? Working on the Area of Rectangles

My third-grade class had been working on the area of rectangles and other shapes. As area tools, the students were using 2″ × 2″ squares of purple construction paper. Students would cover up the given shape and report the area in "purple square units." After being introduced to the square-inch measure, they figured out that a purple square unit was equal to 4 square inches. Next they measured rectangles using different area tools. They made rectangular arrays composed of purple squares and square inches to show equivalent areas. They also recorded in inches the dimensions of the rectangles.

From these explorations, students became quite comfortable with purple squares and square inches. Most of them seemed to connect this work with rectangles and dimension immediately to their work with arrays and multiplication. However, some subtle ideas were at play here. I wasn't sure if my students were making a connection between the linear measures of the sides of a rectangle and its area measure in square units.

To check how these connections were developing, I designed an assessment exercise that would allow me to examine systematically my students' ideas. I distributed three-by-five-inch index cards and asked students to find the area of this card, using whichever method they preferred. I offered purple squares, white one-inch squares (which the class called "square inches"), and rulers. I also gave them a separate worksheet to record their thinking. I knew I would need to watch them as they worked, as well as examine their worksheets carefully.

Eli, Peter, and Cal all used rulers to measure the sides of their index cards. They marked off the sides in inches and drew in the array, using their rulers as straightedges to draw lines across and down the rectangle. On the recording sheet, Eli re-created the array and wrote, "3 × 5 = 15." Peter wrote, "1. I drew the square inches. 2. I knew it was 3 × 5 and 3 times 5 = 15." (See Eli's and Peter's work in fig. 10.5.)

First Cal represented his work as shown in figure 10.6a, without indicating multiplication. After he had done the assignment, he came up with another way to think about it and wrote that up, too (fig. 10.6b).

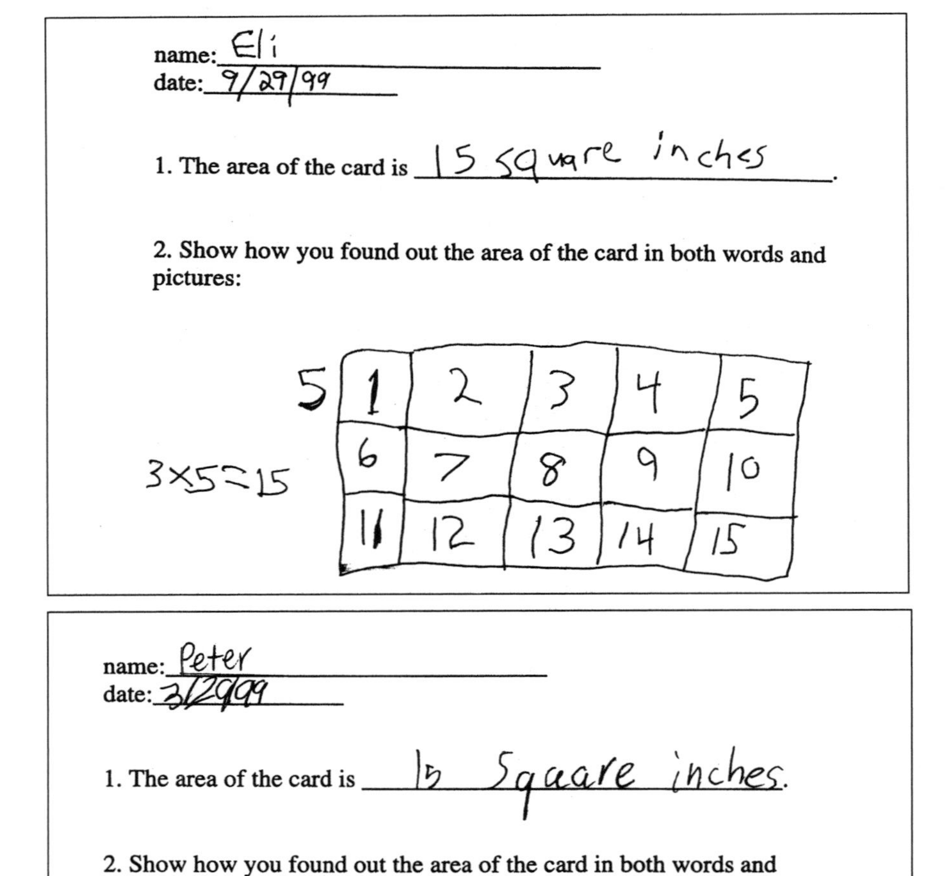

Fig. 10.5. Eli and Peter used rulers to fill in squares and wrote 3 × 5 = 15.

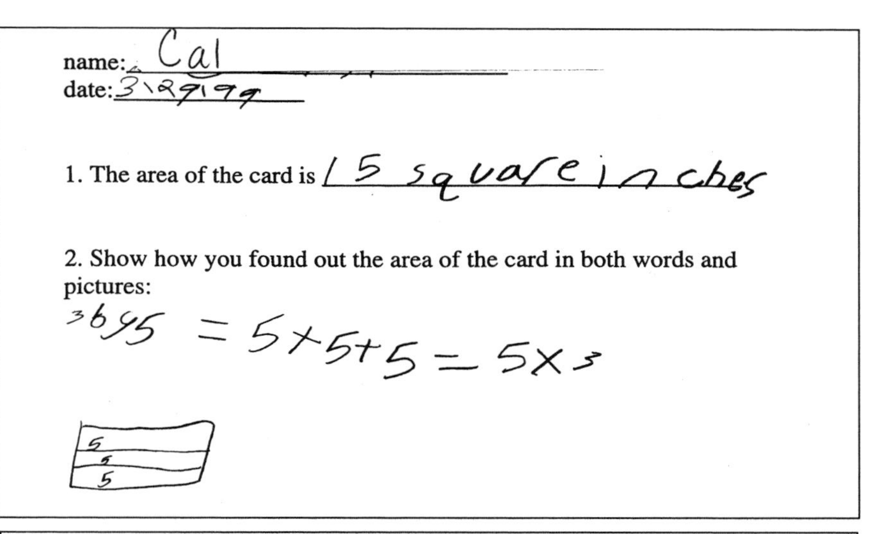
name: Cal
date: 3\29\99

1. The area of the card is 15 square inches

2. Show how you found out the area of the card in both words and pictures:

2. Show how you found out the area of the card in both words and pictures:

5 square inches + 5 square inches + 5 square inches

Fig. 10.6. Cal first drew the rectangle as a 3-by-5 array of squares and then as three rows each labeled 5.

Some children, like Rebecca (fig. 10.7), used a cover-up method and found that it took 15 white square inches to cover the card. On their recording sheets, these children drew in each square individually. Rebecca indicated that she knew the dimensions of the card but didn't seem to apply addition or multiplication to get to 15.

I was intrigued by the group of students who used a collection of purple and white squares to cover the area. After placing two purple squares on the card, they could tell how many more square inches were needed to cover the card completely. Karen wrote, "If the purple square is 4 square inches, then the top is 2 and the bottom [*her arrow points to the side*] is 2. The first line is 5 with two purples + a square inch. Same with the rest, except the bottom line is all square inches" (see fig. 10.8a.).

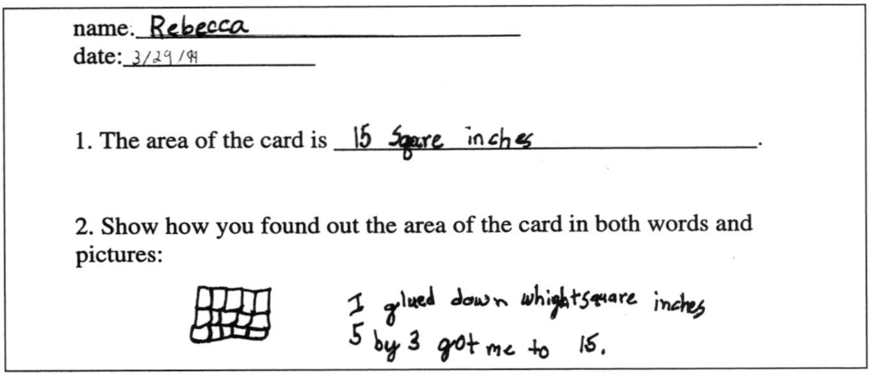

name: Rebecca
date: 3/29/99

1. The area of the card is 15 Square inches.

2. Show how you found out the area of the card in both words and pictures:

I glued down whightsquare inches 5 by 3 got me to 15.

Fig. 10.7. Rebecca explained how she used white square inches to find the area.

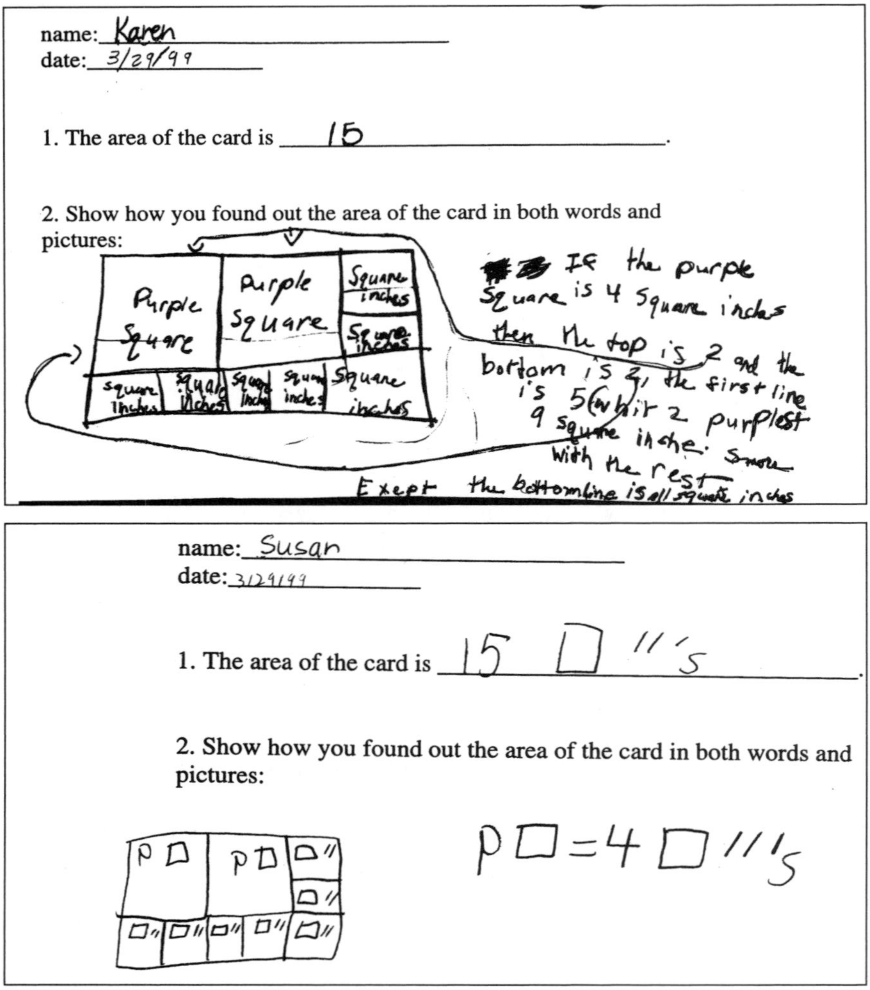

name: Karen
date: 3/29/99

1. The area of the card is 15.

2. Show how you found out the area of the card in both words and pictures:

Purple Square | Purple Square | Square inches | Square inches
Square inches | Square inches | Square inches | Square inches | Square inches

If the purple Square is 4 Square inches then the top is 2 and the bottom is 2, the first line is 5 (whit 2 purpl) 9 square inche. Same with the rest Exept the bottomline is all square inches

name: Susan
date: 3/29/99

1. The area of the card is 15 □ "s.

2. Show how you found out the area of the card in both words and pictures:

P □ | P □ | □" | □"
□" | □" | □" | □" | □"

P□ = 4 □ "'s

Fig. 10.8. Karen and Susan used a composite of different units to measure the area.

As shown in figure 10.8b, Susan also used a collection of different units and explained her representation with a code to show that a purple square is equivalent to four square inches.

I was confused by the work of two students, Anne and Cathy. I had seen that Anne (fig. 10.9) had begun with a ruler and on her recording sheet had drawn a rectangle with sides labeled “5” and “3” and had written, “I [used] square inches. I only [used] 8 because if 5 × 3 = 15, 8 of...” but then didn't finish and crossed out what she had written. For question 1, she did write in the correct answer, “15 square inches.” I would like to know what she was thinking when she wrote, “I only [used] 8.” Was there an idea that she then abandoned? If so, why did she abandon it? What does the “5 × 3 = 15” mean to her, and how does she explain her answer, 15 square inches? Is she making the important connections, or are there some missing links? I wish I had gotten to her sooner to see how she was thinking.

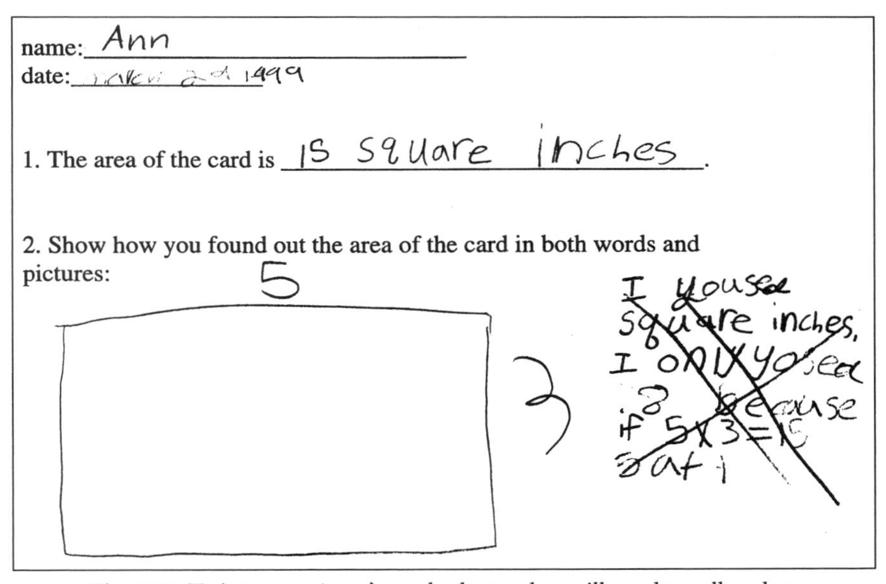

name: Ann

date:

1. The area of the card is 15 square inches.

2. Show how you found out the area of the card in both words and pictures:

Fig. 10.9. To interpret Anne's work, the teacher will need to talk to her.

Cathy also used a ruler to find the dimensions of the sides of the index card. Having done that, she represented the card as shown in figure 10.10. What are the ideas that Cathy still needs to put together?

All in all, I was able to learn a lot about how my students were thinking about measuring area, though I do have more questions. In the days to come, I will investigate how much the cover-up method will support the connection between these physical manipulations and the linear measurements of the dimensions.

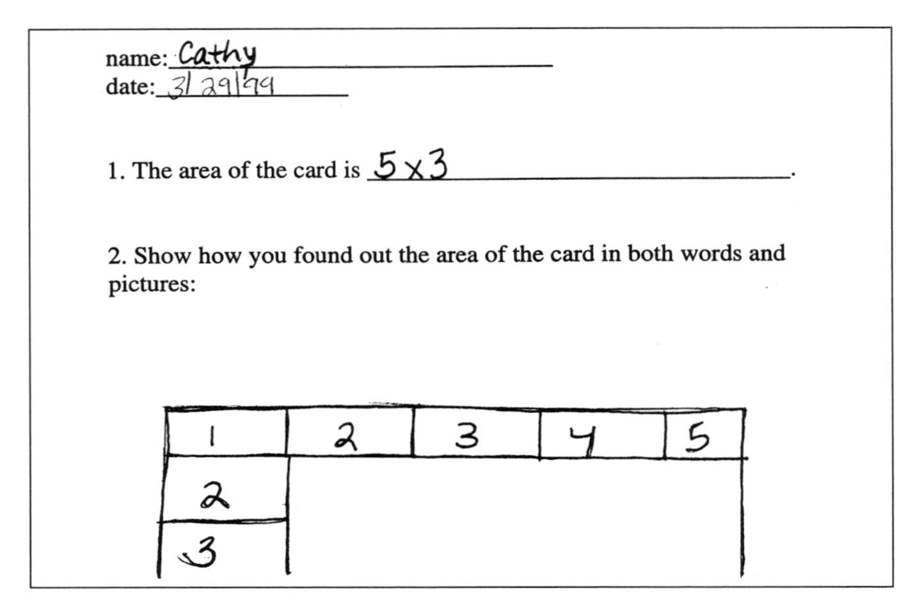

name: Cathy
date: 3/29/99

1. The area of the card is 5 x 3.

2. Show how you found out the area of the card in both words and pictures:

Fig. 10.10. Cathy measured the length and width of the index card and concluded the area is 8.

ELABORATING ON JAN'S CASE

A case like Jan Szymaszek's offers her the opportunity to come together with a group of colleagues to inquire into students' understanding. What are the conceptual issues revealed in this piece of writing? Although Jan has carefully described her students' work, there is much about her students' thinking that bears examination. In this section, we point to some of the issues that could be pursued.

First, Jan's case illustrates what is so common in any class: given a complex mathematical issue, students address relevant problems with a variety of resources, skills, and understandings. As often happens, Jan's assessment has answered some of her questions but has also raised new ones.

Jan set out to address a fairly specific question: how were her students connecting the linear measures of the sides of a rectangle to its area measured in square units? For some of her students, she can confidently answer that they are making important connections. Eli, Peter, and Cal, for example, first measured the index card with rulers to identify the length and width in inches and then drew in the array of 1-inch squares, putting into action the idea that the lengths of the sides in inches determine the number of square inches to be drawn in each row and column. These three children also multiplied the dimensions and came up with the answer of 15 square inches.

There are also differences among these children's work. On his recording sheet, Eli reconstructed the array and counted squares. One can note the order of his count, row by row, in contrast to the count shown in figure 10.1, around the outside and through the middle.

On Cal's recording sheets (fig. 10.6), his two representations of the index card—one as a 3-by-5 array of squares, the other as three rows each labeled 5—show the connections he has made among the array, repeated addition, and multiplication. His second drawing seems to show how he sees the row as a unit of five squares, iterated to cover the rectangle.

Although most of the other children correctly solved the problem, too, they didn't provide evidence that addressed Jan's original concern: Were they connecting the linear measure of the rectangle's sides to its area as measured in square units? These children covered the rectangle with square units, some of them using multiple units, and determined the area by counting up square inches. It is not clear, therefore, whether they see how the number of such units is related to the length and width of the rectangle. They do demonstrate an understanding that the rectangle must be completely covered by the units, without overlaps or gaps, and the units must be aligned in an array with the same number of units in each row and column.

When looking at Karen's and Susan's recording sheets (fig. 10.8), one might consider the work done in earlier years to make sure children understand that measurement entails using the same unit repeatedly. For example, you can't measure the size of a room with paces of different length. It would seem the children here are breaking that rule, using both purple squares and square inches to measure the area of the card. But, of course, they're not, because they understand how the two units are related. Since 1 purple square is equivalent to 4 square inches, they know that 2 purple squares and 7 square inches make 15 square inches. This unit coordination is like saying that 3 nickels and 4 cents are worth 19 cents.

In addition, Karen's writing may suggest ideas about linear dimensions. She explains that since a purple square is 4 square inches, the top and bottom are 2 and the "first line is 5." One might conclude that she is paying attention to one dimension of the rectangle: the length is 5 inches. However, since she has not identified units, we do not know if she is counting inches (the length of the rectangle) or square inches (the number of squares in the top row), or both.

There are several things to notice about Cathy's work (fig. 10.10). She drew in a row of five rectangles and a column of three rectangles (indicating units of area), corresponding to the lengths of the sides measured in inches (linear units), which she found by using a ruler. She also drew the corner rectangle as shared by the row and the column, labeled "1" for the count in each direction. Thus, she seems to understand the relationship between the linear measure of a side of a rectangle and the square unit that measures the rectangle's area. However, she hasn't filled out the array and has written that the area of the card is 8 (perhaps by adding 5 and 3). Cathy's work, along with Anne's, has alerted their teacher to pay particular attention in coming days. Have they yet to see how rows (or columns) of squares are iterated to cover the rectangle, or how rows and columns are coordinated?

In Jan's work with her class on the area of a rectangle, she had identified a key idea: that her students must make a link between the linear dimensions of a rectangle and the number of square units in its rows and columns. Through her assessment question, she discovered that some students appear to have made the link; others, she can't tell. Of those others, most have put together other important ideas: that to measure a rectangle's area, it must be completely covered by the units without gaps or overlaps, and that square units must be aligned with the same number of units in each row and column. Yet her assessment left her with questions about two of her students: what do Anne and Cathy understand?

Even while she—and we—are left with questions about individual students, the worksheets of students in this class provide learning opportunities for us. Their work is not idiosyncratic. Rather, each child's paper points to conceptual issues common among children. By studying the work of Jan's class, other teachers can anticipate the thinking of their own students, thus becoming better prepared to interpret the ideas the children present and better positioned to help them move their ideas forward.

TEACHERS LEARNING MATHEMATICS WHILE STUDYING CHILDREN'S THINKING

The issue of visualizing the structure of rectangles and Jan Szymaszek's case are offered in this article to illustrate what can be learned when teachers write about their students' thinking and meet with colleagues to discuss such writing in a disciplined way. In the projects that spawned Jan's case, teachers work through a set of related concepts, exploring the mathematics for themselves while they examine students' engagement with the same content.

For example, in a seminar on measurement (Schifter, Bastable, and Russell 2002a, 2002b), the structure of a rectangular array is one of several issues explored. Other issues include students' confrontation with the ambiguity of the words *big* and *size* as they come to see that any object has a variety of measurable attributes; what a unit is, and what young children must understand about units in order to measure; ideas of dimensionality, and the different kinds of units used to measure length, area, and volume. The issues of structuring rectangles are extended to the visualization of 3-D arrays.

Through their explorations into these themes of measurement, teachers develop their own understanding of the mathematics. They, too, work on their skills to decompose and recompose shapes, figure out the reasoning behind formulas for area, visualize two- and three-dimensional structure, and investigate the complex relationship between area and perimeter and between volume and surface area.

In such seminars, there is a rich interplay between teachers' discoveries of their own ways of approaching the mathematics and their investigations into students' understanding of related topics. At a general level, through their own learning of mathematics, teachers experience the process, working through confusions, sometimes frustration, persisting in order to reach the satisfaction of new insight. Teachers who themselves experience the process are better able to guide their students through it.

As teachers learn mathematics to develop a rich network of connections among ideas, they are able to identify when a student broaches an important concept. And as they identify conceptual challenges to their students, they also learn how to call on their students' reasoning as a resource for learning. We have seen the potential for this in Jan's case: Cathy's work in figure 10.10 reveals mathematical insight even though she incorrectly concludes that the area of the card is 8. With an awareness of what Cathy *does* understand and how this is related to what Cathy still needs to learn, Jan can help to move her from where she is to where she needs to go. Through examining students' work together, teachers can develop the disposition to find the strengths in each child's mathematical thinking to help fill conceptual gaps.

"Here's something I've learned that I call key," one teacher declared on the last day of a seminar. "I've learned that my students—all of our students—have far more prior knowledge and reasonable ways to think about measurement than I had ever imagined" (Schifter, Bastable, and Russell 2001b, p. 218).

REFERENCES

Battista, Michael T., Douglas H. Clements, Judy Arnoff, Kathryn Battista, and Caroline Van Auken Borrow. "Students' Spatial Structuring of 2D Arrays of Squares." *Journal for Research in Mathematics Education* 29 (November 1998): 503–32.

Outhred, Lynn N., and Michael Mitchelmore. "Representation of Area: A Pictorial Perspective." In *Proceedings of the Sixteenth Psychology in Mathematics Education Conference,* Vol. 2, edited by William Geeslin and Karen Graham, pp. 194–201. Durham, N.H.: Program Committee of the Sixteenth Psychology in Mathematics Education Conference, 1992.

———. "Young Children's Intuitive Understanding of Rectangular Area Measurement." *Journal for Research in Mathematics Education* 31 (March 2000): 144–67.

Schifter, Deborah, Virginia Bastable, and Susan Jo Russell. *Developing Mathematical Ideas.* Parsippany, N.J.: Dale Seymour Publications, 1999a.

———. "Teaching to the Big Ideas." In *The Diagnostic Teacher: Constructing New Approaches to Professional Development,* edited by Mildred Z. Solomon, pp. 22–47.

New York: Teachers College Press, 1999b.

———. *Developing Mathematical Ideas: Measuring Space in One, Two, and Three Dimensions Casebook.* Parsippany, N.J.: Dale Seymour Publications, 2002a.

———. *Developing Mathematical Ideas: Measuring Space in One, Two, and Three Dimensions Facilitator's Guide.* Parsippany, N.J.: Dale Seymour Publications, 2002b.

Simon, Martin A., and Glenn W. Blume. "Building and Understanding Multiplicative Relationships." *Journal for Research in Mathematics Education* 25 (November 1994): 472–94.

11

About Students' Understanding and Learning of the Concept of Surface Area

Cinzia Bonotto

THE RESEARCH Center in Mathematics Education at the University of Padua (Italy) has been working on the ideas expressed by the Italian programs for primary schools and has been developing new approaches to the mathematical concepts that were the objectives of the curriculum. Since 1993–94, it has been implementing certain elementary school activities to capitalize on the knowledge and techniques students usually develop outside school. In brief, we believe that the development of formal mathematical knowledge should lead to algorithms, concepts, and notations that are rooted in a learning history that starts with the students' informal, experientially real knowledge. The idea is not only to motivate students with everyday-life contexts but also to look for contexts that are experientially real for the students and can be used as starting points for developing new mathematical knowledge (Gravemeijer 1999).

Traditional classroom teaching often seems to favor the separation between classroom and real-life experience. This rift is particularly significant with respect to the concept of *surface*. Outside school, any child is capable of recognizing a real surface, in particular a plane surface. The child distinguishes between it and a line or a number, considering it as a two-dimensional space. In school, though, pupils tend to identify a surface [geometric object] with its area, that is, its measure with respect to a predetermined two-dimensional unit of area [numerical object], thus considering surfaces only at a numerical level. We think that one reason for this may lie in the fact that not much importance is given to the concrete act of measurement. With the rectangular surface, the fact that measuring does not take place impedes children's understanding of the meaning and role of area units. It is also an obstacle in comprehending why a formula is used to calculate a surface area, whereas an instrument is used to measure a line.

The importance of taking direct measurements came out clearly in a study carried out by Nunes, Light, and Mason (1993). Children were asked to carry out comparisons of magnitudes (length and area respectively) in situations where simple visual estimate or direct superposition did not suffice. Thus, they had to rely on measurement tools like a ruler or area unit to obtain information about the magnitudes to reach a conclusion. The results showed a strong correlation between the number of correct answers and the application of a measuring strategy based on a count of area units. Not only is the use of an area unit important, it is also important to be able to recognize the way that the units are organized. If the formation is analyzed in rows and columns, one is able to explain the formula that is in fact used to calculate the area of a rectangular figure (Outhred and Mitchelmore 2000).

Many studies stress that both elementary and secondary school students have an inadequate understanding of the area concept. Several authors attribute this poor performance to a tendency to learn the area formula by rote (see, e.g., Tierney, Boyd, and Davis 1990). This may result from the fact that teachers do not allow children enough time to develop an understanding of the multiplicative structure of rectangular arrays.

In the project described here, we compared two methods of teaching the concept of surface area. One was a traditional method, based on the acquisition and application of the formula for the area of a rectangle in a fifth-grade class. The other was a more innovative system that drew on students' out-of-school, real-world experiences, in a fourth-grade class. Whereas the fifth graders simply learned to apply the formula, the fourth graders learned to derive it, for example, by counting how many area units (small squares) there were in a real surface that they could easily manipulate (a sheet of paper). This procedure reflects a strategy—that of evaluating the size of a rectangular surface by the number of small squares it contains—with which the children are familiar; indeed, they often used it outside school.

Background to Instruction

We deem it important to immerse children in a classroom culture that focuses on the importance of activities of realistic mathematical modeling, namely, activities that are both real-world-based and quantitatively constrained. Such activities enhance the relationships between informal out-of-school mathematics and formal in-school mathematics in the development of abstract mathematical knowledge while preserving the focus on everyday meaning. This allows students to become involved with mathematics, to revise their views of it as a remote body of knowledge, and to develop a positive attitude toward school mathematics.

For this to happen, two changes are necessary: one in teachers' attitudes toward mathematics, and one in the norms that guide student-teacher relationships. Our studies, presented as a possible model for others, take into account these factors. We used objects that incorporate mathematical elements that a person encounters in everyday situations—for example, a receipt from a supermarket, bottle and can labels, and so on (Bonotto 2001). Material like this is particularly meaningful because through it students learn to analyze and interpret the reality around them in mathematical terms. In our experiments, pupils were introduced to mathematics in a way that would as to make it easier for them to move from situations in which mathematics is normally used to the underlying mathematical structure and from the mathematical concepts to the real-world situations. Students are expected to approach a problem as a situation to be organized and structured mathematically, not primarily as a situation for application of ready-made solution procedures. This does not mean that the student's knowledge of solution procedures does not play a role, but that the primary objective of the student is to make sense of the problem.

Furthermore, in the classroom activities we applied a variety of instructional techniques, involving both an extensive use of the children's own written descriptions of the methods they adopted and whole-class discussions.

First Study

Subjects, Material, Hypothesis

The first study was carried out in the second quarter of the 1998–1999 school year. It took place in the fifth grade of a primary school of a small school in a village in northeastern Italy, with twenty pupils. The class had already dealt with the concept of surface and had used tangram pieces to construct and to recover some plane figures. (Tangram is an old Chinese game in which a square has been cut up in a particular fashion into seven differently shaped pieces. The aim of the game is to reproduce diverse geometric figures through a combination of those seven pieces.) Square measures had only just been briefly mentioned. The class was accustomed to traditional teaching methods.

Three sessions were led, each lasting about two hours. The teacher was present throughout, along with two observers. Each session used photocopies of the cover of a rectangular ring binder of a kind currently sold in stationery stores. Each pupil was given a photocopy of that cover, which contained the following data: the number of sheets (90) in the binder and the dimensions (21×29.7), which correspond to a standard European Union paper format known as A4 size, and the size of the individual squares in the graph paper of the binder (5 mm).

In the first session, after a brief introductory discussion during which the class learned to read the numerical data supplied on the label of the binder, the children were asked to answer a few questions in writing. Three of the questions were as follows:

1. To you, what does 21×29.7 mean? What unit of measure do you think 21×29.7 is written in?
2. What do you think the surface of each sheet in the binder is?
3. If you lay out, in any way you wish, all 90 sheets, without any overlap, do you think they will cover a surface of 1 m^2?

The aim of these questions was to understand what the pupils meant by surface.

Each child was given the same photocopy one month later. At this point the next two experiments were carried out. In the second experiment, we especially wanted to bring out visual comparison, using the answers the pupils had given to the previous questions as a starting point for a general discussion. Having analyzed the photocopy, the children were given the possibility of reviewing the answers they had given the preceding month and of correcting any mistakes in them.

The last session came one week later. Here we wanted to look at the coverage, or filling in, with area units. The pupils were given the same photocopy and asked to answer one question in writing:

4. How many squares whose sides measure 5 mm can you draw on one sheet? Write how you plan to figure this out.

The objective of the last two sessions was to analyze how and to what extent, in real-life situations, the children managed to apply the knowledge they had acquired through traditional teaching methods based on the acquisition and application of the formula for the area of a rectangle.

We conjectured that a kind of teaching, based substantially on a drawing of a rectangle with the formula $b \times h$ next to it, promotes the children's tendency to identify the concepts of surface and surface area, and that this way of teaching leads pupils to think that the only valid way of determining the size of a surface is applying the formula. Moreover, we suspected that if one presented the size of a surface as the product of two linear dimensions, the meaning and the real representation of square measures would not be made clear.

We therefore expected the fifth graders to consider surface purely at a numerical level, mechanically associating a square unit of measure to that number by applying the formula and mnemonically remembering the rule for going from one square measure to another.

Some Results

The answers given to questions 1, 2, and 3 confirmed our conjectures. The children equate the concepts of surface and surface area; see, for example, the following answers of Maura and Andrea to the first question:

> In my opinion, 21 × 29.7 indicates that if I multiply these two numbers, I can get the surface of the sheet and also its unit of measurement…

> 21 and 29.7 indicate width and length which, multiplied by each other, give the surface of the sheet …

and those of Marina and Fabio to the second question:

> Each sheet will have this surface: 21 × 29.7 = 623.7 cm^2 …

> Every sheet will measure 21 × 29.7 = 623.7 [*he multiplies the figures in columns*], a surface of 623.7 cm^2 for each sheet because surface equals area.

In the answers to the third question, we observed a purely arithmetical procedure, correct and undoubtedly effective, consisting of multiplying the area of one sheet of paper by the total number of sheets, and changing the result to square meters. Some children did this last operation directly, simply by moving the decimal point, like Alessandra:

> If we lay out all the sheets over a surface of 1 m^2, for me we'll cover it, even more. If we multiply 623.7 by 90, we'll get 56133.0 cm^2 [*she multiplies by columns*]. I change 56133.0 cm^2 = 5.6133 m^2.

Other children explicitly wrote the relation between cm^2 and m^2, like Tiffany:

> I think that the 90 sheets of 623.7 cm^2 will manage to cover 1 m^2 because if I multiply 623.7 × 100 the result is 62370 cm^2 and since 1 m^2 corresponds to 10000 cm^2 I can say that the 90 sheets can cover 1 m^2 and even more (they can cover up to about 6 m^2).

The idea of coverage came out clearly in only three children's answers. Among these, the method Andrea adopted is interesting:

> To me, I lay out 90 sheets inside of blocks that can take 6, I use 16 blocks of 6 sheets, and I should get 1 m^2….

Thus, after Andrea noticed that 6 sheets were necessary to cover his desk, he used his desk as a new unit of measure. Following his intuition, he managed to use a strategy that is much closer to that used outside school. There is an intuitive aspect involved in covering one surface area with another, and this can prove to be useful in understanding the role and meaning of square measures.

In the last experiment, as we expected, the pupils had recourse to arithmetic operations, dividing the area of the sheet of paper by the area of the 5 mm × 5 mm square. Only three children took a different path, probably imagining the sheet covered by little squares. These children calculated the number of squares that could be drawn in a line along each of the two sides of the sheet, and then they multiplied the numbers they had obtained. All the children showed difficulty in thinking of the small squares as a sample unit that could be used to arrive at the size of the sheet.

In this class, there was clearly very strong conditioning taking place because of formal procedures. The children regularly had recourse only to arithmetical procedures to solve problems that could have been approached through real-life, practical means. They seemed to identify the concepts of surface and the measurement of surface; the rectangular surface was the product of the lengths of the sides, to which one automatically attributed a square unit of measure, and the square unit was an abstract entity that one obtained by measuring two equal linear units of measure.

From the classroom discussions, which for reasons of brevity we cannot report here, it became evident that there was a difference for the children between the way one evaluates a figure in school and the way one estimates a real surface outside school. This led to an incoherence between the answers they gave when they thought of the sheet as a rectangle and those given when they simply observed a sheet of paper. For these pupils it was very important to arrive at an answer to the problem, and it did not necessarily matter if their answer was coherent with the figure they were studying or not.

SECOND STUDY

Subjects, Material, Hypothesis

This study was also carried out in the second quarter of the 1998–1999 school year. It took place in the fourth grade of a primary school of a small school in a village in northeastern Italy, with twenty-two pupils. In this class the concept of surface area had not yet been dealt with, although the children had played with tangram pieces when the teacher was teaching them about perimeter. It should also be noted that in this class other mathematical concepts had been introduced starting from out-of-school situations. Thus, the children were used to a certain kind of instruction that encourages them to view situations in their daily lives from a mathematical perspective. Four sessions were carried out, at one-week intervals, each lasting two hours. The teacher was present throughout, along with two observers.

In the first session, each child was given a photocopy of the cover of a loose-leaf ring binder containing graph paper, such as those currently sold in stationery stores. This cover contained the following data: the number of sheets (90) in the binder and the dimensions (15×21), which correspond to a standard European Union paper format known as A5. In photocopying the binder label, we deliberately concealed the information pertaining to the size of the graph-paper squares, so as not to complicate the instructions we wanted to give. Having briefly read the label, the children were supposed to answer three questions in writing:

1. To you, what does 15×21 mean?
2. What unit of measure do you think 15×21 is written in?
3. Choose and write the dimension of the little squares, which no longer appears on the label. Then tell me how many of these little squares you

need to fill up one sheet of paper.

The aims of the first session were to analyze how the label was read and to introduce the comparison of surfaces.

The class examined the answers and discussed them. During the discussion, rectangular sheets of different sizes were used. Then the children were asked to do the following exercise for the next week: cover the blank surface of the cover of the binder photocopy with 1-cm squares.

This exercise was to be the basis of the second session, in which we wanted to introduce coverage with area units. In this second experiment, four questions were asked, although not simultaneously; a question was asked only when all the children had answered the preceding one. The first three concerned the sheet that the children had filled with little squares.

1. How many little squares did you use to fill in the page? Write how you counted them and why you did it that way.
2. What did you need the little square for? What does the little square mean to you?
3. To measure the surface of the sheet, could you simply have used a ruler? Explain your reasoning.

The last question concerned another rectangular sheet that had been distributed to each child of the class, filled in only partially with little squares.

4. Quickly tell me the surface area of the part of the sheet that is covered by little squares.

In the third session, the children had to answer three questions about two rectangular sheets they had been given. The sheets were of equal surface area, and they were congruent; one was filled with squares measuring 1 cm × 1 cm, the other with squares of 5 mm × 5 mm.

1. Describe the two sheets.
2. Calculate the surface area of each of the two sheets, look at the results, and comment on them.
3. What does the little square represent in each of the two sheets?

In the fourth session each child was given a sheet of millimetric graph paper, which served as the basis of a general discussion.

In these last three sessions, the aims were (*a*) to observe how the children analyzed the surface, (*b*) to observe how clear the role of the little square as a unit of measure was to the children, and (*c*) to introduce the mathematical way of writing a square measure. We wished to propose an alternative to the standard teaching method. Our method consisted of visually comparing real surfaces that are limited in size and easily manipulable, so as to be able to estimate their areas and then verify the estimate by filling in or covering the surfaces with small squares and counting them. In our opinion, the visual comparison should make the meaning of *surface* clear from the outset. The second step was for the pupils to approach the problem of measuring a surface. Since they could not use a ruler, using area units would allow them to measure a surface directly, without going through the process of measuring linear dimensions, avoiding the related problems involved with this process.

This way, in our opinion, it would become much clearer to the pupils (*a*) that comparing different surfaces does not mean only comparing the numbers one gets by calculating their areas, (*b*) that it is therefore necessary to distinguish between a surface area and measuring that surface, (*c*) that visual comparison can be verified by real measurement of the surfaces, (*d*) what square units of measure are and why they are used, and (*e*) what the real meaning of the formula $b \times h$ is.

Some Results

During the whole-class discussions, the children appeared to understand the concept of surface very clearly; they themselves defined it as the "internal" space of the sheet of paper. They did not have difficulty in estimating size through observation and then verifying their idea by placing one surface on top of the other.

When asked to fill in their photocopy with 1 cm × 1 cm squares, instead of drawing each single square, the children drew two bands of squares along the two perpendicular sides of the sheet, which gave a rectangular array of squares. For the children, counting the squares meant multiplying the number of squares drawn along the two perpendicular lines, since they had already used this procedure when they learned to multiply. Thus, applying it here, they themselves introduce the usual formula for the area of a rectangle.

The third session allowed us to observe the degree of comprehension of the concept of surface area and the role of the square as an area unit. In the preceding sessions in which approximate estimation and numerical calculation were introduced, the surfaces were all filled with 1 cm × 1 cm squares; in this instance, the unit areas were different. We had thought that the use of squares of different sizes would be an obstacle for the pupils in recognizing and verifying equal surfaces. We noted, however, that all the children calculated the sizes of the two surfaces by multiplying the number of squares along the two sides of the respective rectangles, and some children explained that the different results they obtained were due to different unit samples. An example is Caterina's answer:

> The sheets have the same surface area and only the number of little squares changes because the size of the squares is a sample. In the two sheets, there's a different sample: a 5 mm one and a 1 cm one. The surface area of the two sheets is this: The 1 cm one = 80 cm. The 5 mm one = 320 mm. The surface size doesn't

change, only the number of little squares does. The difference is of 240 squares, 320 – 80 = 240.

Next to the number obtained, everyone specified which size of squares they had used, that is, which area unit, to calculate the number. This means that for these children the surface area of a figure is not just a "pure" number but rather it represents the magnitude of some entity.

There is no doubt that the role of the little squares as representative of a sample was clear. In fact, none of the pupils contradicted themselves in the process of observing, first, by superposition, that the sheets had equal surfaces, and then by seeing that there was a difference in the numerical results they had gotten by directly measuring the surfaces. The answers to the third question gave further confirmation that the students had grasped this role of the little squares as a sample. Andrea's answer is an example:

> In each of the two cases the square represents a space that serves to calculate the area surface of the sheet.

Some Considerations and Open Questions

Pupils in the fifth-grade class who had simply learned the formula for calculating surface area viewed the surface of a rectangular figure differently from pupils in the fourth-grade class who had been taught a different procedure to evaluate it. The latter procedure consisted of visual comparison to start with, followed by numerical calculation, that is, covering the surface and counting concrete, square units of measure. Two activities were thus highlighted: visual comparison and direct surface measurement.

All the children showed that they clearly understood the significance of comparison; they first looked at and estimated the surfaces at hand, then verified their estimation by placing one on top of the other. One difference between the two classes, however, did appear. The fifth graders simply compared the sheets as they had been asked to and stopped there, but the fourth graders went further, comparing their sheets with another real surface that they themselves normally used (their report-card notebook). They considered comparison to be an activity that allows one to determine which figure occupies more space. This is undoubtedly due to the kind of teaching they have received since their first year in school, a teaching method that has always taken into consideration the children's daily lives.

Furthermore, there was a difference for the fifth-grade children between the way one evaluates a figure in school and the way one estimates a real surface outside school. This led to differences in the answers they gave when they thought of the sheet as a rectangle and their answers when they simply observed a sheet of paper.

In the fourth grade, where we introduced the concept of surface area by having the children first visually compare real surfaces and then measure them directly, there was never any confusion between the initial evaluation and the final calculation. By observing the two rectangular sheets of graph paper, the children were able to deduce that the sheets had the same surface area. They were also capable, however, of explaining the different numerical results they had gotten when they counted the small squares on each sheet or surface.

Regarding the concept of surface measurement, for the fifth-grade children, who were already familiar with the notion of surface, measuring a surface meant obtaining a number and applying the arithmetical operation defined by the formula, without analyzing or estimating the figure first. This was confirmed in a study we carried out among students in the second year of junior high or middle school, for whom measuring a surface exclusively meant applying a formula; these students then proved to have a number of difficulties related to the use of the small square as an area unit.

For the children of the fourth-grade class, to measure a surface means to count area units, which can change from one surface to another. Square measures have a more concrete meaning for them, who, given a real sheet of paper, could see for themselves that it was not possible to measure its surface with a ruler. In this way the rule for going from one square measure to another is a consequence of the relation between the actual area units used. Furthermore, knowing how the little squares were arranged, the fourth-grade pupils chose the procedure of multiplying the number of squares they had drawn along the two sides of their sheets as a quick way of counting. They not only knew how to get to the formula to calculate the area of a rectangle but also were able to explain it. Therefore, direct measurement proved useful in acquiring the concept of surface area. Furthermore, this is a procedure that the pupils frequently use out of school—for example, to draw a figure of a certain size on a sheet of graph paper.

The use of small squares as a first approach to the formula can, however, create some problems. In our studies, we noted that pupils often tended to express both area and linear measurements by numbers of squares. This could cause confusion between the concepts of square and linear measures, especially if the concepts aren't dealt with in depth.

We deem that instruction could be designed to take into account both direct and indirect measurement, and to establish a connection between these two approaches, where the first one serves to make the second one meaningful (by "direct measurement" we mean the determination of dimensions by counting, e.g., using squares as a first step in approaching the formula, whereas by "indirect measurement" we mean that a formula is applied, e.g., when the size of a figure is derived from the measurement of linear dimensions). These two methods can thus be mutually integrative and complementary.

We believe that our research is also a useful tool to change attitudes with respect to mathematics, on the part of both pupils and teachers (see Bonotto

2001). The usefulness and accessibility of the discipline of mathematics, which many students find difficult and abstract, becomes more apparent when one enables children to draw new mathematical knowledge from the reality around them. One also helps overcome the rift between "schoolroom" and "out-of-school environment," giving greater value to the knowledge and strategies children possess in practice.

References

Bonotto, Cinzia. "How to Connect School Mathematics with Students' Out-of-School Knowledge." *Zentralblatt für Didaktik der Mathematik* 33 (June 2001): 75–84.

Gravemeijer, Koeno. "How Emergent Models May Foster the Constitution of Formal Mathematics." *Mathematical Thinking and Learning: An International Journal* 1 (April 1999): 155–77.

Nunes, Terezinha, Paul Light, and John Mason. "Tools for Thought: The Measurement of Length and Area." *Learning and Instruction* 3 (January 1993): 39–54.

Outhred, Lynne, and Michael C. Mitchelmore. "Young Children's Intuitive Understanding of Rectangular Area Measurement." *Journal for Research in Mathematics Education* 31(March 2000): 144–67.

Tierney, Cornelia, Christina Boyd, and Gary Davis. "Prospective Primary Teachers' Conception of Area." In *Proceedings of the XIV Annual Conference of the International Group for the Psychology of Mathematics Education*, vol. 2, edited by George Booker, Paul Cobb, and Teresa N. de Mendicuti, pp. 307–15. Oaxtepex, Mexico: International Group for Psychology of Mathematics Education, North American Chapter, 1990. (ERIC Document #ED411138)

12

The Measurement of Time: Transitivity, Unit Iteration, and the Conservation of Speed

Constance Kamii

Kathy Long

Authors of textbooks seem to believe that the measurement of time can be taught by teaching children how to read the clock. However, this teaching involves only social-conventional knowledge, which is often arbitrary. For example, the fact that "1" means "1" for the short hand and "5" for the long hand is a social convention that seems arbitrary to young children. Furthermore, unlike length and volume, time is not visible. How do children become able to measure time, which they cannot even see?

Piaget ([1946] 1969) conceptualized two aspects of measurement—namely, transitive reasoning and unit iteration. These ideas can most clearly be understood in his description of the "towers" experiment about the measurement of length (Piaget, Inhelder, and Szeminska ([1948] 1960). In individual interviews, he asked children to build a tower having the same height as a model 80 cm tall. The child's tower was to be built on a table 90 cm lower than the base of the model. The child was given smaller blocks than the ones used in the model so that one-to-one correspondence could not be used. Long strips of paper as well as a ruler and three sticks were provided—one stick 80 cm long, one longer, and one shorter than 80 cm.

Before the age of about seven, children had no use for the sticks, ruler, or strips of paper. Three-year-olds only glanced at the model, but four- and five-year-olds attempted to make more precise comparisons by looking alternately many times at the model and the copy. When they later wanted to make an even more precise comparison, children tried to bring the two towers together. The interviewer did not allow this direct comparison, and the children used various body parts in an attempt to compare the two towers as precisely as possible.

Finally around the age of seven, children began to use a stick as a third term. (If tower A was the model and tower B was the child's copy, C here was a stick—the third term.) Piaget (1970) explained this use of a stick by the phrase *transitive reasoning*, which becomes possible as the child becomes able to coordinate mental relationships. Transitivity refers to the ability to deduce a third relationship from two or more other relationships of equality or inequality. For example, we can show sticks A and C (see fig. 12.1a) to a four-year-old and ask, "Which one is longer?" Four-year-olds have no trouble replying that A is longer. The interviewer then announces, "I'm going to hide the longer one (A)," brings stick B out, and inquires whether B or C (fig. 12.1b) is longer. Young children again have no difficulty answering that C is longer. The interviewer then asks the crucial question: "Is this stick (B) as long as the one I've hidden (A)? Or is this one (B) longer, or is the one I've hidden (A) longer?"

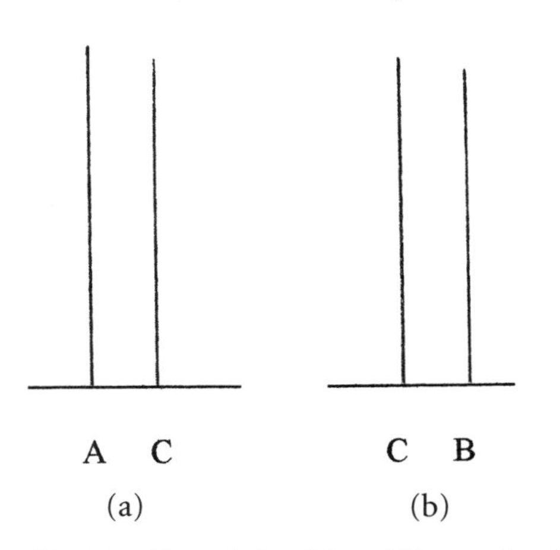

Fig. 12.1. Two relationships of "longer than"

Four-year-olds usually answer that they cannot know because they did not see A and B together. When they later become able to deduce, usually around the age of seven or eight, that A is longer (without looking at A), they are said to have constructed the logic of transitive reasoning (Piaget 1970). If direct comparison of two lengths, such as A and B, is impossible, it is necessary to use a third (C in this situation) to find out if A is equal to, longer than, or shorter than B. Sticks and rulers function as a third term, which enables us to compare lengths that cannot be compared directly. For children who cannot yet reason transitively, sticks and rulers are useless for comparing two lengths that cannot be placed next to each other.

In the towers task, Piaget, Inhelder, and Szeminska went on to show a small block to the child and asked if it could be used to compare the height of the two towers. The children who could reason transitively divided themselves into two groups: the less advanced group said that the small block was too small to be of any use, but the more advanced group used it as a unit to iterate and count. These children placed the small block at the bottom of the tower, marked the upper end of the block on the tower, moved the block up until its lower end was exactly on the mark, and repeated the same procedure upward to the end of the tower, without any gaps or overlappings. These actions indicated that the children thought about the length of the block as a part of the height of the tower. For children who cannot think of the length of a block as a part as well as a unit of the whole length, the block is completely useless.

To summarize, the two cognitive abilities necessary for children to measure length are (*a*) transitive reasoning and (*b*) unit iteration. Children construct unit iteration out of transitivity. (We say this because unit iteration appears later than transitivity, and all the children who demonstrated unit iteration also demonstrated transitivity according to our research [Kamii and Clark 1997; Long and Kamii 2001; Reece and Kamii 2001].) Transitive reasoning involves comparing the whole of A (fig. 12.1) with the whole of B, by means of the whole of a third term, C. Unit iteration involves making a part-whole relationship within each whole. Measurement in a strict sense involves the numerical quantification of units. In a broad sense, however, Piaget sometimes used the term *to measure* to refer to more qualitative comparisons with transitive reasoning only.

To explain children's development of transitivity and unit iteration, it is necessary to review the fundamental distinction Piaget ([1967] 1971, [1945] 1951) made among three kinds of knowledge according to their ultimate sources: physical, social (conventional), and logicomathematical knowledge.

THREE KINDS OF KNOWLEDGE

Physical knowledge is the knowledge of objects in external reality. Our knowledge of the color, shape, and weight of a stick or any other object is an example of physical knowledge. These properties are in objects and can be known empirically by observation.

Examples of *social (conventional) knowledge* are our knowledge of holidays and conventional units such as meters and minutes. Social knowledge has a source in conventions made by people, and physical knowledge has a source in objects.

Whereas physical and social knowledge have sources outside the individual, the source of *logicomathematical knowledge* is in each person's mind. Examples of logicomathematical knowledge are mental relationships we create between and among objects, such as *different* and *similar*. With respect to sticks A and B in figure 12.1, for example, it is just as true to say that the two are different as it is to say that they are similar, since the source of these relationships is our minds. If we think about them as being different, they become different, and if we think about them as being similar, they become similar. The reader must have noted that transitivity and unit iteration are examples of logicomathematical knowledge, not empirical knowledge (based on observation or experience). These relationships are not observable or directly teachable; they must be constructed, or made, by each child.

In a recent study involving 383 children in grades 1–5, Kamii and Clark (1997) found, in the measurement of length, that transitivity is usually constructed by grade 2 (29 percent in grade 1 and 72 percent in grade 2) and that unit iteration is demonstrated in grade 3 (55 percent). In the measurement of volume, Reece and Kamii (2001) found with 257 children in grades 2–5 that the majority (51 percent) demonstrated transitivity in grade 3 and that the majority (56 percent) showed unit iteration in grade 4.

Whereas the measurement of length and volume requires transitive reasoning and unit iteration, the measurement of time requires the construction of a third aspect of logic: the child must be able to think that the instrument used to measure time functions at a constant speed (referred to as *the conservation of speed* in the study that follows). To determine the grade levels when transitivity, unit iteration, and the conservation of speed appear with respect to measuring time, 120 students in a suburb of San Francisco were interviewed. This total consisted of 30 students each from kindergarten, second grade, and fourth grade at an elementary school and 30 students from sixth grade at a nearby middle school. Equal numbers of boys and girls were included at each grade level, and the students were mostly from middle- to upper-middle-class homes. Further details can be found in Long and Kamii (2001). We summarize below the findings related to transitivity, unit iteration, and the conservation of speed.

TRANSITIVITY IN THE MEASUREMENT OF TIME

In the transitivity task, children were asked to listen to and compare the lengths of two pieces of recorded music (both from *The Nutcracker* by Tchaikovsky)—one thirty seconds long, the other forty seconds long. They were then shown a set of materials and asked if they could use them to prove (or make sure) that one piece was longer than the other (or whatever they had said). The materials consisted of a supply of water (in a one-liter bottle), an empty one-liter clear-plastic bottle, a funnel made with an empty bottle (8 cm in diameter and 23 cm in height) with its bottom removed and a clear-plastic tube attached to its mouth, a large tray to catch any spilled water, and a water-soluble, fine-point marker (see fig. 12.2). These materials allowed the children to run water through the funnel into the empty bottle for the length of time the first piece of music played, mark the water level when the music stopped, and then refill the bottle and follow the same procedure for the second piece. If the water level was higher at the end of the second piece, children could deduce that the second piece was longer.

If the children had no idea how to use the materials, the interviewer demonstrated plugging the bottom tip of the funnel with a finger, filling the tube with water, then removing the finger so that the water was free to run into the empty bottle. They were then given a second chance to think of a way to use the materials to determine which piece of music was longer.

Fig. 12.2. Performing the transitivity task

Criteria Used to Categorize Children into Three Groups

Children were categorized at Level 3 (the highest level) if they started the water flow when the music started each time, stopped the flow when the music ended each time, and marked the bottle to be able to remember exactly where the water level was at the end of the first piece of music. These children were categorized at the highest level because they knew from the beginning that they could deduce that $A = B$ if $C_A = C_B$, (where A stands for the duration of the first melody, B for the duration of the second melody, and C_A and C_B, respectively, for the duration of the water flow). If $C_B > C_A$, however, they would deduce that $B > A$. Children were also categorized at Level 3 if they could immediately articulate a plan that clearly used transitive reasoning or could explain their plan after they were shown how the water could be run through the funnel and into the bottle. If children did not mark the bottle after the first piece but gave some clear indication that they remembered where the water level had been, either on the bottle the water ran into or on the top funnel the water was running out of, they were placed in this category.

The children who could not think of any way to use the materials to compare the lengths of the music, even after they had been shown how the water could be run through the funnel and into the bottle, were categorized at Level 1. If they showed an effort to make an indirect comparison but were not careful about starting and stopping the flow of water precisely at the right moment, or if they were not concerned about the exact level of water in either the bottle or the funnel, they were categorized at Level 2, an intermediate level.

Findings

It can be seen in table 12.1 that the number of children who demonstrated transitive reasoning increased with each grade level. There is a quick increase from 0 percent in kindergarten to 53 percent in second grade and to almost 100 percent in fourth and sixth grades.

TABLE 12.1

Percent of Students at Level 3 for Each Measurement-of-Time Task by Grade Level

	Grade Level			
	K	2	4	6
Task	$n = 30$	$n = 30$	$n = 30$	$n = 30$
Transitivity	0	53	90	96
Unit iteration (water)	0	13	36	70
Unit iteration (sand)	0	16	50	83
Conservation of speed	0	10	30	83
Equal intervals (sand)	0	20	56	76

UNIT ITERATION

Two unit-iteration tasks were given. The first one required children to find another way to prove (or make sure) that one of the pieces from *The Nutcracker* was longer than the other (or whatever the child believed at that moment). In this task they were given two small vials, each with a hole in the bottom, and were told that they could plug the bottom and fill up one vial with water before the music started. When water was allowed to run through the hole at the bottom of each vial, the water took about ten seconds to flow from one vial to the other. Thus the duration of the water running through the vial could be considered a unit of time, and a part of the whole duration of the music. Children demonstrated unit iteration by repeatedly transferring water from one vial to the other, counting the number of times a vial became empty, and using the number of times a vial was emptied to compare the lengths of the two pieces.

Children who could not think of a way to use the vials were given a demonstration by the interviewer. They were shown how to let the water run from one vial to the other, using fingers to plug the holes at the bottoms of the vials. They were then asked again if they could think of a way to use the vials. The interviewer also offered help with manipulating the vials if the child had difficulty, stipulating that the child must tell the interviewer exactly what to do.

The second unit-iteration task involved a commercially made ten-second sand timer. We gave this second task because we wanted to know if unit iteration was easier with a familiar instrument than with vials. Children were first asked if they could use the sand timer to know for sure which piece of music was longer. To demonstrate unit iteration with this device, they needed to start the timer when the music started and flip it each time the sand ran out until the end of the music. They also had to count how many flips were required to time each piece of music and judge which was longer.

Criteria Used to Categorize Children into Three Groups

Children who could articulate and demonstrate a reasonable plan for the first unit-iteration task, either immediately or after they were shown how water could run from one vial to the other, were categorized at Level 3. This plan required letting water run from one vial to an empty one, starting with the music, then switching the positions of the vials as soon as the water had run out, and repeating this process as many times as necessary to the end of the music. When children demonstrated their plan, they needed to show every intention of switching the positions of the vials as quickly as possible so that there would be no gaps between the units.

Children who had no idea what to do with the vials were categorized at Level 1. Those who let the water run from one vial to the other but were surprised that the music continued and had no idea what else to do were also placed at Level 1.

Those who demonstrated some idea of unit iteration were categorized at Level 2. For example, some children lost some water while transferring it from one vial to the other and said it was not necessary to have the same amount of water in the vial each time. If they repeatedly started the water running before the music started or continued after the music stopped and saw no problem with this procedure, they were also placed at Level 2.

If children were categorized at Level 3 using the two vials and could clearly describe what they would do with the sand timer, they were also placed at Level 3 on the second unit-iteration task. If they had not been placed at Level 3 on the first task, they were asked to demonstrate their plan with the sand timer, even if their verbal explanations were adequate.

Children were categorized at Level 1 if they had no idea what to do with the sand timer or if they turned it over once at the start of the music but did not know what to do after the sand had run through and the music was still playing. Those who could describe something resembling unit iteration but, for example, did not think that it was important for all the sand to go down before flipping it over were placed at Level 2.

Findings

A steady increase in proficiency can be seen in table 12.1 throughout the grades tested for both unit-iteration tasks. The water-and-vials task showed a large increase in successful performance between fourth and sixth grades (from 36 percent to 70 percent). The sand-timer task shows a slightly better performance. This difference was not unexpected given the fact that the sand timer was a familiar instrument.

It is interesting to note the thinking of three kindergartners who were categorized at Level 2 of the unit-iteration task with the sand timer. These children had all used a sand timer in class or at home to play games. They started the sand timer with the music, flipped it over when the sand ran out, and counted the number of times the sand timer was flipped. However, they did not relate the number of flips to the duration of the music. When asked why she kept flipping over the sand timer, one of them said, "Because I needed to keep pretending there was more sand." Clearly, she did not think about each run of the sand as a unit of time; she just needed to keep the sand going as long as the music was playing.

THE CONSERVATION OF SPEED

Children were shown two identical plastic bowls, one containing thirty black marbles and the other one empty. They were told that their task was to move the marbles from one bowl to the other as *slowly* as possible. The timing device that would indicate to children when to start and when to stop consisted of two clear plastic cylinders (3.5 cm in diameter and 15 cm in height). One of the cylinders was filled with water and had a hole in the bottom (initially plugged by the interviewer's finger). The other cylinder had a red line taped 9 cm from its bottom (see fig. 12.3). Children were told to begin moving the marbles when the interviewer unplugged the hole and let the water run from the top cylinder to the bottom cylinder, and to stop when the water level reached the red line (approximately twenty-five seconds). When the signal was given, the children started to move the marbles slowly from one bowl to the other. If they began to speed up their actions, the interviewer gently reminded them to move as slowly as they possibly could. When the water level reached the red line, the children stopped. If they were not paying attention to the water level, the interviewer pointed out that it was time to stop.

Fig. 12.3. Performing the conservation-of-speed task

The children were then asked to return the marbles to the original bowl while the interviewer poured the water back into the first cylinder. This time, the children were instructed to move the marbles as *fast* as they could. They were told to watch the water rising to the red mark again, to know when to stop.

When the task was completed and the marbles had been moved two times, the children were asked, "Did the container fill up at the same speed both times, or did it fill up faster the first time or faster the second time?" If the children answered that it filled up faster the first time or faster the second time, the interviewer clarified this answer by elaborating it. For example, if a child said the cylinder filled up faster the first time, the interviewer responded, "So the water filled up faster when you were going more slowly." The child confirmed this statement or clarified his or her answers.

If the children stated that the cylinder filled up at the same speed both times, they were asked, "How do you know?" When the children could not give an answer that included some sort of logical deduction, a countersuggestion was made: "Another boy (or girl) told me yesterday that the water went faster when he was moving the marbles faster. Do you think he was right or you are right? What would you tell him to convince him that you are right?" This procedure was followed to determine how firmly the child held the view that the speed was the same both times.

Criteria Used to Categorize Children into Three Groups

When children said the container filled up at the same speed both times and could justify their answers, they were categorized at Level 3. The children's justification needed to include a logical rationale, such as "The water goes down in the same way no matter what people are doing," or "You used the same containers and the same water, and it went through the same hole, so it had to go at the same speed both times."

Children were placed at Level 1 if they said that the water filled up faster when they moved the marbles slowly or that the water filled up faster when they moved the marbles faster. If they said it was the same but could not explain their answers or were easily swayed by the countersuggestion, they were also placed at Level 1. Children who said that the speed was the same and could not be swayed by the countersuggestion but could give only a partial explanation were placed at Level 2.

Findings

The conservation-of-speed task proved to be surprisingly difficult and appeared later than Piaget's findings suggested (Piaget [1946] 1969). The percent in table 12.1 rises very slowly from kindergarten to fourth grade and then makes a dramatic rise to 83 percent in sixth grade. These data suggest that it is not until sometime between fourth and sixth grades (an increase of 53 percent) that most children construct the notion that the speed of the instrument must be constant and is independent of their own actions.

Equal-Intervals Task

Before students began the unit-iteration task with the sand timer, a question was posed to check whether they believed the deductive necessity for the speed of the timing instrument to be constant. The question posed was whether or not it took the same amount of time for all the sand to run through one way as the other. If students said it took the same amount of time either way, they were asked how they knew that was so.

The children were categorized at Level 3 if they said that the interval was the same and logically explained why that had to be true. For example, some children stated that the same amount of sand went through the hole each time; so it had to be the same amount of time. Children were categorized at Level 1 if they said that one way took longer than the other. If their only means of comparing was to count, they were also placed at Level 1. Children who were placed at Level 2 were unable to justify their correct answer or initially said one way took longer than the other but later changed their minds.

As can be seen in table 12.1, more children responded at Level 3 at an earlier age with the sand timer than with the marbles and water. It was nevertheless not until fourth grade that a scant majority of students came to believe in the deductive necessity of equal intervals. (When children construct higher-level logic, they begin to feel that the sand has to take the same amount of time going down in one direction as in the other.) The conservation-of-speed issue separates the measurement of time from that of length and volume. A unit of time is not an observable object like a ruler or a measuring cup. A unit of time must be created out of the motion of an object and the logicomathematical notion that the speed of that motion is (and must be) constant. Children must be aware of the deductive necessity of the conservation of speed to compare intervals of time accurately. Therefore, their understanding of measuring time cannot be considered complete until all three aspects are present: transitive reasoning, unit iteration, and the conservation of speed. It is not surprising that, according to our three studies (Kamii and Clark 1997; Long and Kamii 2001; and Reece and Kamii 2001), the logic required in the measurement of time is constructed later than the measurement of length or volume.

EDUCATIONAL IMPLICATIONS

Piaget's research and theory show that logicomathematical knowledge such as transitive reasoning and unit iteration is deeply rooted in children's thinking and that concrete operations develop one after another out of an undifferentiated whole between the ages of about 6 and 11. For example, as stated earlier, transitivity and unit iteration first appear with respect to length, then with respect to volume, and then with respect to time. In other words, length is easier to "logicomathematize" than volume, and volume in turn is easier to logicomathematize than time. We can therefore not say that if we teach "this," children will acquire transitivity, or that if we teach "that," students will be able to express unit iteration in the measurement of time. Accordingly, students may benefit from having experiences with time throughout the day rather than engaging in specific activities that focus narrowly on specific mental relationships. Several examples that may encourage children's construction of time follow.

Making Decisions Related to Time

The teacher can ask students to make some of the moment-to-moment decisions that occur each day in the classroom. For example, teachers can ask students, "If we have to go to lunch at 12:00, do we have time to read the story and celebrate Allison's birthday?" Students can also help their teachers make decisions that involve class presentations. "How much time would it take for five children a day to give a daily news report if each spoke for ten minutes? If that's too much time to sit and listen, how long should the reports be?" Decision making can occur on a daily or weekly basis. "If the spelling test is going to be on Friday, what should you be doing each day this week to get 100 percent?" Other decision-making opportunities arise when students are asked to consider long-term projects. "If the science fair is scheduled for February 21 (six weeks from now), what has to be done by when?" In general, if students know what their assignments are and when they are due, they can take on the responsibility and begin to tell the teacher what they are planning to do each night.

Resolving Conflicts Related to Time

A teacher introduces a new math game that everyone is excited to play. The next day, one child puts it on his desk as soon as he arrives to reserve it for game time. Another child insists that she should have it first because she has signed her name on a signup sheet that some of the students invented the day before. In this situation, the teacher can turn to the class asking for a fair way to resolve the conflict and for a rule to prevent this kind of problem in the future. Children think hard when there is a conflict to resolve.

Making Timing Devices

Children in primary grades often face the problem of limiting the time their classmates use the computer, swings, or other equipment. Since most children can reason transitively by second grade, the teacher might suggest that the children find a way to create a timer to limit the time fairly. The teacher could make a variety of containers available such as plastic soda bottles and empty milk cartons as well as materials such as water and sand, but it would be up to the students to find a way to use them to measure time. They might also want to create timing devices for a variety of activities, such as to determine who in the class can jump rope the longest or keep a glider in the air the longest.

According to Piaget's theory, logicomathematical knowledge develops in an interrelated way by differentiation and coordination. An example of differentiation is distinguishing between spatial and temporal relationships. When young children finish a race behind someone, for instance, they say that the winner ran for a longer time, even if the two started at the same time and stopped at the same time. This is an example of basing a temporal judgment on a spatial criterion. As children coordinate spatial relationships with spatial relationships—and temporal relationships with temporal relationships—they gradually separate space from time and begin to understand that the interval (or time) was the same. For children's development of the logic necessary to understand the measurement of time, it is therefore desirable to encourage them to think hard not only about time but also about all kinds of objects, people, and events all day long. A good way to encourage thinking is to put children in situations that require debate, negotiations, and decision making.

REFERENCES

Kamii, Constance, and Faye B. Clark. "Measurement of Length: The Need for a Better Approach to Teaching." *School Science and Mathematics* 97 (March 1997): 116–21; (October 1997): 299–300.

Long, Kathy, and Constance Kamii. "The Measurement of Time: Children's Construction of Transitivity, Unit Iteration, and Conservation of Speed." *School Science and Mathematics* 101 (March 2001): 1–8.

Piaget, Jean. *Play, Dreams, and Imitation in Childhood.* 1945. Translation, New York: Norton, 1951.

———. *The Child's Conception of Time.* 1946. Translation, London: Routledge & Kegan Paul, 1969.

———. *Genetic Epistemology.* New York: Columbia University Press, 1970.

———. *Biology and Knowledge.* 1967. Translation, Chicago: University of Chicago Press, 1971.

Piaget, Jean, Bärbel Inhelder, and Alina Szeminska. *The Child's Conception of Geometry.* 1948. Translation, London: Routledge & Kegan Paul, 1960.

Reece, Charlotte, and Constance Kamii. "The Measurement of Volume: Why Do Young Children Measure Inaccurately?" *School Science and Mathematics* 101 (November 2001): 356–61.

13

Bridging Yup'ik Ways of Measuring to Western Mathematics

Jerry Lipka

Tod Shockey

Barbara Adams

The Yup'ik Eskimo people of southwest Alaska, like many indigenous peoples throughout the world, have much to offer mainstream society. Their perceptual abilities have been noted by explorers (Nelson 1899) to anthropologists (Carpenter 1973) to educators (Palliscio, Allaire, and Mongeau 1993). In the rather unpredictable environment of southwest Alaska, which to many outsiders may appear featureless and barren, the Yup'ik have learned to thrive as they travel across the tundra. Visualizing and estimating distances, developing a rich vocabulary for describing spatial relations, and using their own system of body measures all stem from centuries of living, hunting, gathering, and moving on and across this land. This deep knowledge of the environment and keen observation of natural phenomena are ideal topics for mainstream schooling, particularly elementary-level mathematics.

The past decades have seen numerous reports calling for educational programs for American Indian/Alaska Native (AI/AN) schools and communities that are based on local culture and employ a group's vernacular language

---

Without the involvement of Yup'ik elders and teachers, this paper would not be possible. There are far too many to name. In particular, however, we want to dedicate this paper to Mary and Frederick George of Akiachak, Alaska; Evelyn Yanez of Togiak, Alaska; and Ferdinand and Nancy Sharp, Henry Alakayak, and Anecia Lomack of Manokotak, Alaska. This work has been partially funded by the Alaska Schools Research Fund, University of Alaska Fairbanks, and the National Science Foundation, Instructional Materials Development, award #9618099, "Adapting Yup'ik Elders' Knowledge: Pre-K-to-6th Grade Math Instructional Materials Development." Opinions expressed are those of the authors and not necessarily those of the foundation, the Alaska Schools Research Fund, or the Yup'ik elders and teachers.

(Deyhle and Swisher 1997; Indian Nations at Risk Task Force 1991; Pavel 2001; Tippeconnic 1999; Yazzie 1999). This call is due, in part, to the nationwide, persistent gap between the academic performance of AI/AN students and their nonnative peers. AI/AN students continue to underperform in core academic subjects, particularly mathematics (see the NCES Web site, nces.ed.gov/nationsreportcard/mathematics/results/scale-ethnic.asp). Yup'ik students in rural southwest Alaska appear to score even lower on standardized tests than other AN groups (see the Alaska Department of Education Web site, www.eed.state.ak.us/tls/assessment/).

We present some of the results of a long-term collaborative project by University of Alaska Fairbanks faculty, Alaska teachers, and Yup'ik elders, funded by the National Science Foundation.[1] Titled "Adapting Yup'ik Knowledge Pre-K to 6th Grade Math and Instructional Materials Development," the project strives to confront the challenges outlined above by connecting everyday experience and knowledge of Yup'ik people to elementary school mathematics. The fifteen supplemental modules we have developed cover a wide range of both Yup'ik tasks and mathematics concepts. In this article, we will focus on the richly textured system of Yup'ik measurement and translate portions of this cultural wealth into lessons involving prealgebraic thinking and fractions, ratios, and proportions that are used in Western classrooms.

Much of the information presented in this article originated from Mary George, a Yup'ik teacher and lifelong resident of Akiachak, Alaska, and her husband Frederick George, a respected elder with substantial knowledge of star navigation, hunting, fishing, and preserving food.

Traditional Yup'ik culture uses measuring, estimating, spatial reasoning, and proportional thinking for many tasks, most of which are closely linked to subsistence (hunting, gathering, and fishing) skills. The measures used are often related to the human body. In this article, we first describe examples of everyday Yup'ik practice that contain embedded mathematics. Second, we show how these cultural practices can address prealgebraic thinking. Specifically, we concentrate on linear measures and their relationship to a larger system of measures, and show how concrete representation expressed through proportional thinking can aid students' understanding of mathematics.

BACKGROUND TO THE YUP'IK CONTEXT AND YUP'IK CULTURE

The Yup'ik people of southwest Alaska are a people living in two cultures. At the start of the new millennium, seasonal rounds of subsistence hunting,

---

1. The grant is titled "Adapting Yup'ik Elders' Knowledge Pre-K to 6th grade Math and Instructional Materials Development." The University of Alaska Fairbanks, College of Rural Alaska's Bristol Bay Curriculum Project, and the School of Education's Alaska Schools Research Fund also fund this work.

fishing, and gathering of wild foods remain a mainstay of their economy. Villages typically range in size from between 75 to 600 people. Most villages are located off the road system, and transportation links between villages typically consist of local air taxi services, boat, and snowmachine. Extended families play a role in the village's social structure, and elders are respected leaders in the community.

However, most villages have running water, phones, computers, and other modern amenities. Mainstream U.S. culture has made major inroads through satellite TV, videos, and easy air transportation to larger towns, the city of Anchorage, and beyond. As Alaskan residents, all children are required to attend school and take state and national standardized exams. Unfortunately, test scores are generally low, and Yup'ik children are underrepresented in college or professional work. Teaching seems to be one of the few professional careers that dominate. Some children do not see the relevance of school mathematics, or most academic subjects for that matter, to a successful life. This is a concern for elders, who would like young people to participate more fully in both the larger economy and in the subsistence economy. Subsistence living, coupled with mainstream North American life, provides rural Yup'ik students with ample opportunities to have the best of both worlds. Only recently have some schools tried to bridge the gap between Yup'ik culture and mainstream schooling by incorporating some of their traditional practices and wisdom into academic subjects (Barnhardt and Kawagley 1999).

Many Yup'ik cultural practices connect well to school-based mathematics. For example, the entire process of catching, preserving, and distributing salmon uses a variety of units of capacity. A net full of fish will be put into wheelbarrows and moved to where the fish can be processed and hung to dry. A "fish rack full of drying salmon" becomes a unit of measure, and expert adults can reasonably estimate the number of fish racks full of salmon that they will need to fill their smokehouse, a small wooden structure at the bottom of which is a fire to smoke the fish, a method of preservation. From this, they accurately estimate the number of times their smokehouse must be filled to provide enough salmon to last the winter. Later, they distribute the salmon to other members of extended families, dividing it by using another unit of measure, a "tote." This process of establishing units of measures and converting from one unit to another occurs in other subsistence activities as well. When berries are picked, they are first collected in a scoop, then transferred to buckets, then to plastic bags, then placed in a freezer. Each of these containers represents a unit of measure.

METHODOLOGY

In developing the mathematics curriculum, each module followed a particular Yup'ik cultural storyline. The mathematics embedded within the cultural activity may have a direct relationship with school mathematics such as in the Estimating, Measuring, Counting, and Grouping module, which introduces the Yup'ik numeration system (base 20 and subbase 5), or it may require a high degree of interpretation to connect it to school mathematics, such as the Elastic Geometry module (see Lipka et al. 2001).

The processes and outcome of our research were framed by the sentiment: "How can this work help the next generation of schoolchildren?" (Lipka, Mohatt, and the Cuilistet 1998) This long-term research was aimed at classroom and schoolwide reform. To accomplish this, we collected data by interviewing elders and Yup'ik teachers as we were engaged in a variety of everyday Yup'ik activities and in more formal, face-to-face meetings. Most meetings were conducted in Yup'ik and were translated into English. We videotaped elders and Yup'ik teachers as they taught us specific skills. We followed the Yup'ik master-apprentice model of learning, and engaged in subsistence and other activities as apprentices. Through this, we gained an understanding of Yup'ik ways that allowed us to begin the process of adapting their knowledge to schooling in a Western-style classroom.

Transforming the data into curriculum was a complex iterative process involving the Yup'ik elders and teachers, mathematicians, and educators. The elders sometimes developed the mathematics activities themselves, but more often it was done by the academic part of the team. We would then have the elders and teachers perform the activities. They would provide feedback, both of their personal understanding of how their knowledge was transformed for schooling, as well as for pedagogical insights. Through a series of trials and errors, we gained an intuitive sense of the connections between the embedded mathematics and the school mathematics. The examples cited in this article arise from this process.

Modules have been piloted in villages and larger urban sites in Alaska, as well as in selected sites in California and New York. In fact, the findings of a preliminary study using an experimental design (random assignment of teachers to treatment and control groups) for the Design and Build It module found that all treatment students outperformed all control group students on pre- and posttest score differences at a statistically significant difference (Lipka and Adams 2001).

EXAMPLES FROM YUP'IK CULTURE—MEASURING AND PROPORTIONAL THINKING

One of the most esoteric and interesting Yup'ik practices of measurement, as related by Frederick George, is star navigation. He has spent considerable time teaching project staff how he observes the Big Dipper, the North Star, and Cassiopea in their relationship to one another and how they can be used to orient oneself when traveling. Frederick's practice of star navigation includes using hand measures to approximate the height of these stars and constellations in the night sky. He also uses their nightly cycle as a means of telling (measuring) time and measures (15 degrees per hour) the change in his course that he must make to compensate for the perceived movement of the night sky. This wonderful system of navigation is the essence of one of our modules called Star Navigation.

The Yup'ik have developed a system of measures taken from the human body. For example, *yagneq* is the measurement from fingertip to fingertip of opposite hands while the arms are held outward (an approximation of an individual's height). *Taluyaneq* is the distance from the middle of the chest to the fingertip of an outstretched arm. *Tallinin* is the distance from the armpit to the end of a clenched fist of an outstretched arm. This way of measuring is obviously indispensable for such activities as traveling, hunting, and fishing as the measures are always "on hand" when needed.

Also, many times, it allows a structure to be built to fit perfectly the person or people using it. A kayak, for example, needs to fit the kayaker as best as possible. It is like a shoe—the better the fit, the more closely connected a person is to the earth and to the movement of walking (or for the kayak, the more closely connected the person is to the sea and to the movement of paddling). The craft is in harmony with his body. It would be difficult to create such a perfect fit with standard measures that would have the length be fifteen feet and the width, two feet, for example.

But with Yup'ik body measures, a kayak can be tailor-made to the builder. A Yup'ik kayak is two *yagneq*, one *taluyaneq*, and one *tallinin* in length, or approximately three times the paddler's height. The distance from the bow to the front rim of the cockpit is *yagneq* plus *taluyaneq*. The diameter of the cockpit is *tallinin*, and the distance from the rear rim of the cockpit to the stern is *yagneq*. The depth of the kayak is one *ikugamek*, the distance from the elbow to the end of a clenched fist (see fig. 13.1).

In addition, the Yup'ik kayak is constructed so that the paddler sits closer to the stern (back). This allows the bow to rise slightly out of the water to avoid smashing into potentially damaging ice floes or other dangers. Instead, the boat will rise over them. The longer bow also provides room for storing "tools" on the bow deck that are too long to be stored in the kayak, such as an ice prod and a spear. When a Yup'ik kayak is used for salmon fishing, more of the catch is loaded at the stern than at the bow, thereby maintaining the slightly raised bow position.

Another task that requires the flexibilty of the Yup'ik system of body measures is house building. John Pauk from Manokotak, Alaska, at a meeting held in Fairbanks with Yup'ik elders and teachers, explained how he built his house. He decided to use his own system of measures, rather than standard or regular cuts of lumber, such as two-by-fours. His system of measures was

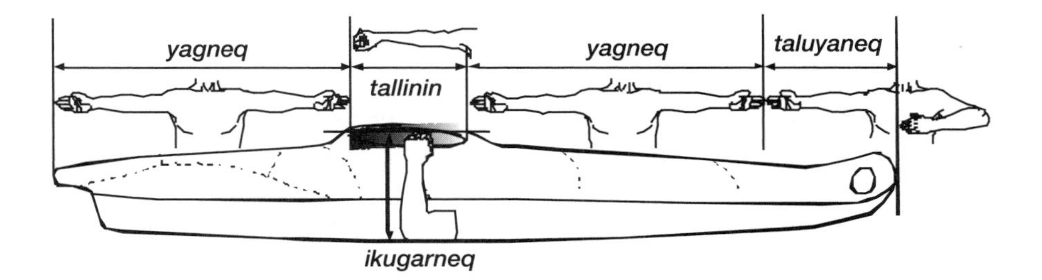

Fig. 13.1. Kayak illustration with body measures

related to his body size, the material he was using, and the nature of his task (building a house). Once he established this system, every single piece of building material was in proportion to this measure, as well as in proportion to his body. The result is a tailor-made house.

Another way the Yup'ik measure is by "eyeballing" or visually estimating the size of an object. Take, for example, making clothes. At another meeting with Yup'ik teachers, elders, and university faculty, Paul's wife, Lillie Gamechuk (Pauk) asked Anecia Lomack, a Yup'ik teacher from the same village, to stand up. Lillie was across the room from her. Anecia stood up for a minute, and then Lillie asked her to sit down. The measuring was complete. Lillie never physically touched her or used any kind of measuring instrument, such as string or measuring tape. Lillie then used scissors and paper to make out of paper a *qaspeq* (a hooded Yup'ik dress or long shirt), hat, and belt for Anecia.

A short time later, Anecia tried on her paper garments. Each item fit. As Lillie later explained, she uses a "normal" body type, familiar to her from years of observing and measuring women for making clothing. From this "norm" she makes adjustments, all in her mind's eye. Her visual approach to measuring is quite common among many Yup'ik elders.

Building Fish Racks: Connecting Yup'ik Measurement to Standard Mathematics

Another involved system of Yup'ik measurement has to do with catching salmon. During the short Alaskan summer, most coastal people are busy fishing, primarily for salmon, which constitutes a large portion of the year's food supply. The need to harvest a lot in a short time (three to six weeks) necessitates efficient methods. For example, when people build a fish rack, they often use locally available materials such as driftwood, freshly cut poles, or scrap lumber from other projects. New, store-bought lumber is rarely used, primarily because of its high cost and lack of availability. Other factors that influence the appearance and strength of the structure include location (near the ocean, exposure to strong winds), and the amount of fish it must accommodate. Depending on these conditions, and on the amount of salmon typically harvested, people build their fish racks accordingly. Mary George has developed a system of body measures to measure the length of her fish rack and its poles, the length of the different types of salmon, and the space between the poles needed to fit the different types of salmon cuts. For example, a king salmon cut is approximately the same measurement as Mary George's shoulder-to-shoulder measurement. To determine the length of a fish rack pole, Mary measures the distance from the middle of her chest to the end of one outstretched arm, five times (see fig. 13.2).

Many body measures are proportional to another one, for example, one *yagnek* equals two *taluyaneq*, making the system easily applicable to the Western concepts of ratios, proportions, and equivalencies that are the heart of the sixth-grade module, Measuring, Prealgebra, and Proportional Thinking: Drying Salmon.

In this module, students simulate building a fish rack and apply Mary George's system of body measures to determine the length of their respective fish poles (horizontal poles from which the salmon are hung) and the height at which they are placed, using their own body as the unit of measure. As they finish, students observe that although they each used the same part of

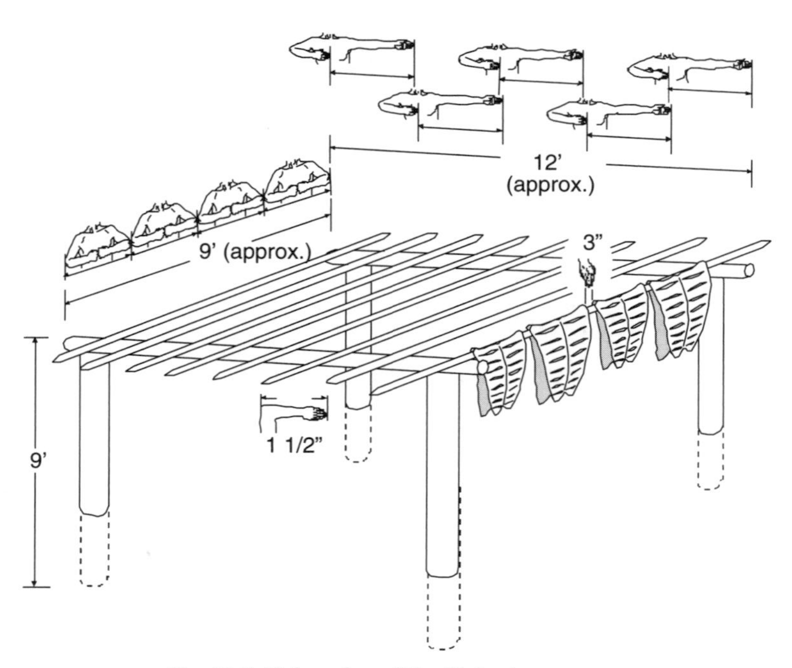

Fig. 13.2. Fish rack and Yup'ik body measures

their body as a measure, the distance between two outstretched hands, say, the length of each pole and the height at which it rests differs from person to person. Building a fish rack lays the groundwork for exploring and understanding the differences between standard measures and body measures. Students learn to organize data, create tables, and graph the data. They are also introduced to, and use, a set of proportional two-dimensional fraction bars that represent Yup'ik body measures (see figs. 13.3 and 13.4) and also show students how body measures could develop into standard measures.

The fraction bars represent a shift from the actual body measures to mathematical representation. Mathematics relies heavily on the concept of representation. Abstracting from the concrete to symbolic representation provides freedom for mathematical manipulation. In this module, students have a strong understanding of measurements because they have visualized them and physically manipulated them. However, as the students become competent at forming algebraic equations and expressions, the physical measures only provide limitations. Students are challenged to write out sentences that express a relationship, such as "1 *yagneq* equals 2 *taluyaneq*." They discover that it is easier to use symbols, such as $1Y = 2T$.

Students learn that these symbols often represent variables, which represent different body measures, different lengths of salmon cuts, and different lengths of poles. For example, although Y represents *yagneq*, which refers to the same part of the body for everyone (the distance between two outstretched hands), Y represents a length that may differ from person to person.

Students learn to express a variety of relationships as equations such as $Y = T + 2I$ or $Y = T + 1I + 2N$ where I represents *ikusegneq*, the measure from the elbow to the fingertips, and N represents *naparneq*, the measure of the hand with thumb outstretched. From here, they move to learning about ratios and expressions of constant relationships. For example, T/Y and Y/T relate numerically to 1/2 and 2/1, respectively. As students express body relations as ratios, they discover that the same relationships exist for a variety of body measures. For example, the ratio of Y/T is the same as the ratio of T/I. This leads to understanding proportion as a constant. We ask students to explore two equivalent ratios by posing the question "How are these the same and how are they different?" During a trial lesson, students answered this question in the following ways: "They look same on the outside but are different on the inside." "Both are halves, but they are different sized halves." "They become different after you look at half of what." "Each of them is half of a different thing." When asked what a variable is, students gave the following responses: "It is a mystery number." "It can vary." "It is a letter and the letter stands for something." As an additional test, we asked them to write a multiplication equation. This is the most challenging case, as it represents area and is written with exponents, which may be a new concept for many students at this level. For example, $T(T) = T^2$.

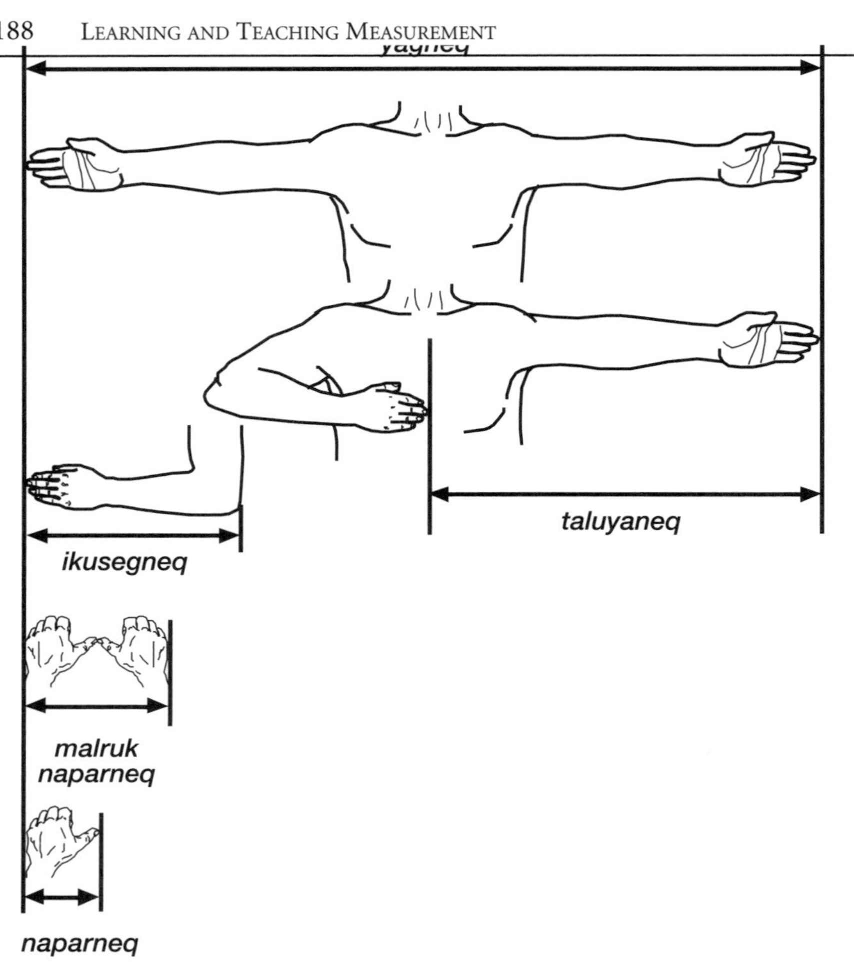

Fig. 13.3. Yup'ik body measures

<table>
<tr><td colspan="12">Y</td></tr>
<tr><td colspan="6">T</td><td colspan="6">T</td></tr>
<tr><td colspan="3">I</td><td colspan="3">I</td><td colspan="3">I</td><td colspan="3">I</td></tr>
<tr><td colspan="2">M</td><td colspan="2">M</td><td colspan="2">M</td><td colspan="2">M</td><td colspan="2">M</td><td colspan="2">M</td></tr>
<tr><td>N</td><td>N</td><td>N</td><td>N</td><td>N</td><td>N</td><td>N</td><td>N</td><td>N</td><td>N</td><td>N</td><td>N</td></tr>
</table>

Fig. 13.4. Fraction bars–based on Yup'ik body measures

The Mathematics and the Pedagogy

Throughout the module Measuring, Prealgebra, and Proportional Thinking: Drying Salmon, we attempt to move students from thinking arithmetically to thinking algebraically. Algebraic thinking involves finding patterns and general relationships instead of finding one answer, the foundation of arithmetic thinking.

A second important concept in algebra is representation. We have incorporated a variety of representations throughout the module, including pictures, sentences, symbols, graphs, and tables. Often we will introduce a concept using pictures or a hands-on manipulative, then encourage students to move from these devices to symbols, graphs, and tables. Writing descriptive sentences benefits both the students and the teacher by illuminating the students' thinking process.

Pedagogically, students are presented with open-ended challenges that require them to solve real-world, Yup'ik-based problems, providing them with the opportunity to learn about another culture, as well as learn new cognitive strategies for solving practical problems. Although the problems are open-ended, they are constrained by the use of body measures and other materials that limit the possibilities for exploration. This creates a constructivist-oriented classroom in which students discover the answers to the problems given them.

Applying Everyday Mathematics to Your Neighborhood

To be able to bridge everyday or ethnomathematics knowledge from the community to the school, the Dutch program Realistic Mathematics Education (RME), the Funds of Knowledge project in Arizona (Civil and Khan 2001; Gonzalez et al. 2001), and the work in Alaska with the Yup'ik (Lipka, Mohatt, and the Ciulistet Group 1998) first identify everyday situations that are rich in mathematical possibilities. Freudenthal (1973) believes that young people should be introduced to mathematics by using their experiencing and drawing the mathematics out of those experiences. Then one needs to assess potential activities for their authenticity both to cultural practices and to mathematics.

Every culture is replete with multiple instances of people using standard and nonstandard measures that can be used in the classroom to connect students' everyday knowledge (including their intuitive knowledge) to school mathematics. Measuring is a way of organizing space—involving linear measures of length, two dimensions and area, and three dimensions and volume. Measuring in everyday contexts also includes approximating angles and coordinating angles and distance in locating.

Gravemeijer (1998), in an innovative ethnomathematics project, uses the students' knowledge of shadows, that is, how the length of a shadow changes as the position of a person changes in relation to the light source (moving closer or farther from the light source). Similarly, we have students measure the length of their shadows at different times of the day so that they can observe the movement of the shadows as well as their varying lengths. Then, the students create a sundial that allows them to measure lengths, angles, and movement (15 degrees an hour). In other modules, we emphasize the important role measuring plays in the Yup'ik community—in making clothing, navigating and locating, and building structures such as fish racks, homes, and kayaks—incorporating these activities into mathematics problems.

Another example of connecting community to schooling is presented by Civil (Civil and Khan 2001) in her work in Hispanic communities in Arizona. She employed the concept of "funds of knowledge" (Gonzalez et al. 2001). In this instance, the researchers found that many of the families they interviewed were engaged in gardening. Civil and her colleagues developed one mathematics unit that used the theme of gardening and connected it to measurement in one, two, and three dimensions. The curriculum included such activities as maximizing garden space given a specific perimeter and exploring the amount of soil required for each plot (volume). In these examples measuring lengths, areas, and volume of a garden plot bridged everyday knowledge to school knowledge.

Gerdes's (1999) attention to the accumulated wisdom of cultural practices of African artisans has brought to awareness the mathematical ideas encoded within material objects. Each cultural group has their own unique ways of organizing reality and giving it meaning. Gerdes suggests paying attention to the cultural cognitive aspects as well as to reflection on everyday practice, since each way can bring out mathematical ideas. Through such work African teachers may recognize the contributions of their culture or Yup'ik teachers can appreciate the fact that their elders "know mathematics," and elementary school students may better appreciate that what they know outside school may count in school.

CONCLUSION

From the keen observations of Yup'ik Eskimo elders of the seemingly barren tundra to students observing their shadows being cast and to everyday activities such as gardening, rich opportunities abound that can connect intuitive understandings to school mathematics. Connecting everyday mathematics to school mathematics increases the possibility of improving students' mathematical understandings.

In fact, research conducted on one of our modules has attempted to determine if a curriculum based on students' everyday knowledge and community knowledge can improve their mathematics performance. The results from this initial study (Lipka and Adams 2001) have shown that the treatment group (those that used the project's curriculum) outperformed the control group by a statistically significant degree. Thus, this initial study suggests that bridging school and community can improve students' mathematical understandings while simultaneously respecting the students' culture and everyday knowledge. The approach of this project and its initial studies show promise for altering the long-term gap in mathematics achievement between American Indian/Alaska Native students and nonnative students. The work of this project continues, and more research can be expected in the future to further tease out the many factors that affect students' mathematics performance.[2]

REFERENCES

Barnhardt, Raymond, and Angayuquq Oscar Kawagley. "Culture, Chaos and Complexity: Catalysts for Change in Indigenous Education." Alaska Native Knowledge Network, May 1999. www.ankn.uaf.edu/ccc.html.

Carpenter, Edmund. *Eskimo Realities.* New York: Holt, Rinehart & Winston, 1973.

Civil, Marta, and Leslie Khan. "Mathematics Instruction Developed from a Garden Theme." *Teaching Children Mathematics* 7 (March 2001): 400–405.

Deyhle, Donna, and Karen Swisher. "Research in American Indian and Alaska Native Education: From Assimilation to Self-Determination." In *Review of Research in Education,* vol. 22, edited by Michael Apple, pp.122–94. Washington, D.C.: American Educational Research Association, 1997.

Freudenthal, Hans. *Mathematics as an Educational Task.* Dordrecht, Netherlands: Reidel, 1973.

Gerdes, Paulus. *Geometry from Africa.* Washington, D.C.: Mathematical Association of America, 1999.

Gonzalez, Norma, Rosi Andrade, Marta Civil, and Luis Moll. "Bridging Funds of Distributed Knowledge: Creating Zones of Practices in Mathematics." *Journal of Education for Students Placed at Risk* 6, no.1–2 (2001): 115–32.

Gravemeijer, Koeno. "From a Different Perspective: Building on Students' Informal Knowledge." In *Designing Learning Environments for Developing Understanding of Geometry and Space,* edited by Richard Lehrer and Daniel Chazan, pp. 45–66. Mahwah, N.J.: Lawrence Erlbaum Associates, 1998.

Indian Nations at Risk Task Force. *Indian Nations at Risk: An Educational Strategy for Action.* Final report of the Indian Nations at Risk Task Force. Washington, D.C.: U.S. Department of Education, 1991.

---

2. The effects of a culturally based mathematics curriculum on Alaska Native students' academic performance is a three-year study funded by the U.S. Department of Education, Office of Educational Research and Improvement.

Lipka, Jerry. "Culturally Negotiated Schooling: Toward a Yup'ik Mathematics." *Journal of American Indian Education* 33 (spring 1994): 14–30.

Lipka, Jerry, and Barbara Adams. "Improving Rural and Urban Students' Mathematical Understanding of Perimeter and Area." Paper to Alaska Schools Research Fund—School of Education, University of Alaska, Fairbanks, 2001.

Lipka, Jerry, Sandra Wildfeuer, Nastasia Wahlberg, Mary George, and Dafna Ezran. "Elastic Geometry and Storyknifing: A Yup'ik Eskimo Example." *Teaching Children Mathematics* 7 (February 2001): 337–43.

Lipka, Jerry, with Gerald Mohatt and the Ciulistet Group. *Transforming the Culture of Schools: Yup'ik Eskimo Examples.* Mahwah, N.J.: Lawrence Erlbaum Associates, 1998.

Nelson, Edward. *Eskimos about the Bering Strait.* Washington, D.C.: Smithsonian, 1899.

Pallascio, Richard, Richard Allaire, and Pierre Mongeau. "Spatial Representation of Geometric Objects: A North-South Comparison." *Inuit Studies* 17 (1993): 113–25.

Pavel, Michael. "Schools, Principals, and Teachers Serving American Indian and Alaska Native Students." *ERIC Digest.* Charleston, W.Va.: ERIC Clearinghouse on Rural Education and Small Schools, 1999. (ERIC Document Reproduction no. ED 425 895)

Tippeconnic, John. "Tribal Control of American Indian Education." In *Next Steps: Research and Practice to Advance Indian Education,* edited by Karen Swisher and John Tippeconnic, pp. 33–52. Washington, D.C.: Office of Educational Research and Improvement, 1999. (ERIC Document Reproduction no. ED 427 902)

Yazzie, Tarajean. "Culturally Appropriate Curriculum: A Research-Based Rationale." In *Next Steps: Research and Practice to Advance Indian Education,* edited by Karen Swisher and John Tippeconnic, pp. 83–106. Washington, D.C.: Office of Educational Research and Improvement, 1999. (ERIC Document no. ED 427 902)

Introduction

Students' experiences with measurement on the secondary school level shape their facility with measurement as an adult. Dossey (1997) observed, "Measurement plays an important role in the everyday quantitative actions of citizens. It is perhaps the most visible, but least considered, aspect of quantitative literacy…." (p. 180). In grades 6 through 8, measurement is a strand that easily can be found in the curriculum, but it is seldom mentioned in high school curricula. One of the differences between *Principles and Standards for School Mathematics* and the NCTM's 1989 *Standards* is the extension of the measurement strand to grades 9–12. Measurement can be used to develop and tie together important ideas of algebra, data analysis, and geometry. Measurement naturally occurs in high school in science and technology classes. The articles in this section contain suggestions not only for instruction but also for assessment.

The articles provide ideas for using literature, technology, and manipulatives to enhance the mathematical experience of students. Some of the ideas presented by Hartmann, Choppin, and Enderson involve using technology such as the TI-92, Dynamic Geometry software, and other software with which some teachers may or may not be familiar to develop concepts about measurement. Austin uses literature to develop concepts and explore links with other disciplines. Geoboards are common on the elementary school level but can also be employed effectively on the secondary school level. Burns and Brade use geoboards to develop area formulas in middle grades and the circular geoboards to explore properties of a circle. Lubienski and Strutchens analyze U.S. students' performance on measurement items on NAEP and present questions and suggestions for classroom instruction and pedagogy. Two articles relate measurement experiences of adults: Adams and Harrell detail the use of estimation by adults, and Steinback focuses on the instruction of measurement in adult education.

Measurement is a common skill needed to study science and mathematics. The articles for the middle grades follow along with the suggestions of NCTM and other mathematical organizations concerning the type of experiences students should have to develop a strong understanding of measurement concepts. Hodgson uses successive approximations to find the area of irregularly shaped regions, an article by Burke using the TI-92 software to explore irrational numbers, and the advantages of problem-centered instruction are detailed by Boston and Smith.

On the secondary school level, we may add to our students' confusion. We give students a large number of problems that have no context and are without units. As observed by Jean Taylor, "Many students think of mathematics as a set of precise rules yielding exact answers and are uncomfortable with the idea of imprecise answers, especially when the degree of precision in the estimate depends on the context and is not itself given by a rule" (Taylor 1998, p. 31). There are other concepts of measurement that are not addressed in this volume such as derived units, the use of different scales, and the analysis of precision and error in measurement. Throughout this section a common theme is the student being able to build his or her own understanding of measurement and to have a variety of measurement experiences.

Veronica Meeks and Robert Wheeler

References

Dossey, John, "Defining and Measuring Quantitative Literacy." In *Why Numbers Count: Quantitative Literacy for Tomorrow's America*, edited by Lynn A. Steen, pp. 173–86. New York: College Board, 1997.

Taylor, Jean. "The Importance of Workplace and Everyday Mathematics." In *High School Mathematics at Work: Essays and Examples for the Education of All Students*, edited by the Mathematical Sciences Education Board, pp. 30–34. Washington, D.C.: National Academy Press, 1998.

14

What Students Know about Measurement: Perspectives from the National Assessment of Educational Progress

Marilyn E. Strutchens
W. Gary Martin
Patricia Ann Kenney

MEASUREMENT topics appear on commonly used standardized tests and large-scale assessments, one of the most visible of which is the National Assessment of Educational Progress (NAEP). In the current NAEP framework (National Assessment Governing Board 2000), measurement is one of five content strands. This strand includes topics such as selecting and using appropriate measurement instruments (for example, rulers and protractors) and appropriate units of measurement in both standard and metric systems; converting from one measurement unit to another within the same system; and computing and comparing perimeter, area, surface area, and volume. In this paper, we will discuss what U.S. students know about measurement on the basis of the results of the NAEP assessment.

The assessment is given at grades 4, 8, and 12, and topics are marked as appropriate for a grade level according to when it is introduced in the school curriculum. For example, at grade 4 area may be assessed using a figure superimposed on a grid. By grade 8, it is assumed that students can use formulas to compute area. The assessment includes multiple-choice items, constructed-response (short answer) items, and extended contructed-response items in which students are expected to provide explanations. In addition to

This article is based on work done for the NCTM NAEP Interpretive Report Project supported by a grant to NCTM from the National Science Foundation (NSF), grant no. RED-943189. Any opinions expressed herein are those of the authors and do not necessarily reflect the views of NCTM or NSF.

overall results for measurement, NAEP also provides results on individual items, some of which are publicly released, as well as samples of students' work related to these items. Thus, an analysis of NAEP can furnish useful insights into the performance of U.S. students on measurement concepts, including what they know and which topics are particularly difficult. In addition, since the NAEP mathematics assessment has been given about every four years since 1972, we can trace achievement over a period of time.

Past administrations of NAEP have shown that measurement is a difficult topic for students to master. Figure 14.1 shows average scale scores (based on a 0–500 point scale) for measurement from the 1990, 1992, 1996, and 2000 assessments. As expected, performance across grade levels has increased steadily between assessments. The more interesting results appear within the grade levels themselves. At all grade levels, performance in measurement in 2000 was significantly improved from 1990; at grades 4 and 8 performance was also significantly improved from 1992. When 1990 and 2000 are compared, the differences ranged from 14 scale-score points at grade 8 to 10 and 8 points at grades 4 and 12, respectively. Although these results suggest that students became more proficient in measurement topics over this period, the number of students performing at the higher end of the scale is still low.

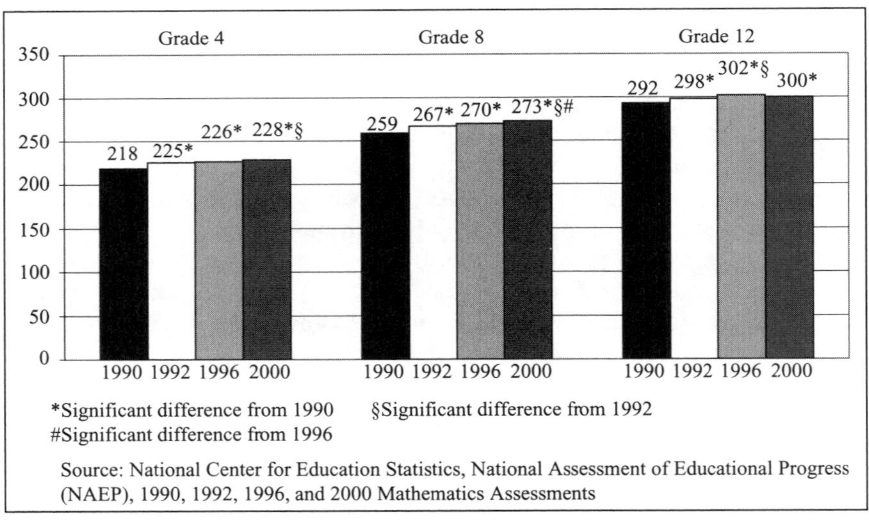

Fig. 14.1. NAEP scale scores for the measurement content strand: 1990–2000

Detailed analyses of NAEP results, items, and students' work related to measurement have consistently shown major deficiencies in this area (Martin and Strutchens 2000; Kenney and Kouba 1997; Lindquist and Kouba

1989). Most researchers attribute this low performance to the limited experiences students have with measurement, primarily using standard measurement instruments and calculating measures with formulas (Nitabach and Lehrer 1996). Wilson and Rowland (1993) suggest the following emphases in students' learning of measurement: (1) identifying attributes to be measured, (2) making comparisons according to the attributes, (3) establishing appropriate units of measurement based on the attributes, (4) recognizing the need for standard units of measurement, and (5) creating formulas to help count units.

In addition, Nitabach and Lehrer (1996) suggest that a common core of measurement assumptions should guide teachers when they are creating problems for students to solve and making decisions about how to interact with students during problem solving: (*a*) units of measure should be identical, (*b*) measurement involves iteration, (*c*) a scale has a zero point, and (*d*) measurement is characterized by additivity. These perspectives can serve as a lens to examine students' understanding of measurement based on NAEP items and to suggest strategies for teaching that may improve students' development of measurement concepts and skills.

PERSPECTIVES ON MEASURING LENGTH

Recent administrations of NAEP imply that students have basic knowledge of measuring length, but that their knowledge tends to be superficial. For example, in the 1996 NAEP the overwhelming majority of fourth graders could identify an instrument appropriate for measuring length and could choose an appropriate unit for measuring length in English units; eighth- and twelfth-grade students were very successful in choosing an appropriate unit for measuring plant growth. However, fourth-grade students had difficulty with items in which they were asked to use a ruler to either measure an object or produce an object with a given length, with correct response rates below 30 percent.

Of particular interest is an item in which students were asked to determine the length of an object pictured above a ruler where the ends of the object and ruler were not aligned. This item (or an item much like it) has been given in several assessments; a similar, publicly released item from an earlier assessment is shown in figure 14.2. Fewer than one-fourth of fourth-grade students in the 1996 sample correctly identified the length of the object. Although the success rate in 1996 rose to nearly two-thirds at grade 8 and more than four-fifths at grade 12, this still implies a rather startling lack of understanding of measurement. In the assessment in which the particular item in figure 14.2 was used, many students either selected the number at the end of the object (8 in the problem in fig. 14.2) or counted the number of lines instead of the number of units (6 in the problem in fig. 14.2).

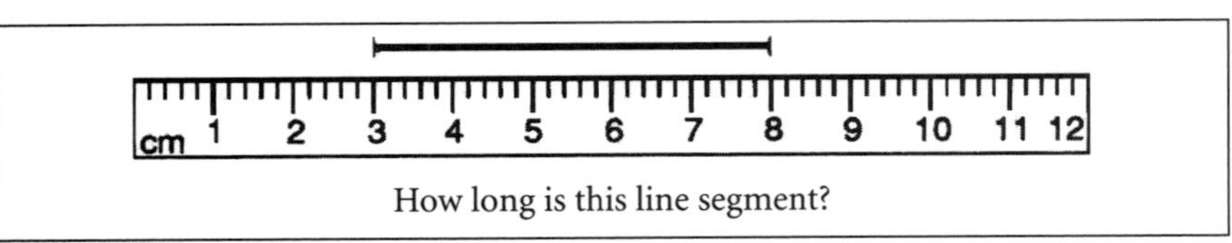

Fig. 14.2. Task from 1986 NAEP assessment in which students are asked to determine the length of an object pictured above a ruler, where the ends of the object and ruler are not aligned (Linquist and Kouba 1989)

These responses suggest that although students have a general understanding of the attribute of length and appropriate units for measuring length, they do not adequately understand the role of the unit in measuring length. Wilson and Rowland's (1993) analysis suggests that students may benefit from experiences in which they are asked to measure the length of objects using nonstandard units. For example, they might be asked to use different sizes of paper clips to measure the length of their desk, thus focusing attention on what a unit is and how units are combined to measure objects. Further, they might be asked to design their own ruler using a nonstandard unit, thus focusing attention on the role of units in the measuring instrument (Van de Walle 2001). With the knowledge built in such activities, they may be more likely to be able to use rulers reliably and flexibly.

PERSPECTIVES ON PERIMETER

Perimeter is a special application of length that measures the distance around a region. Given our previous observations regarding students' difficulties with NAEP items involving length, it should not be surprising that they also have substantial difficulties working with perimeter. For example, fewer than half of fourth graders in the 1996 NAEP were able to find the length of one side of a geometric figure given its perimeter, and fewer than 20 percent were able to produce a figure with a given perimeter.

Although correct responses to this item were substantially higher at grades 8 (about 2/3 correct) and grade 12 (over 3/4 correct), responses to another task (see fig. 14.3) suggest major limitations of these older students' conceptions of perimeter. In this item, students were given paper shapes labeled *N*, *P*, and *Q* and asked to determine the shape with the longest perimeter. Students were reminded that perimeter is the distance around a figure, but they did not have access to a ruler. Only 6 percent of the eighth-grade students and 12 percent of the twelfth-grade students answered correctly, where a correct response had to include both the conclusion that *P* has the longest perimeter and an explanation of how the student decided that *P* was the correct choice. More than half of the students said that *P* had the longest perimeter, but they did not give an adequate explanation or provided no explanation at all.

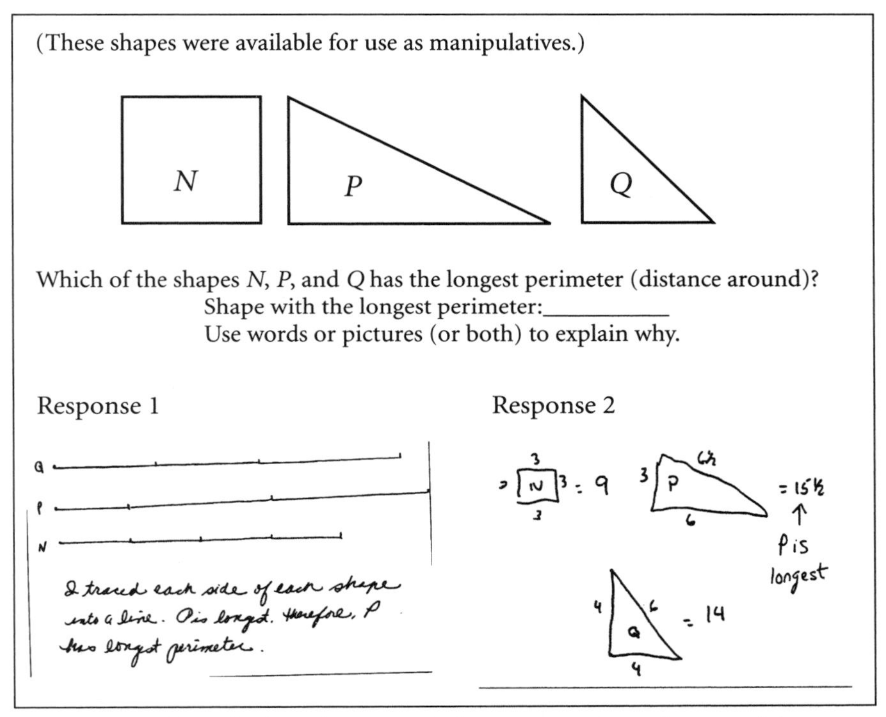

Fig. 14.3. Task in which students are asked to compare the perimeters of three figures, along with sample responses

Martin and Strutchens (2000) analyzed a small set of responses to gain insights into the kinds of responses given. In their analysis, they pointed out that some students simply wrote that *P* "looked like" it had longer sides, whereas others attempted to measure all the sides of each figure using a self-determined scale but did not state exactly how they determined that *P* had the largest perimeter.

It may be instructive to look at some of the correct response strategies they identified, since these may suggest productive ways of approaching perimeter. One common approach is shown in Response 1 of figure 14.3, an actual response given by a twelfth-grade student. In this approach, which relies on an empirical comparison, the student drew a line segment for each shape by tracing the length of each side of the figure. It is then obvious that *P* must have the largest perimeter. This tactic of "unrolling" the sides of a figure allows the use of a comparison strategy, suggesting that the student has a solid grasp of the basic concept of perimeter. Teachers might want to encourage this strategy. May (1999) suggests a related approach in which students make shapes using strips of heavy paper connected at their endpoints by brads; removing one brad allows the students to do this same "unrolling" strategy. Alternatively, teachers might have their students stretch a piece of string around the sides of a figure, noting (with their fingers or with a mark) the beginning and ending points. Pulling on the ends of the string so that it is taut then demonstrates the total perimeter.

A somewhat more sophisticated approach is shown in Response 2 of figure 14.3, given by an eighth-grade student. The student used a measurement system apparently based on assigning a length of 3 units to a side of square *N*. The student then assigned values to the sides of shapes *P* and *Q* on the basis of how they compared to a side of *N*, or 3 units. For example, the student noted that one leg of *P* is the same length as a side of *N*, that the other leg is about twice as long, and that the hypotenuse is a little more than twice, so the perimeter of *P* is about 3 + 6 + 6 1/2 units, or 15 1/2 units. Likewise, the legs of *Q* are a little larger than a side of *N* (say about 4 units) and the hypotenuse is about double, so the perimeter of *Q* is about 4 + 4 + 6, or 14 units. Although the student mistakenly gave the perimeter of *N* as 9 instead of 12, the student correctly concluded that *P* must have the largest perimeter. This approach again draws on a comparison strategy but uses a self-assigned unit, allowing a numerical comparison. This flexible use of units suggests a good understanding of units in relation to the measurement of perimeter.

To encourage this viewpoint, teachers may want to have their students use nonstandard units (such as paper clips) laid end to end to find the perimeter of various shapes. Robertson (1999) describes an activity in which students are asked to estimate and then actually count how many "footsteps" or paces it will take to walk around the playground. In addition to the association of units with perimeter, this activity can also highlight the need to have standard units, since the answers may vary substantially depending on the students' shoe sizes. Students should be able to build on such informal experiences to develop more formal calculations of perimeter using a ruler to measure the sides of a figure and then adding them up, eventually leading to the formulas commonly used to calculate perimeter.

PERSPECTIVES ON AREA

Area is measurement in two dimensions seeking to determine how much surface is enclosed within a region. One can determine the area of an enclosed region by tiling it with congruent regions serving as the unit of measure. However, students' initial experiences with area are often limited to the use of a formula (such as area = length × width). The following sections demonstrate the limited understanding students have, both in comparing the areas of shapes and in conceptualizing the units of measurement.

Comparing Areas

One of the first steps in developing an understanding of area is for students to make comparisons of shapes with different-sized areas to focus on the attribute being measured. Figure 14.4 shows a publicly released item featured in the 1992 and 1996 NAEP mathematics assessments that asked students to evaluate statements comparing the areas of shapes *N* (a square) and *P* (a right triangle), where they were given a pair of each shape to use as manipulatives. Students did not do well on this item at any of the grade levels on the 1996 or 1992 assessments. However, as one might expect, twelfth-grade students performed better than those in the other two grade levels, one major problem being that (as in the perimeter problem in fig. 14.3) students did not give adequate explanations.

Samples of students' responses at grades 4 and 8 are shown in figure 14.4 to illustrate correct ways in which students solved the problem. The most intuitive way to solve the problem was just to experiment with the manipulatives by placing the different shapes on top of each other to compare their areas, as shown in Response 1. This fourth-grade student's response is particularly intriguing, since it uses an animated style of drawing to demonstrate that the two shapes have the same area. Here, the two figures are superimposed, and part of shape *P* is moved so that it fits over the part of shape *N* not already covered by shape *P*.

Response 2, produced by an eighth grade student, illustrated that one could use two of each shape to form a rectangle of the same area, implying that one of each must also be the same area. The student used both words and pictures to justify the choice of Bob's claim as correct.

Response 3 was produced by an eighth-grader, but similar responses were seen in the twelfth-grade set of responses. This response is an example of solving the problem through a comparison of the dimensions of the figures and the use of formulas—the most abstract method. Because students did not have access to a ruler, they had to compare the shapes to see that a side of square *N* is the same length as one base of the triangle *P* and that the length of one of the bases of *P* was twice that of its other base. Once the students established these relationships, a solution to the task using a formula became possible—as long as students knew the formulas for the areas of a square and a triangle.

Students' responses to this task revealed that students understood area at different levels, and that students who had a conceptual understanding of area could solve the problem without resorting to formulas. Moreover,

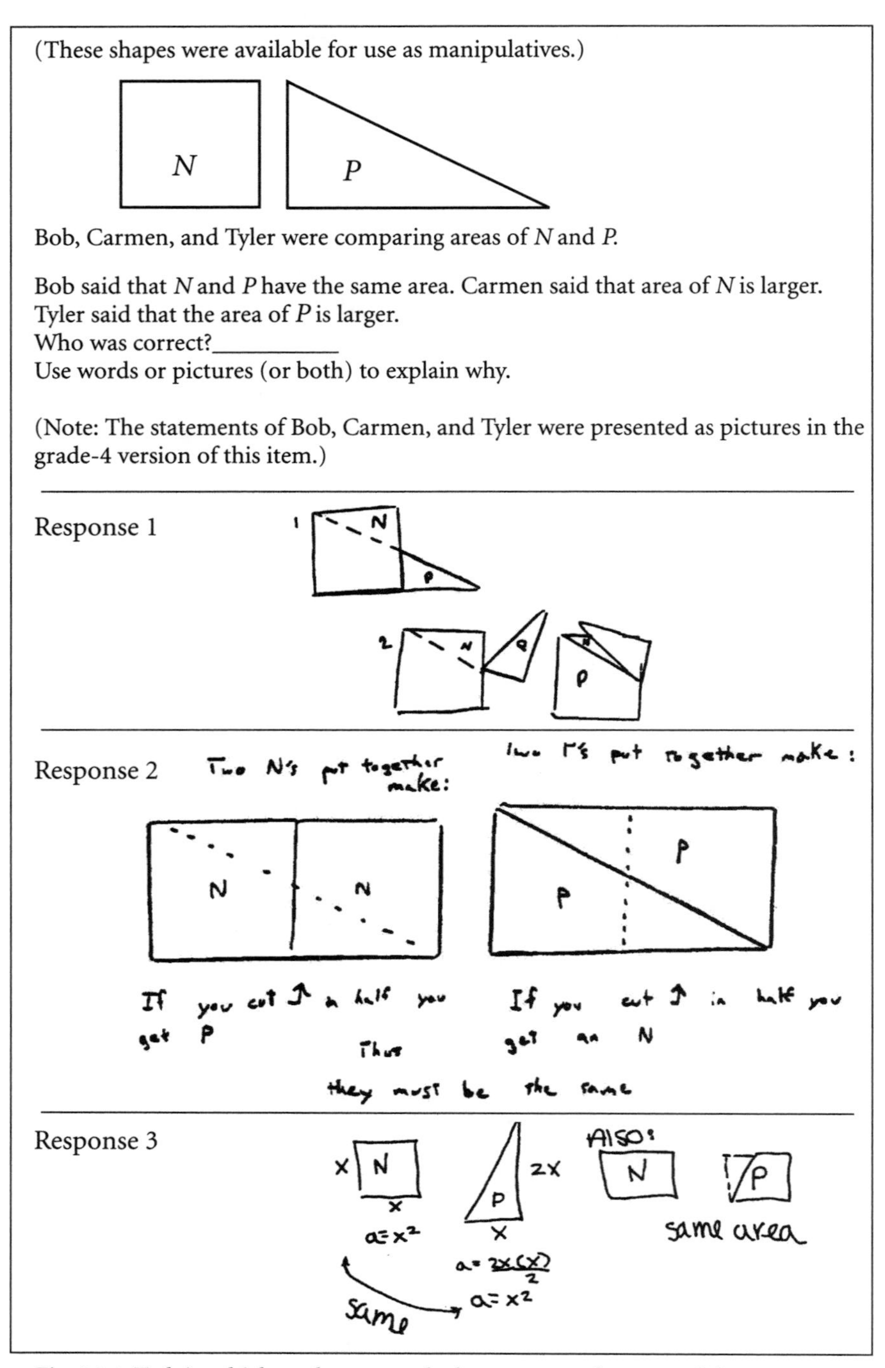

Fig. 14.4. Task in which students are asked to compare the areas of three figures, along with sample responses

manipulating the shapes to determine how their areas compared may have been more efficient for students who may have difficulties remembering the formulas used to find the areas of the shapes. (See Strutchens, Harris, and Martin 2001 for more examples of students' responses.)

Students' first experiences with area should include comparing the areas of different shapes in a variety of contexts. Nitabach and Lehrer (1996) reported an activity in which a teacher gave first- and second-grade students three rectangles made of construction paper. Each rectangle was made of twelve square units (1×12, 2×6, and 4×3), but they did not have any demarcation lines on them; the children were not told the relationships among the rectangles. The teacher asked the students to determine how much space each of the shapes covered in the context of making a quilt. By the end of the lesson the students discovered that the shapes covered the same amount of space by folding the shapes and laying them on top of each other, as they tried to determine if one shape could be cut apart to fill another shape. The teacher posed pivotal questions, such as "How can we find out which shape covers the most space?" and "How are you going to prove to someone what you think might be true?" to help students discover that each shape was made of twelve square units.

Understanding Area Units and Area Formulas

Counting square units to determine the area of a region and drawing a shape when given an area in square units are tasks that show whether students understand area units. For the 1996 NAEP mathematics assessment fewer than half of the fourth-grade students (41 percent) and a majority of eighth-grade students (78 percent) were able to choose the area of a figure embedded in a centimeter grid. Moreover, an item that was given in the 1992 assessment asked fourth- and eighth-grade students to draw a rectangle with an area of 12 square units. About two-fifths of the fourth-grade students and two-thirds of the eighth-grade students successfully completed this task (Kenney and Kouba 1997). One would expect students to do well on both items because they do not require a formula, only a conceptual understanding of area.

Although the 1996 NAEP mathematics assessment did not include many items involving area formulas, only 44 percent of the eighth-grade students and 60 percent of the twelfth-grade students were able to choose the correct numerical expression of the area for a given situation in a given item. Other analyses by Martin and Strutchens (2000) revealed that students were not successful on items that required them to determine the number of subregions it would take to fill a larger region or determine the areas of irregular shapes. These results indicate that students need more meaningful experiences with area, such as counting the number of square units it would take to cover a region. Rather than memorize particular formulas for certain shapes, they need to understand why the formulas work. Finally, they need experiences that require them to apply the area formulas in a variety of contexts rather than do straightforward computations with the formulas.

Perspectives on Surface Area and Volume

Recent NAEP analyses have included relatively few items involving surface area and volume. However, responses to the few items that were included suggest that students have great difficulty with these concepts. Correct response rates did not surpass 40 percent on any of three items involving surface area on the 1996 NAEP, none of which have been released to the public. The difficulties students have with area (as discussed in the previous section) make their difficulties with surface area understandable. Likewise, fewer than 40 percent of the students were successful on three items involving volume included in the 1996 assessment. For example, fourth graders were given a diagram of a cube made up of smaller cubes and asked to determine its volume by determining the number of smaller cubes in the larger cube; eighth- and twelfth-grade students were asked to compute the volume of a cylindrical container given its height and the radius. These findings suggest that students need to have deeper experiences with surface area and volume, instead of just computing with formulas. As with area, they would likely benefit from experiences with the respective attributes and with counting the appropriate units needed to tile (for surface area) or fill (for volume) a solid.

Perspectives on Measuring Angles

Over the past several administrations of NAEP, students have consistently had difficulties with angle measure. For example, in the 1996 NAEP, fewer than one-third of fourth graders could identify angles smaller than a right angle, and fewer than 15 percent could sketch an angle larger than a given measure. By the eighth grade, students' performance had risen somewhat, but only two-thirds were able to correctly order a set of angles, even though this item could be easily answered by visual inspection without using a protractor. Since this NAEP item has not been released to the public, figure 14.5 gives an example of a similar task. Without solid experiences with angle measure, students will experience increasing difficulties as they explore the properties of more-complex geometric figures. Indeed, fewer than one-fourth of twelfth graders were able to draw a perpendicular line to a given segment and to then measure the angle formed by that line with another given line.

A closer look at some NAEP items suggests some approaches that may be useful for teachers to consider. First, students need to have more experiences in qualitatively comparing the size of angles instead of immediately focusing

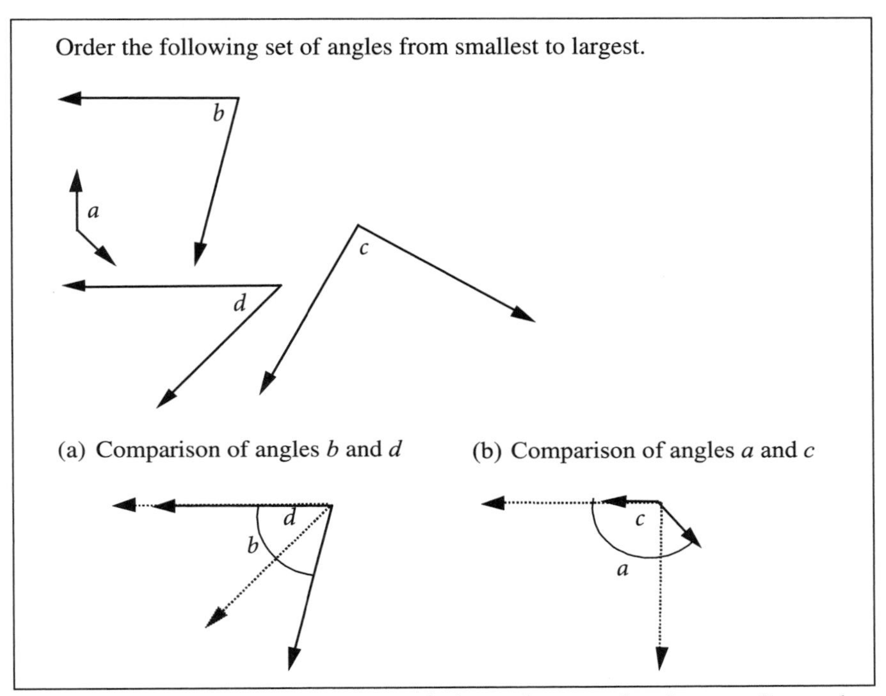

Fig. 14.5. Task in which students are asked to order a set of angles according to the angle measure

on the use of a measuring instrument. For example, to successfully complete the task given in figure 14.5, students need to be able to visualize how much an angle is "bent" or how "pointy" it is. They could trace an angle and superimpose it on top of others in order to see the relationship better, as seen in figure 14.5a, in which angle *b* can be seen to be larger than angle *d*. Furthermore, it is crucial that they understand that the lengths used to represent the sides of the angle do not affect its angle measure, so that angle *a* in figure 14.5 is actually larger than angle *c* as shown in figure 14.5b, even though the sides of angle *c* appear to be longer. The use of benchmark angles, like a right angle or a 45-degree angle, may enhance both their ability to compare angles and their underlying concepts of angle measure. For example, for the item in figure 14.5, a student could determine that angle *b* is smaller than angle *a* without the use of a protractor simply by observing that angle *b* is less than a right angle and angle *a* is more than a right angle.

Wilson and Adams (1992) suggest that students may benefit from experiences with nonstandard units of angle measure, such as iterating wedges to measure angles, before they begin using protractors. In this way, they can see how the unit is iterated to "fill" the angle. However, unlike length, the unit is rotated so that successive units are placed edge-to-edge (as in fig. 14.6b) rather than placed so that the bottoms of the wedges are lined up (as in fig. 14.6a). Building on their experiences with benchmarks and iteration, students might estimate angle *a* in figure 14.4 to be 135 degrees, since it is a right angle and about half of a second right angle.

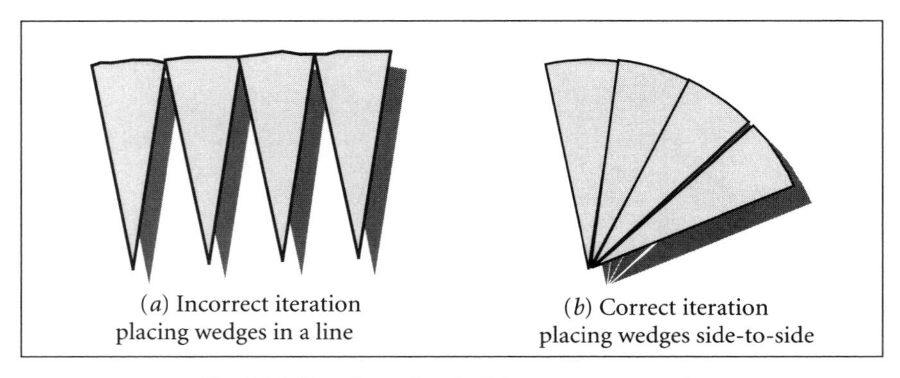

Fig. 14.6. Iterating a "wedge" to measure an angle

Rich, informal experiences in which students explore the attribute of angle size through comparison and through the iteration of a unit of measure can set the stage for students to understand standard units of measurement for angles (usually the degree), to be able to use the common tool for measuring (the protractor), and to be able to make good judgments about the reasonableness of their answers. The use of activities that exemplify the angle as an action (turning) may promote insight into the nature of angles. With this knowledge in hand, students will be ready to tackle more-challenging situations involving angles.

CONCLUSION

Students in the United States continue to have major difficulties with concepts of measurement. Despite years of negative results in the NAEP mathematics assessment, little improvement has been seen. As has been pointed out in virtually every section of this article, these difficulties can often be explained at least in part by the framework identified in the beginning of the article. That is, students lack experience with attributes and units of measurement before embarking on the use of standard measuring devices and formulas. Without a concerted effort to give students experiences that build their conceptual understanding of these different types of measurement, performance with measurement will likely continue to lag in the coming years.

REFERENCES

Kenney, Patricia Ann, and Vicky L. Kouba. "What Do Students Know about Measurement?" In *Results from the Sixth Mathematics Assessment of the National Assessment of Educational Progress*, edited by Patricia Ann Kenney and Edward A. Silver, pp. 141–63. Reston, Va.: National Council of Teachers of Mathematics, 1997.

Lindquist, Mary M., and Vicky L. Kouba. "Geometry." In *Results from the Fourth Mathematics Assessment of the National Assessment of Educational Progress*, edited by Mary M. Lindquist, pp. 44–54. Reston, Va.: National Council of Teachers of Mathematics, 1989.

Martin, W. Gary, and Marilyn E. Strutchens. "Geometry and Measurement." In *Results from the Seventh Mathematics Assessment of the National Assessment of Educational Progress*, edited by Edward A. Silver and Patricia Ann Kenney, pp. 193–234. Reston, Va.: National Council of Teachers of Mathematics, 2000.

May, Lola. "Creating Good Models." *Teaching Pre-K–8* 29 (January 1999): 24, 1.

National Assessment Governing Board (NAGB). *Mathematics Framework for the 1996 and 2000 National Assessment of Educational Progress.* Washington, D.C.: NAGB, 2000.

Nitabach, Elizabeth, and Richard Lehrer. "Developing Spatial Sense through Area Measurement." *Teaching Children Mathematics* 2 (April 1996): 473–76.

Robertson, Stuart P. "Getting Students Actively Involved in Geometry." *Teaching Children Mathematics* 5 (May 1999): 526–29.

Strutchens, Marilyn E., Kimberly A. Harris, and W. Gary Martin. "Assessing Geometric and Measurement Understanding Using Manipulatives." *Mathematics Teaching in the Middle School* 6 (March 2001): 402–5.

Van de Walle, John A. *Elementary and Middle School Mathematics: Teaching Developmentally.* 4th ed. New York: Addison Wesley Longman, 2001.

Wilson, Patricia S., and Ruth E. Rowland. "Teaching Measurement." In *Research Ideas for the Classroom: Early Childhood Mathematics*, edited by Robert J. Jensen, pp.171–94. New York: Macmillan, 1993.

Wilson, Patricia S., and Verna M. Adams. "A Dynamic Way to Teach Angle and Angle Measure." *Arithmetic Teacher* 39 (January 1992): 6–13.

15

Providing Opportunities for Students and Teachers to "Measure Up"

Melissa Boston

Margaret S. Smith

By the conclusion of middle school, students should understand perimeter, area, surface area, and volume and should have developed strategies and general formulas to measure these attributes in selected figures (National Council of Teachers of Mathematics [NCTM] 2000). To realize these expectations, the middle school curriculum should include measurement tasks that provide opportunities for problem-centered instruction and promote students' understanding of measurement concepts. Situations in which one attribute of a geometric figure is allowed to vary while another attribute remains constant can provide rich problem-solving opportunities that meet both these objectives for middle school students. The two examples discussed here involve maximizing the area of a rectangle for a fixed perimeter and minimizing the surface area of a rectangular prism for a fixed volume. After establishing a rationale for including such tasks in the middle school curriculum, we provide vignettes and discussion surrounding the opportunities for problem-centered instruction and for students' learning. We conclude by presenting opportunities for teachers' professional development centered on such tasks.

Rationale

Middle school students' experiences in learning perimeter, area, surface area, and volume have traditionally been limited to memorizing and applying formulas (Bright and Hoeffner 1993). This instructional approach has not proven effective, as noted by research and standardized testing that indicate that mid-

The information in this article is based on work supported by National Science Foundation (NSF) grant #9731428. Any opinions, findings, conclusions, or recommendations expressed in this article are those of the authors and do not necessarily reflect the views of the NSF.

dle school students harbor fundamental misconceptions about measurement attributes. Students have difficulty drawing a figure of a given perimeter or area (Chappell and Thompson 1999), and they frequently confuse perimeter and area (Chappell and Thompson 1999; Bright and Hoeffner 1993) as well as surface area and volume (Martin and Strutchens 2000). In addition, middle school students seem unaware that area can vary for rectangles of a fixed perimeter and experience great difficulty in explaining how two different rectangles can have the same area but different perimeters (Chappell and Thompson 1999). Results from the National Assessment of Educational Progress in 1996 show that only 1 percent of eighth-grade students were able to provide at least a satisfactory explanation of how to maximize the area of a rectangular dog pen given a specific amount of fence (Kenney and Lindquist 2000).

In contrast, instruction in measurement that goes beyond application of formulas can strengthen students' understanding of perimeter, area, surface area, and volume (Bright and Hoeffner 1993). Incorporating tasks into the middle school curriculum that examine variations in one attribute while another is kept constant can create opportunities to center instruction on genuine problems that facilitate middle school students' understanding of measurement concepts. These tasks can also provide students with opportunities to relate numeric and geometric patterns, use technological tools, and create powerful generalizations (Hersberger and Frederick 1995).

ENACTING MEASUREMENT TASKS IN THE CLASSROOM

The following vignettes feature teachers and students who participated in QUASAR, a mathematics reform project that sought to improve the instruction and learning of mathematics in urban middle schools (Silver and Stein 1996). These vignettes illustrate how two specific tasks provided opportunities for problem-centered instruction that allowed students to develop and strengthen their understanding of measurement concepts. The first vignette focuses on the explorations of one group of students as they struggle to maximize the area of a rabbit pen. The second vignette shows the mathematical potential of using stacking cubes to create rectangular prisms.

DESIGNING RABBIT PENS

The Fencing Problem

Mrs. Marlowe, a seventh-grade teacher in an urban middle school, presented her students with the following task: "Each of the seventh-grade classes at Franklin Middle School will raise rabbits for their spring science fair. Each class will use one side of the building as one of the sides of their rectangular rabbit pen, and each class wants their rabbits to have as much room as possible. How do you think each class should design its pen?"

Mrs. Marlowe had several objectives in posing this problem to students. First, she wanted students to explore a situation in which they would need to make group decisions and create numerical data. She hoped that students would make conjectures and recognize the need to gather evidence to support or refute these conjectures. Second, she expected students to investigate the different rectangular pens that could be made with a specific amount of fence and to organize their data in a way that illustrated which pen yielded the most area. If different groups chose different amounts of fence, the class could look across reports and generalize that the most room is afforded when the length (the side parallel to the school building) is twice the width.

Day 1: The Power of Students' Convictions

Mrs. Marlowe distributed the Fencing Problem, asking students to "try to get a picture of what this might look like." She made it clear that the problem would cause some frustration, but provided no other guidance on how to proceed: "This is going to cause some disequilibrium with you, where you are not going to know what to do. You will have questions, and you will have to make decisions as a group. I'd like you to get started. Talk to each other and see what you are going to do."

In a group consisting of three boys (Darren, James, and Michael), Darren immediately assumed the role of leader. As soon as Mrs. Marlowe finished speaking, he raised his hand and asked, "How do we know how big the wall is?" The teacher reiterated, "Make a decision," and began to circulate to other groups. After some discussion and negotiation, the boys were confident that they had found a solution. A visit by Mrs. Marlowe, however, convinced Darren's group to continue thinking. They eventually decided that a 100-by-100-yard square (fig. 15.1), which had an area of 10,000 square yards and used 300 yards of fence, was "the" solution and held steadfastly to this belief.

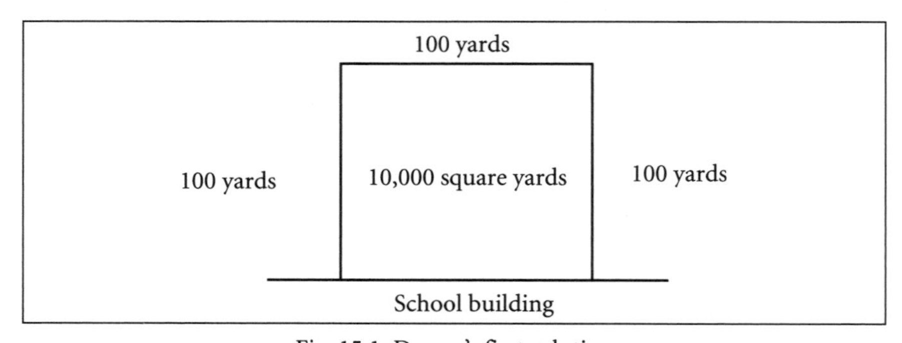

Fig. 15.1. Darren's first solution

Mrs. Marlowe's repeated queries, "How do you know that this is the greatest area possible for that rectangle? How do you know that for that amount of fence, that particular rectangle has as much room as possible?" had little impact on the group. The boys remained convinced that the area would remain constant regardless of the configuration of the rectangle. Even when Mrs. Marlowe told the group that she was not convinced that they had found the pen with the most area and suggested that they test some different dimensions, Darren continued to insist that, "If the amount of fence is the same, then the area will be the same. It has to be." The other boys in the group appeared content to acquiesce to Darren's leadership.

Day 2: Darren's Epiphany

The next day began with Mrs. Marlowe indicating that paper was available if any groups were ready to create a poster depicting their strategies and conclusions. Darren's group decided that they were "definitely" ready to make their poster. Darren was just beginning to draw when Mrs. Marlowe approached the group. She reiterated that the boys needed to prove that their pen enclosed the most area for 300 yards of fence, stating, "I would like to see some reasoning that this is the most area with this amount of fence. I am not convinced." Leaving behind this challenge, Mrs. Marlowe moved on to other groups.

Darren emphatically asserted to the group, "I swear it's the same. If you always have 300 yards of fence, it's the same no matter what shape you put it in." James suggested that they write that down "for proof." Underneath the diagram of a 100-by-100-yard square attached to a school building, Darren wrote, "We think that no matter how you shaped the pens it will be the same area if you keep the same amount of fence every time." Once again, the boys appeared satisfied with their solution. When Mrs. Marlowe returned, she read the statement on the poster and asked, "Where's your data to show that?" Darren responded that they did not need data because they "just knew." Mrs. Marlowe repeated the need to prove this assertion, and again left the boys with the suggestion of building different rectangles with 300 yards of fence.

Darren began to sulk, but James returned to the 125-by-50-yard rectangle he had suggested the previous day. He entered "125 × 50" on the calculator and excitedly told the group, "Look, we were wrong!" Darren looked at the display of "6250" and insisted that it could not be right. He told James to repeat the calculation, adamant that, "It has to be 10,000. It has to be the same as the pen that's 100 by 100." Darren then took over the calculator himself, obviously bothered by the discrepancy and still muttering, "I know it has to be the same." Michael then stepped in to support Darren's convictions by using the diagram on the poster to show that the area would remain the same because "even if you want to make it way longer, you can cut this off and put it here and it still equals 10,000." Darren challenged, "But how can you tell that for sure?" Darren then appeared to have a breakthrough. He quickly drew a diagram of a 125-by-50-yard-rectangle on the graph paper next to the 100-by-100-yard square (see fig. 15.2).

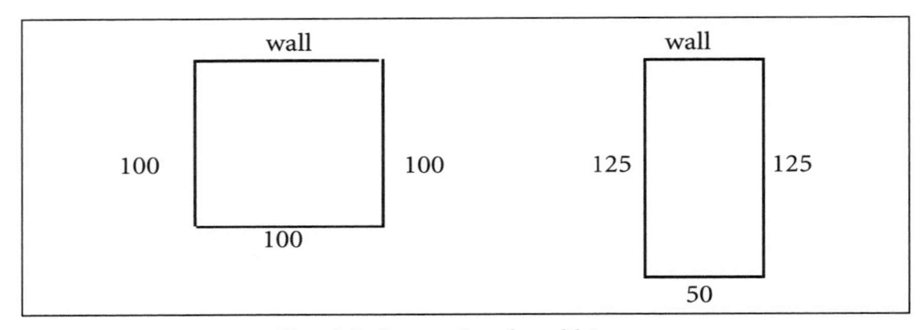

Fig. 15.2. Comparing the rabbit pens

He compared the two for a while, and then asked the other boys, "Doesn't it look smaller? I don't know... it should give an area of 10,000 but...." Darren stopped midsentence and suddenly asked Mrs. Marlowe if his group could step out into the hallway.

Once in the hallway, Darren positioned James and Michael as the corners of a 100-by-100-yard rabbit pen with one side adjacent to the wall, using each of the square floor tiles to represent a 10-by-10-yard square. He then paced off a rectangle that was 12 1/2-by-5 square tiles to represent the 125-by-50-yard rectangle and positioned himself in one corner and a classroom observer in the other. He looked at the two rectangles and shook his head in disbelief saying, "The area is definitely smaller. We were wrong about everything!" The other boys agreed, and they raced back into the classroom to amend their poster. Mrs. Marlowe asked the boys what they had discovered, and Darren excitedly told her that the two rectangles had different areas. As the bell rang, she instructed the boys to continue thinking about this at home and pull their ideas together for tomorrow.

Discussion

This vignette is interesting for several reasons. Pedagogically, it depicts a teacher who has centered instruction on a challenging task that asked students to consider a genuine problem situation. Mrs. Marlowe recognized that the Fencing Problem would cause her students "disequilibrium," a term she had introduced previously to give her students a label for the frustration they often felt in problem-solving situations. Even when the group of boys appeared to be struggling with the problem, she let them make important mathematical discoveries without direct instruction. Each day in her initial encounter with the boys, Mrs. Marlowe asked questions intended to focus their ideas on the main goals of the problem. She began by leaving her suggestions open but became more directive in subsequent interactions by proposing a strategy that would challenge Darren's convictions. Mrs. Marlowe did not refute Darren's ideas herself but instead insisted that the boys determine whether their conclusions were valid. The main mathematical ideas of the problem thus remained open for exploration and discovery.

Furthermore, the vignette provides a glimpse at students' thinking that illuminates the boys' conceptions of perimeter and area, how these attributes are measured, and the relationships between them. All three boys appeared to have a solid understanding of perimeter as the distance around the rectangle and of area as the amount of space inside the rectangle. They used the "length times width" formula to determine the area of a rectangle and were able to determine dimensions for a rectangular pen with a specified amount of fencing. The boys easily conversed about units of measurement for both perimeter and area and seemed comfortable in using graph paper and floor tiles to represent different rectangles. The vignette also illustrates the value of visual diagrams in supporting students' thinking. Darren approached the task with the conviction that area would not vary for rectangles created with a fixed amount of fence. The discrepancy between Darren's ideas and the results of the area calculation were enough to spur his thinking, but numerical data alone did not convince Darren that his convictions were false. Only through visually representing the two rectangles on graph paper and with floor tiles did Darren come to realize his misconception. The openness and accessibility of the task, combined with its implementation as a problem-solving activity, allowed the boys to engage in an exploration that lead to the discovery of interesting mathematical relationships.

CUBE COVERS

The Space Armor Task

Students in several QUASAR classrooms explored surface area and volume by using stacking cubes to create rectangular prisms of different volumes. Teachers initiated the task by asking students to consider the following situation: "For space travel, one day's supply of food is concentrated in this cube, which we will call a food pellet. To prevent the food from contamination during the flight we must wrap it in very special paper called space armor" (Shroyer and Fitzgerald 1986).

In this scenario, the number of food pellets being packaged or covered corresponds to volume, and the cost of covering the package, at $1 per square unit of special paper, corresponds to surface area. The exploration of space armor lasts several days. Students are presented with the task of covering rectangular prisms of different volumes (i.e., packages of 1, 2, 3, and so forth, food pellets) and of minimizing the amount of space armor needed to wrap these "food packages." The exploration is intended to help students (1) conceptualize surface area and volume and the units used to measure these attributes; (2) develop, or develop meaning for, the common formulas for determining surface area and volume of rectangular prisms; (3) realize that surface area can vary for rectangular prisms of a fixed volume; and (4) generalize that for different rectangular prisms with the same volume, elongated prisms have a greater surface area than more compact prisms (or conversely, that surface area decreases as the prisms become more cubelike). The lesson from Mr. Holbert's class, featured in the following vignette, is near the end of the exploration.

Mr. Holbert's Class

In the previous lesson, students in Mr. Holbert's class had built rectangular prisms with 24 cubes and entered the dimensions, surface area, and volume into a chart that the teacher had introduced earlier (see fig. 15.3).

Volume in Cubic Inches (# of cubes)	LENGTH (Front Edge)	WIDTH (Side Edge)	Height	Surface Area in Square Inches (Cost)
24	24	1	1	98
24	12	2	1	76
24	8	3	1	70
24	6	4	1	68
24	6	2	2	56
24	4	3	2	52

Fig. 15.3. Table of data for 24 cubes

At this point, students and teacher had abandoned the space armor context and were incorporating mathematical terminology into their discussions. Students easily connected the dimensions of the rectangular prisms with the formula for volume. Most were able to generate different dimensions for rectangular prisms with a volume of 24 cubes by grouping the factors of 24 in different ways instead of working directly with the cubes. Students still appeared to be using the cubes to verify their calculations for surface area, though most had created a procedure very similar to the common formula by realizing that opposing sides of the prism had the same area.

Homework was to enter the data for a 36-cube rectangular prism into the chart in order to compare the characteristics of 24- and 36-cube solids that seemed to affect surface area. Mr. Holbert hoped that students would generalize that compactness minimized surface area and that the most compact rectangular prism would be a cube. He began class by asking students to take a few minutes to write in their journals about the following: "Which 36-cube rectangular prism has the largest surface area? The smallest? Give reasons for your answers."

Before discussing students' responses, Mr. Holbert asked students to "make generalizations about the relationship between dimensions and surface area" by comparing the information in the chart for the 24- and 36-cube prisms. As he visited each of the groups, he often pointed to a sign on the wall that read, "Group first, teacher last!" to indicate that he would not respond to a question until students had wrestled with it sufficiently themselves. Sometimes he simply wrote the question on a side chalkboard and directed students' attention back to it whenever they were capable of answering it themselves.

After about ten minutes, Mr. Holbert called the class together and asked for ideas. Markus offered, "The largest surface areas have the most sides showing, and the smaller surface areas have the least sides showing." The class seemed to agree, so Mr. Holbert asked, "How does that relate to the dimensions?" Janelle suggested, "The larger the dimensions, the greater the surface area." Mr. Holbert hesitated, and another student rephrased Janelle's comment as, "The longer the one dimension, the greater the surface area. The more compact ones have a smaller surface area." Most of the class again appeared content with this observation. Mr. Holbert continued, "What shape would give the smallest possible surface area?" Jake replied, "Each side would be a square, it would be a cube." The teacher responded, "How does that relate to the dimensions? What would the dimensions be?" Lauren replied, "They would all be the same. Only for a volume of 24, you couldn't do that with cubes (holding up a stacking cube); $2 \times 2 \times 2$ is too small and $3 \times 3 \times 3$ is just a little too big."

The class decided that a cube with a volume of 24 units and one with a volume of 36 units would not be much different; both would have linear dimensions close to 3 units. Mr. Holbert then directed students to work in groups to consider what would happen to the surface area if they "cut the top layer off a $6 \times 2 \times 2$ and placed it end-to-end to create a $12 \times 2 \times 1$" and continued to repeat this process of doubling the length while cutting the height in half (see fig. 15.4). He clarified his directions with the aid of a diagram on the chalkboard and also indicated that the next rectangular prisms in the series would have dimensions $24 \times 2 \times 1/2$, then $48 \times 2 \times 1/4$.

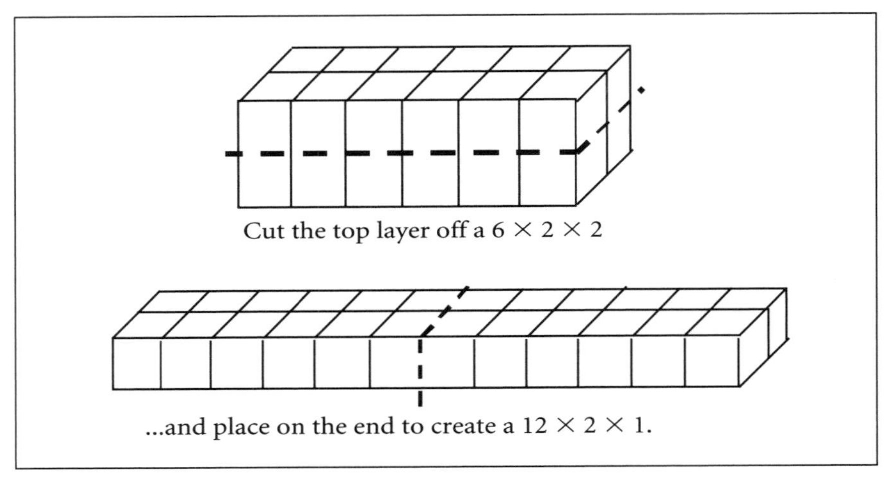

Fig. 15.4. Making a 6 × 2 × 2 into a 12 × 2 × 1

When students reconvened to discuss this as a class, they first negotiated why the surface area would not double each time, as initially proposed by a student. They concluded that the surface area almost doubles, "minus the area of the ends that are stuck together." Mr. Holbert pursued this a little further, getting students to generalize that as the process continues, the ends get thinner and thinner and the surface area gets closer and closer to actually doubling. He summed up the discussion by reiterating that, "surface area gets increasingly larger as the rectangular prism becomes more and more elongated, or as the one dimension becomes increasingly larger than the others." To close the lesson, Mr. Holbert asked students to "write about what would happen to a 6 × 2 × 2 if I rearrange it into a 2 × 3 × 4—if I make it more compact" and compare this to the journal entry they wrote at the beginning of class.

Discussion

Mr. Holbert implemented a problem-solving task that allowed for exploration and discovery. In the role of facilitator, he provided the opportunity for students to grapple with mathematical ideas and eventually come to their own understanding. He supported students' thinking by asking questions that guided students toward the mathematical goals of the lesson and by providing appropriate scaffolding, such as introducing a chart to organize data, on which students could build increasingly sophisticated ideas. The teacher refrained from assuming the mathematical authority in the classroom and allowed students to pursue answers to their own questions and to address the questions of their fellow classmates.

Mathematically, the vignette portrays a task that provides a concrete representation of surface area and volume, allowing students to develop meaning and understanding for each of these attributes and of the formulas and units with which they are measured. Research suggests that concrete experiences in using stacking cubes to create rectangular prisms helps to develop middle school students' understanding of surface area and volume (Battista and Clements 1996). In addition, connecting various mathematical representations—tables of data, formulas, and rectangular prisms made of cubes—helped students to make sense of mathematical ideas and to communicate their ideas to others.

The problem is rich mathematically, providing students with the opportunity to identify patterns in numerical data and use these observations to make generalizations. Students also identified patterns in the characteristics of the rectangular prisms that always seem to have the greatest surface area and those that always seem to have the least surface area. By allowing students to approach the task using a variety of representations, Mr. Holbert provided multiple entry points for students. The task does not prescribe a direct solution method, nor can students simply apply a predetermined formula or procedure. Instead, students must rely on their understanding of surface area and volume to form conceptions about how these quantities are measured and related.

Together, the lessons from Mrs. Marlowe's and Mr. Holbert's classrooms show the potential of incorporating challenging tasks and problem-centered instruction into the middle school classroom. Students develop meaning for measurement attributes and discover interesting mathematical relationships through their own problem-solving efforts as the teacher facilitates and guides these efforts toward the mathematical goals of the lesson. As illustrated in the vignettes, providing students with the opportunity to explore situations in which one attribute of a geometric figure is allowed to vary while another remains fixed encourage instruction and learning of measurement concepts that reflect the expectations set forth by the Measurement Standard (NCTM 2000). Such tasks could serve as a powerful addition to the middle school mathematics curriculum.

OPPORTUNITIES FOR TEACHERS' PROFESSIONAL DEVELOPMENT

Tasks such as those used by Mrs. Marlowe and Mr. Holbert can also provide opportunities for teachers to deepen their understanding of measurement concepts. Engaging with these types of tasks can also provide teachers with the opportunity to consider multiple solution strategies, to develop meaning for concepts and procedures, to make mathematical discoveries and connections, and to use a variety of representations. Such experiences highlight the opportunities for students' learning that can be afforded by incorporating challenging tasks into the classroom.

In addition, vignettes or cases such as those that feature the classrooms of Mrs. Marlowe and Mr. Holbert can provide valuable opportunities for teachers to reflect critically on new forms of instruction (Smith et al. forthcoming). By analyzing classroom episodes in which teachers are attempting to implement challenging mathematical tasks, teachers can begin to identify factors that support students' high-level engagement. In this way, teachers engage in critique, inquiry, and investigation into the practice of teaching through the analysis of actual classroom situations. Such "practice-based" professional development experiences provide opportunities for teachers to construct knowledge central to teaching by engaging in activities that are at the heart of a teacher's daily work (Smith 2001).

REFERENCES

Battista, Michael, and Douglas Clements. "Students' Understanding of Three-Dimensional Rectangular Arrays of Cubes." *Journal of Research in Mathematics Education* 27 (May 1996): 258–92.

Bright, George W., and Karl Hoeffner. "Measurement, Probability, Statistics, and Graphing." In *Research Ideas for the Classroom: Middle Grades Mathematics*, edited by Douglas T. Owens, pp. 78–98. New York: Macmillan, 1993.

Chappell, Michaele F., and Denisse R. Thompson. "Perimeter or Area? Which Measure Is It?" *Mathematics Teaching in the Middle School* 5 (September 1999): 20–23.

Hersberger, James, and Bill Frederick. "Flower Beds and Landscape Consultants: Making Connections in Middle School Mathematics." *Mathematics Teaching in the Middle School* 1 (April–May 1995): 364–67.

Kenney, Patricia Ann, and Mary M. Lindquist. "Students' Performance on Thematically Related NAEP Tasks." In *Results from the Seventh Mathematics Assessment of the National Assessment of Educational Progress*, edited by Edward A. Silver and Patricia Ann Kenney, pp. 343–76. Reston, Va.: National Council of Teachers of Mathematics, 2000.

Martin, W. Gary, and Marilyn E. Strutchens. "Geometry and Measurement." In *Results from the Seventh Mathematics Assessment of the National Assessment of Educational Progress*, edited by Edward A. Silver and Patricia Ann Kenney, pp. 193–234. Reston, Va.: National Council of Teachers of Mathematics, 2000.

National Council of Teachers of Mathematics (NCTM). *Principles and Standards for School Mathematics.* Reston, Va.: NCTM, 2000.

Silver, Edward, and Mary Kay Stein. "The QUASAR Project: The 'Revolution of the Possible' in Mathematics Instructional Reform in Urban Middle Schools." *Urban Education* 30 (January 1996): 476–521.

Shroyer, Janet, and William Fitzgerald. *Middle Grades Mathematics Project: The Mouse and the Elephant.* Menlo Park, Calif.: Addison Wesley Publishing Co., 1986.

Smith, Margaret Schwan. *Practice-Based Professional Development for Teachers of Mathematics.* Reston, Va.: National Council of Teachers of Mathematics, 2001.

Smith, Margaret S., Edward A. Silver, Mary Kay Stein, Marjorie Henningsen, and Melissa Boston. *Cases of Mathematics Instruction to Enhance Teaching: Geometry and Measurement.* New York: Teachers' College Press, forthcoming.

16

Measuring Montana: An Episode in Estimation

Ted Hodgson

Linda Simonsen

Jennifer Luebeck

Lyle Andersen

Background

According to the Montana Department of Transportation (MDT), the state of Montana has an area of 147,138 square miles. One look at the state, however, and it's difficult to know how the MDT arrived at that figure (see fig. 16.1). The borders of the eastern two-thirds of the state follow lines of latitude and longitude, so that its shape is approximately rectangular. The western boundary of Montana, however, and overall shape of the western third of the state, are highly irregular. Because of these irregularities, there are no readily available formulas for determining the area of the state.

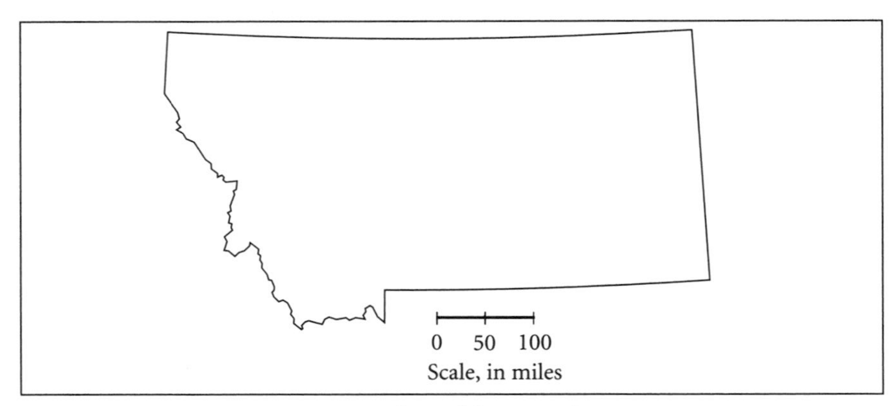

Fig. 16.1. An outline of the state of Montana

In a recent class of preservice elementary school teachers, the topics of measurement and irregular figures were raised. Posting a large map of Montana on the whiteboard, we informed the students of the MDT claims regarding the area of the state.

"How could it be done?" one student asked.

"We can't use squares," a second student added, "since no square would cover the western side of the state."

"Could we ever reach an exact answer?" a third student questioned.

One of the most wonderful and satisfying moments in teaching is the teachable moment. The students in our class, who had been dutifully pondering common area formulas and why they worked, faced a dilemma: How does one find the area of an irregular figure if all that one has available are formulas for regular shapes? Clearly, the students were interested in the question and desired an answer, yet possessed no readily available solution strategy. In other words, what students faced was a true problem! Our challenge as teachers was to direct their interest in such a way that students solved the problem using (and learning) sound mathematics.

In its Measurement Standard for grades 6 through 8, *Principles and Standards for School Mathematics* (National Council of Teachers of Mathematics [NCTM] 2000, p. 240) states that all students should

> develop and use formulas to determine the circumference of circles and the area of triangles, parallelograms, trapezoids, and circles *and develop strategies to find the area of more-complex shapes* (italics added).

When the third student in the class questioned whether an exact answer could be found, we reminded students that the MDT had arrived at an answer, so it must be possible. However, we added, a more pertinent question was "Could the class develop a strategy for finding the area of Montana?" The remainder of this paper traces students' efforts to develop an appropriate strategy and offers several observations about the use of estimation in the mathematics classroom.

STUDENTS' SOLUTION STRATEGIES

Although the area of Montana is known, the question facing students was one of process. By what method could the area of the state be determined? Faced with this task, students were encouraged first to find a range of reasonable answers. Although subtle, this charge—to find upper and lower bounds on the area of the state—had a major impact on students' explorations. Specifically, rather than seek an exact answer, students focused on estimation. In what range of values does the area of Montana fall?

Students' initial investigations focused on the rectangular nature of Montana. Using the map in figure 16.1, students determined that Montana is 550 miles wide at its widest point. From north to south, the greatest distance is 320 miles. Accounting for minor measurement errors (students obtained those figures using rulers and the given scale to estimate the height and width), we found that the state can be bounded by a rectangle that is 600 miles wide by 350 miles high. Likewise, a rectangle with dimensions 400 miles by 250 miles fits easily within the state, as in figure 16.2. Together, these facts lead to the inequality 100,000 mi^2 $< A <$ 210,000 mi^2, where A represents the area of the state.

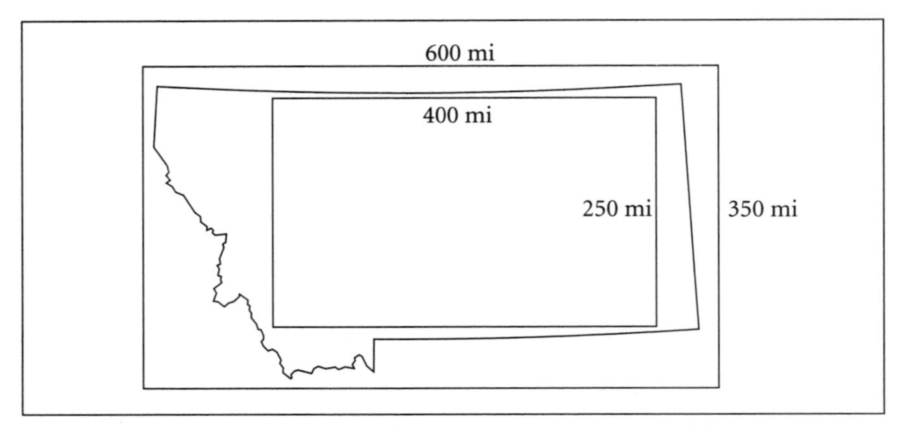

Fig. 16.2. Initial upper and lower bounds on the area of Montana

While providing bounds on the area of the state (as well as knowing that Montana has an area that is greater than that of at least 41 other states), students were uncomfortable with the wide range of possibilities and the amount of error in each estimate. For instance, the rectangle representing the upper bound contains all the state but also includes area from surrounding states and Canada. Likewise, the lower bound fails to include much of the western part of the state. One student suggested that the upper and lower bounds could be averaged, which leads to the surprisingly accurate estimate of 155,000 square miles. However, the student could provide no mathematical justification for averaging the upper and lower limits, and others in the class quickly discarded the suggestion. Nonetheless, students were pressed to find more accurate estimates of the area. Did these initial rectangles provide the best upper and lower bounds, or would other rectangles reduce the estimate error?

By comparing their estimates with others, students quickly determined that although a single rectangle provides an estimate, there is a limit to the accuracy of the estimate. Regardless of the rectangle that is used as an overestimate, for instance, one must include areas outside the borders of the state. Likewise, the underestimate can never completely conform to the boundaries of the state. Moreover, some estimates were better than others. Although we did not emphasize the fact, students seemed to realize that there is an optimal size and position for the rectangles that serve as the upper and lower bound.

The need to reduce measurement error, and the limited accuracy of students' initial estimates, led to the next iteration of the estimation process. In particular, one student observed aloud that the lines of latitude and longitude that were visible on our map partitioned the eastern portion of the State into rectangles.

"If we knew the area of each rectangle," the student reasoned, "we could add the areas to estimate the area of eastern Montana."

Although initially excited about the idea, the students quickly recognized several flaws in this plan. First, the lines of north latitude (from 45 degrees to 49 degrees, in whole-number increments) and west longitude (from 104 degrees to 116 degrees, in whole-number increments) that appeared on our map seemed to form a grid of congruent rectangles, but closer inspection revealed that the rectangles were not rectangles at all. On our two-dimensional rendering of Montana, for instance, the lines of latitude were actually arcs of a circle. Thus, the computations needed to find the area of each "rectangle" were not readily apparent. Second, as one moves from north to south, the "rectangles" composing the grid increase in size, adding another layer of complexity. Lastly, the use of the grid fails to resolve the issue of the irregular western border. How should one determine the area, when the western border of the state fail to conform to a rectangular grid?

Nevertheless, the idea that a grid of small, congruent rectangles would improve the accuracy of their estimate represented a significant step forward. Using the scale that accompanied the map, students superimposed a grid of 100-mile-by-100-mile squares on the map of Montana, as in figure 16.3. Still unresolved, however, was the manner in which to use the grid to estimate the area of the western portion of the state. Several students suggested that we use the grid to obtain upper and lower bounds:

"As an upper bound," one student recommended, "let's count the number of squares that contain some portion of Montana and multiply by 10,000."

"Yeah," a second student added, "and we could add the area of the squares that within the state to find the lower bound."

The problem with this technique, a third student noted, is that it still involves substantial error.

"Look at those squares," the student countered, pointing to several squares in the bottom row of the grid, " There's only a small part of Montana in each, but we're counting the entire square."

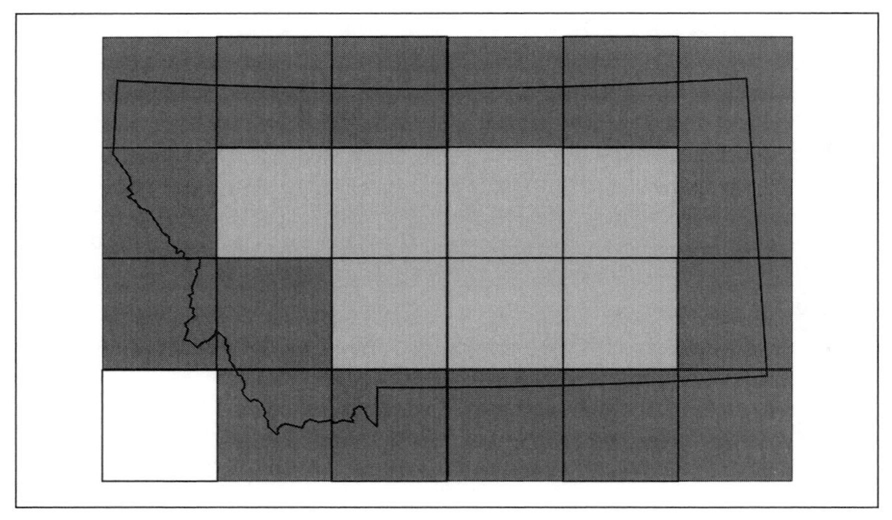

Fig. 16.3. Estimating the area of Montana using a grid of 100-mile-by-100-mile squares

This latter comment represented another teachable moment. Should one simply count the number of squares that contain some portion of Montana and multiply by 10,000 (100 × 100), or should one develop some process for estimating the portion of the state that was contained in each square? In considering these questions, we noted, students had recognized several important concepts underlying estimation. All estimates involve some degree of error, and the best estimates are those that minimize error. Moreover, estimates often vary with regard to simplicity. Although it would be more accurate to estimate the portion of the state contained in each square, the former technique (which we eventually adopted) was much simpler. As is represented by the dark- and light-gray squares in figure 16.3, 23 of the 24 squares in the grid contain some portion of Montana, which yielded an upper bound of 230,000 square miles. Likewise, the light-gray squares were completely contained by the state, yielding a lower bound of 70,000 square miles.

Note that the use of the grid actually yielded worse estimates than the simple rectangular estimates. However, the students seemed to recognize the potential of the grid technique. When asked why the second upper bound was greater than the first, the students responded that the width of the grid was 400 miles, or 50 miles wider than the simple rectangular estimate. Thus, the grid contained more of the area surrounding Montana. To improve the estimate, though, the students responded that one could use smaller squares. Smaller squares, they argued, would include less of the regions outside Montana.

As a second iteration, the students used grids of size 50 miles by 50 miles,

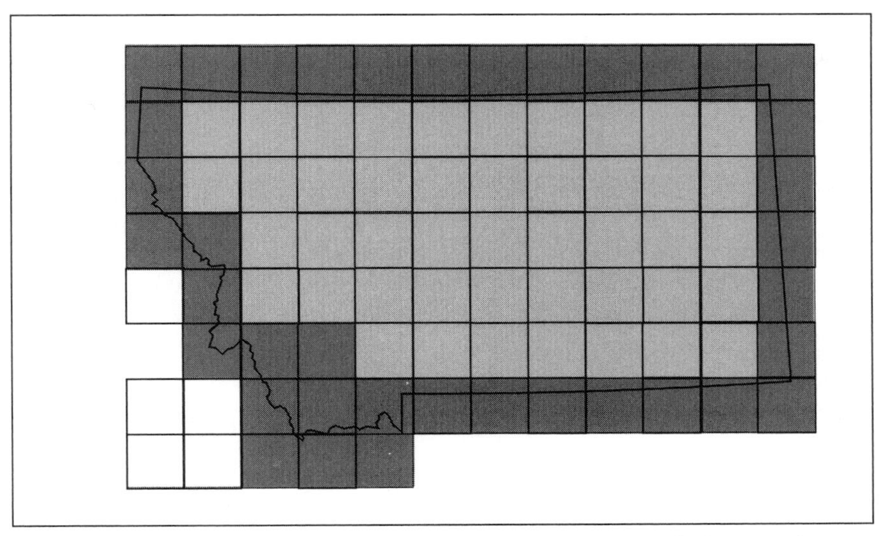

Fig. 16.4. Estimating the area of Montana, using a grid of 50-mile-by-50-mile squares

as in figure 16.4. Using the criteria that were established for the 100-mile grid, students obtained an upper bound on the area by counting the number of squares that contained some portion of Montana and multiplying that number by 2500 (50×50). For a lower bound, students counted only those squares that were entirely within the state and multiplied the total by 2500. The second application of the grid estimation process, with the use of 50-mile squares, yielded the interval $115{,}000 \text{ mi}^2 < A < 207{,}500 \text{ mi}^2$.

As a final estimate, the students used a grid of 20-mile-by-20-mile squares. Using the same procedure as before, and multiplying the number of squares by the area that each square represents (400 square miles), students obtained the following interval for their estimates: $125{,}600 \text{ mi}^2 < A < 166{,}400 \text{ mi}^2$. Although not perfect, the estimate represents a significant improvement over the previous estimation efforts. More important, without any formal training on the concept of limits, our students recognized that as the grid becomes increasingly fine, the range in which the actual area falls becomes increasingly narrow. In doing so, they had developed an intuitive, yet mathematically sophisticated, strategy for finding the area of Montana.

"So, if I take smaller and smaller squares," one student reflected, "my estimates will get closer and closer to the actual area."

"Would it be possible to ever find the actual area?" we asked.

"Well," the student responded, "I'm not sure if you would ever find it exactly, but I think each limit [the upper and lower limits] would grow closer and closer and closer."

DISCUSSION

In the classroom episode described above, students developed a strategy for finding the area of an irregular figure—the state of Montana. As we reflected on the lessons learned during this episode, however, it became clear that it was more than an exercise in measurement and estimation. In particular, the vignette illustrates several important principles about estimation and its use in the classroom.

Principle 1: Estimation Activities Can Promote an Understanding of Mathematics

To some, estimation is nothing more than guesswork. A good estimate, they would argue, is simply a good guess. To label estimation as pure guesswork, however, is to ignore the reasoning underlying the guess. The development of an accurate estimate often requires the use of sophisticated problem-solving strategies and the application of sound mathematics.

As an example, consider the mathematical reasoning that was elicited by the Measuring Montana activity. At the most basic level, the activity requires the use of scale factors. Given a linear scale (1 inch on our map represented 100 miles in the real world), students determined the area represented by rectangles and squares. Although the task did not pose a problem for our preservice teachers, it might be problematic for students in the middle grades.

To develop their estimates, students considered the concepts of upper and lower bounds. Intuitively, the students in our class recognized that counting the number of squares that contained some portion of the state represented an overestimate because "the squares on the edges also contained regions that were not in the state." Similarly, counting only those squares contained entirely within the state represented an underestimate, since "they [the squares] didn't contain all portions of the state."

Students' understanding of bounds was further enhanced by the fact that each placement of a grid yields a unique upper and lower bound. For instance, the use of 100-mile-by-100-mile squares (fig. 16.3) led to the conclusion that the area of Montana is less than 230,000 square miles. However, other placements of the grid can yield other estimates. If the grid is shifted up, for example, as in figure 16.5, one concludes that the area of Montana is less than 200,000 square miles. Both these numbers represent upper bounds on the area of Montana, yet the latter estimate is more accurate than the former. Thus, the activity led our students to consider the existence of "best" upper and lower bounds.

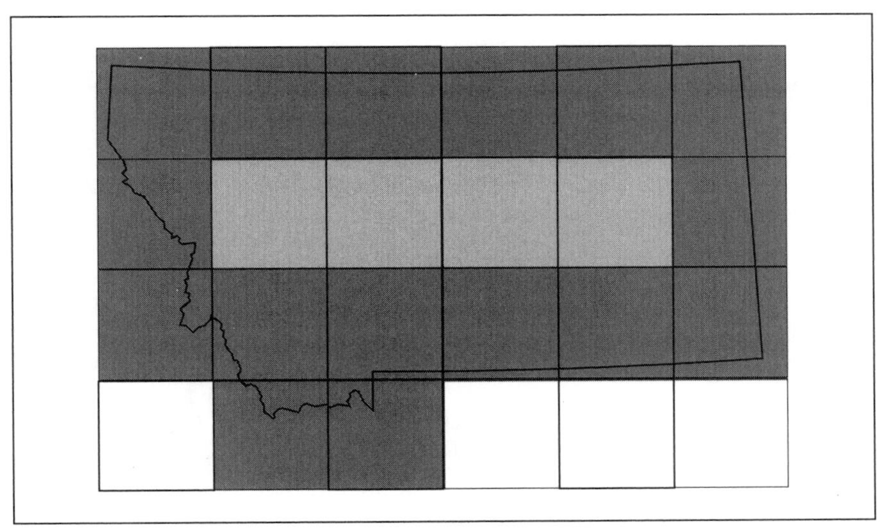

Fig. 16.5. An alternative placement of the 100-mile-by-100-mile grid

Finally, the use of smaller and smaller squares to estimate the area represents a fundamental tool of calculus—the Riemann sum. Although we did not directly mention limits, the students seemed to understand that the upper and lower limits on the area would eventually converge to the actual area. In a high school calculus setting, this sort of activity might be used to introduce Riemann sums and encourage the need to find the area of irregular regions.

Principle 2: Estimation Activities Can Reveal Students' Thinking

In addition to serving as contexts for instruction, estimation activities can serve as contexts for assessment. As an illustration, consider the initial upper and lower bounds on the area of Montana. By enclosing the state with a rectangle, students determined that the area of Montana is at most 210,000 square miles. Additionally, by drawing a rectangle that is completely enclosed by the outline of the state, students determined that the area is at least 100,000 square miles. One student suggested that an estimate of the area of the state could be found by averaging the upper and lower limits but was unable to provide a mathematical justification for the process. Although we would need to probe the student's thinking further to be certain, it appears that the student possesses a procedural understanding of the mean but lacks a conceptual understanding. A common conception of the mean is as the balancing point (Uccellini 1996). By averaging the upper and lower bounds, the student unknowingly balanced the error of the underestimate with that of the overestimate.

Does the student understand the mathematics? Is he or she able to apply it? Does the student understand the benefits of a particular process or technique? Does he or she understand the limitations? Estimation activities can uncover the answers to these questions.

Principle 3: Estimation Can Serve as a Vital Link Between the Real and Mathematical Worlds

We are believers in the use of the real world—the world of students' experience—as a tool for facilitating the study of mathematics. Although not every mathematical task illustrates a real-world phenomenon, we have found that linking the two worlds increases students' excitement about, and interest, in mathematics.

The students in our class lived in Montana, and most had grown up in the state. By focusing on a question of personal interest (How does one find the area of Montana?), the Measuring Montana activity generated students' interest in the concept of area. Additionally, the activity offered them an opportunity to use the measurement tools that had been studied in the class within a context that was personally meaningful.

Throughout *Principles and Standards for School Mathematics* (NCTM 2000), references are made to the benefits of appealing to real-world contexts. Real-world contexts, especially those that are drawn from the students' world of experience, add value and richness to the classroom. Estimation is an integral part of everyday life. By using estimation activities in the classroom and embedding them in real-world contexts, we teach an appreciation for mathematics and mathematical reasoning.

CONCLUSION

We have found the Measuring Montana activity to be an entertaining and fundamentally sound mathematical task. Moreover, the activity is easily adaptable to any state or country. We present it, though, not simply as an interesting activity but also to illustrate some principles about estimation. Estimation is an integral part of the lives of students, can lead to mathematical exploration and discovery, and is an effective assessment tool. It deserves a place in every mathematics classroom.

REFERENCES

National Council of Teachers of Mathematics (NCTM). *Principles and Standards for School Mathematics.* Reston, Va.: NCTM, 2000.

Uccellini, John C. "Teaching the Mean Meaningfully." *Mathematics Teaching in the Middle School* 2 (November–December 1996): 112–15.

17

Estimation at Work

Thomasenia Lott Adams

Gregory Harrell

We develop judgments and perceptions, in the natural and everyday sense regarding the quantities in situations we encounter daily. That is, we estimate. To estimate a measurement means to make a judgment or develop an opinion about a particular measurement attribute (e.g., length, width, area). The process of estimating measurements offers rich opportunities for students to make judgments and develop opinions that can have rewards for their lifelong measurement tasks. Students can use meaningful and useful estimates of measurement attributes to solve problems, validate formal measurements, and make decisions related to measurement (e.g., On the basis of my estimate, should I repeat the measurement?).

Our goal for this project was to collect data on real-life and current applications of estimation. We decided to interview professionals to obtain information about how people use estimation in the workplace. Initially, we made some assumptions about professionals who might frequently employ estimation to complete job tasks. On the basis of those assumptions and, subsequently, a list of professions, we sought people in these professional fields. Descriptions of the eighteen professionals interviewed are given in table 17.1.

For the purpose of meeting our goal, we asked each professional the following questions:

1. For what kinds of tasks do you frequently engage in estimation?
2. Why do you choose to estimate instead of using a tool to obtain a measurement?
3. Why do you choose to use a tool to obtain a measurement instead of estimate?

Frequent Estimation Tasks

Some of the professionals' frequent estimation tasks have been summarized in table 17.2. Several themes regarding the professionals' frequent estimation tasks emerged from these interviews.

TABLE 17.1

Professionals Estimating at Work

Profession (Year of Experience)	Frequent Estimation Task(s)
Traffic patrol officer (4)	• Speed of moving vehicles
Ice cream parlor owner (1)	• Amount of ice cream dipped for a scoop
Pediatrician (17)	• Amount of water loss as an indicator for dehydration • Changes in a child's body weight, temperature, or both
Timber sales consultant (28)	• Feet of lumber in a tract of trees • Size of the tract in acres • Diameter and height of individual trees
Basketball official or referee (8.5)	• Time for players to stay in the lane or shoot the ball
Basketball official or referee (15)	• Time for players to stay in the lane or shoot the ball
Road Traffic Supervisor (21)	• Minimum speed of a vehicle at the time of an accident
Meat Cutter or Butcher (18)	• Thickness and weight of meat • Number of people that a piece of meat will feed
Department store seamstress (32.5)	• Amount of cloth to make or alter a garment • Body dimensions (e.g., width of shoulders)
Contractor or Carpenter (25)	• Dimensions of a room (e.g., area, ceiling height) • Length of ceiling joists and bracing in an attic • Price range for a building project
Land Surveyor (13)	• Time and cost to survey a tract of land • Size of land tracts (e.g., quarter acre, half acre)
Cosmetologist (12)	• Amount of bleach, color, mousse, or gel to use • Length of hair to cut • Length of time required for each appointment
Farmer (peanuts, cotton, wheat) (30)	• Number of seeds for planting an acre • Pounds of fertilizer needed per acre • Fuel needed for tractor to plow a field • Application rate of pesticides and herbicides • Potential annual harvest (e.g., tons of peanuts) • Time required to gather a harvest
Automotive Technician or Mechanic (18)	• Amount of mileage left on brake pads and tires • Temperature and pressure of units (e.g., radiator) • Length of spark plug gaps • Torque value for tightening bolts
Landscaper (8)	• Volume of and distances for laying mulch • Costs to customer for a landscaping project
Licensed Practical Nurse (15)	• Number of supplies to order (e.g., syringes) • Child's temperature and weight • Conversions from pounds to kilograms
Restaurant Kitchen Manager or Cook (28)	• Amount of food needed for the daily menu • Recipe measures (e.g., teaspoon, ounce, cup)
Postal Clerk (7)	• Weight of a package to be mailed • Cost of mailing a package

Theme #1: Professionals Often Use More than One Sensory Skill to Develop an Estimate

The traffic patrol officer uses visual and auditory senses to estimate the speed of moving vehicles. A "sight" estimate of how long it takes for a moving vehicle to travel from one point to another is one factor in determining the probable speed of a vehicle. The patrol officer also uses hearing to obtain an estimated speed for a moving vehicle. The frequency of the audio radar in the patrol vehicle increases as the speed of the target vehicle increases. A steady tone is a signal of a constant speed. A higher pitch is given for higher speeds. After being in the patrol car and monitoring several moving vehicles, the author-interviewer could distinguish the pitch of vehicles in the 30–35-mph range from those in the 40-mph range. A very low speed sounds like a growl, and the sound becomes clearer and intensifies as the vehicle moves faster.

The restaurant kitchen manager or cook uses taste and touch to develop estimates for a "pinch of this" or a "dash of that." It's a set meaning, but it depends on the cook's interpretation. "With a pinch, you just grab between your two fingers [thumb and index]. Of course, someone may have really big hands. His pinch might be twice as much as mine. For a dash, you have a shaker and just give one quick shake." He also makes visual estimates. He knows what a certain amount of water looks like in a container. To demonstrate, he poured water into a large pitcher that had no measurement markers on it to estimate 64 ounces of water. He was very close to the actual measurement (subsequently obtained). He credits his estimation skills to using containers that he is familiar with to make estimates.

Theme #2: Professionals Often Use External Sources to Build or Support Their Estimations

The pediatrician makes estimates about the percent of body weight a child has lost when he hypothesizes that a child is dehydrated. Unless the child is in the hospital or has recently been examined by the pediatrician, he does not have the child's usual weight. He must then turn to other sources to develop an estimate. In particular, he listens to the parent or caretaker of the child to see if any possible causes of dehydration (e.g., vomiting, diarrhea) are present. He also looks at the primary source, the child, for indicators of dehydration (e.g., dry mouth, lethargy) when he estimates weight loss.

The seamstress listens to the customer who wants a garment made or alterations made to a garment to develop estimates. Her goal is to make some initial determinations about how much cloth is needed to create a piece of clothing. For example, she "eyes" the customer's shoulder width to

Table 17.2

Why or Why Not Estimate?

Profession	Reason(s) to Estimate	Reason(s) Not to Estimate
Traffic patrol officer	• It is mandated by state law for writing ticket and is a required skill for officer certification. • It can verify that the radar is working but also prevents drivers from detecting radar surveillance.	• Not applicable: Estimation is required.
Ice-cream parlor owner	• It is impractical to weigh every scoop of ice cream.	• When a customer makes a precise order (e.g., a pint of ice cream) the ice cream must be weighed.
Pediatrician	• It can replace missing data (e.g., normal weight); verify, support, or throw out existing data; and help determine the need for further testing.	• There are consequences for being wrong.
Timber sales consultant	• It saves time.	• There may be uncertainty about classification of trees.
Basketball officials or referees	• There is too much going on in a game to watch the clock.	• [No reason given]
Road traffic supervisor	• No real-time measuring device can be used.	• When dealing with a person's life, future, driver's license, and insurance costs, guessing is not a good thing.
Meat Cutter or Butcher	• It saves time.	• There may be uncertainty about the weight of the meat.

Department store seamstress	• Customers may not want to try on a garment. • Measuring certain body parts (e.g., crotch) may make a customer uncomfortable.	• [No reason given]
Contractor or carpenter	• It saves time.	• Real measurements determine the contract price for the job.
Land surveyor	• The price needs to be given to the customer in advance.	• "Decent estimates" may not be possible.
Cosmetologist	• It is quicker.	• A customer may need a precise coloring mix.
Farmer	• It is a major part of farming.	• [No reason given]
Automotive technician or mechanic	• In this business, time is money.	• It may be an unfamiliar task.
Landscaper	• It saves time and allows more productivity.	• Serious customers need precise amounts.
Licensed practical nurse	• For time, it is not always practical to measure. • On occasion, the patient is not present for measurements.	• Some things cannot be estimated (e.g., blood pressure), or it may not be necessary (e.g., patient is present). • It may be misleading (e.g., "a tall 20 pound baby feels a lot lighter than a short 20-pound baby").
Restaurant kitchen manager or cook	• There are time savings.	• "One must stay faithful in the measurement to have a consistent product all the time."
Postal clerk	• It makes everything go a little faster. • It ensures that the scales are working correctly and is necessary when the scales or computer are down.	• With a computerized system, the scale weight is used to print the price label that is placed on the package.

determine what she is working with when a customer requests her to sew a suit, dress, blouse, or shirt. She states, "If someone wants the pants or dress a certain length, then I measure the person in my mind. I then ask myself, 'Is six inches enough, or should I take off a little more?'" Requests of the customer guide an estimate, and in the end, customer satisfaction defines a good estimate for the seamstress.

Seamstress showing estimate for shoulder width of about 22"

Theme #3: Some Professionals Are Very Concerned about Either Underestimating or Overestimating

The landscaper said, "Money is a factor when I estimate a job. You never want to be under. You always want to be a little over and give a discount, but not be under." When income is a factor, underestimating can cause a person to lose money on a job.

However, overestimating was a concern for others. The restaurant kitchen manager or cook said, "If you are going to guess [estimate], then do it under. If you overguess—say, on a spice—then it will blow your dish away. You can always add [ingredients] to a dish. It's a lot harder and a little more technical when you have to take away." So although the landscaper sees it as important to never be under for an estimate, the restaurant kitchen manager or cook's perception is that underestimating is the best route for his profession.

The contractor or carpenter is concerned about both overestimating and underestimating the costs of projects. If a potential customer asks him to estimate the cost for enclosing a patio, for instance, he first estimates the dimensions of the patio to develop estimates for costs. Before he estimates a contract price for the customer, he measures the patio exactly to ensure that the cost estimate is reasonable. In his words, "If you overprice, you won't get any work. If you underprice and get the job, then you'll lose money."

The cosmetologist has to deal with overestimating more than underestimating. For example, if she overestimates the length of hair to cut for a customer and cuts off too much hair, the customer may be very displeased with her service. If she underestimates and does not cut off enough hair, she can always adjust the cut according to the customer's wishes.

Cosmetologist estimating length of hair to cut

Theme #4: Real-Life Estimation Tasks Are Often Multifaceted

The basketball officials or referees have at least five time estimates they must keep track of during a game: three-second limit for a player staying in the lane, five-second limit for the player to throw the ball in-bounds, a ten-second limit to get the ball passed half court, a ten-second limit for the player to shoot the ball once the ball is in a certain area on the court, and the time limit for the entire game (e.g., 40 minutes for college games). Throughout the game, the referees are constantly switching from one type of monitoring of time limit to another and must be able to focus on other variables of the game while they make these time estimates.

When asked about how much food would be needed for a certain number of people, the restaurant kitchen manager or cook responded, "Say we are going to have 50 people coming to a buffet. How much roast beef am I going to need? I figure 3–4 ounces per person. You're looking at 200 ounces. Basically, that's 12–13 pounds of roast beef. Now, that's end weight—after you've cooked. With roast beef, you're probably going to lose anywhere from 15 to 20 percent of your product, depending on the cut of meat, after you cook it. Trimming off the fat [and] the muscle shrinking from cooking [also contribute], so you estimate from there how much roast beef you are going to need." In this instance, there is a three-step estimation process: one for the raw meat, one for the cooking process, and one for the cooked meat.

WHY ESTIMATE?

Each professional gave a variety of reasons for estimating measurements. Some specific reasons are listed in table 17.2.

Reason #1: The Majority of the Professionals Stated That Estimation Saves Time

Saving time was often connected to professions where large or efficient productions were the key to success in the profession. Think about it from the customer's perspective: just how long do you want to wait for a meal in a restaurant? How long do you want to leave your vehicle at the mechanic shop? Do you want to see someone dip some ice cream, put it in a dish, put the dish on a scale, and then put the scoop (remember it is melting!) on your cone? As customers, it is to our benefit that various professionals use estimation as a skill in the workplace. Even during the interview, this was apparent for the meat cutter or butcher. "This store does a lot of special cuts. In the last thirty minutes, we've had two requests for two-inch sirloins. Everybody back here can just grab their knife and cut two inches." The volume of customers' specific requests indicates a need for developing estimation skills related to those requests.

Reason #2: Estimation Helps Verify the Validity of Measuring Tools and Methods

If the patrol officer sees a car that looks as if it is "running" 60 mph (his first estimate of speed); activates the radar, which says 22 mph, and doesn't hear a high-pitched audible tone; then he concludes that either something is wrong with the radar or his estimate is off. He would suppose that something is wrong with the radar, turn it off, and activate it again to test it.

The postal clerk said that when he grabs a box that he estimates to be two and a half pounds and places it on the scale and the scale shows three pounds, he takes it off the scale. He hypothesizes that the scale is probably wrong. "I've found the scales to be wrong before."

Reason #3: In Some Situations, Precise Measurements May Not Be Possible

In investigating accidents, the road traffic supervisor cannot measure the speed of a vehicle after it has been in an accident—that is, after the vehicle has come to a rest. He can obtain minimum vehicle speed only on the basis of the direct measurement of the skid marks and the drag factor. For example, a 50 percent drag factor with 200 feet of skid marks determines a minimum speed of 60 mph. "These results are based on scientific principle, and the phrase 'at least' is upheld in courts around the country." He added, "We can prove a minimum speed, but it's almost impossible to prove a maximum speed." Therefore, he can use the drag factor and skid marks to say confidently that a vehicle was going at least a certain speed, but he can use this information only to estimate the highest speed the vehicle may have been traveling at the time of an accident.

Road traffic supervisor measuring a drag-factor to determine minimum speed and estimate maximum speed

Reason #4: It Is Required by the Nature of the Job

Some professionals use estimation as a way of life on the job. The peanut, cotton, and wheat farmer stated, "When you have to work with Mother

Nature, the conditions really do make a difference in how you estimate." In his opinion, weather, the market, and government programs are unpredictable variables that require him to estimate. "Your ability to estimate and plan ahead has a lot to do with how successful you are going to be. Estimation is a major part of a farmer's life. Lying in bed at night, I have to try to figure out what the outcome might be under the circumstances. I have to estimate what I can produce on an acre of land."

Reason #5: Customers Often Want Specific Tasks Completed and Estimates for the Cost of the Tasks

Some professions involve providing customers with individualized products or results. These customers most often want to know what the product or result will cost before the professional begins the task. In these instances, the professional makes a judgment on the cost of the task on the basis of a variety of factors (e.g., how much material is needed, how much time it will take to complete the task, how complex the task will be). According to the land surveyor, who bills by the hour, she must first estimate how much time it will take to do a job—that is, survey a tract of property—in order to give a customer a price estimate.

Reason #6: Estimation Is Enjoyable

This was a unique reason for estimating given only by the postal clerk. He said that he estimates even though he does not have to as part of his job. He estimates mainly because he enjoys it, and he finds it to be a challenge. "I

Postal clerk estimating the weight of a package to be mailed

just feel the weight in my hand from playing with it a lot. All through growing up, I grabbed stuff. Being curious, I guess." Now he has to "grab stuff" all the time as he accepts packages from customers and makes estimates about the weight of the packages.

Why Not Estimate?

We hypothesized that there must be reasons for these professionals to decide not to estimate, so we asked each person why he or she would choose not to estimate but opt for a precise measurement instead. All but four professionals had a variety of reasons that estimation may not be employed for a task.

Reason #1: It Is Not Possible

Some measurements cannot be estimated but must be obtained by using an appropriate instrument or methodology. For example, the licensed practical nurse and pediatrician can estimate attributes such as weight and temperature but cannot estimate blood pressure. Therefore, the specific attribute to be measured might dictate whether or not an estimate can be developed.

Reason #2: The Customer Wants a Precise or Consistent Product

When working toward the goal of consistency and pleasing the consumer, the restaurant kitchen manager or cook must consider the dish being prepared. "If I'm using a recipe to make lobster thermidor for the first time, then I would measure. Is the flavor what I am looking for in the dish? Is the texture there? Is it something that would appeal to the customer? I do not want to be hit-or-miss the first time." Also, "For a highly seasoned spice, I really want to measure. Cayenne pepper, red pepper flakes—if you are a little low, it's not too bad. But say you're estimating a teaspoon and you put two teaspoons of cayenne pepper in a dish that calls for one. You're going to notice it. It's going to be hot!"

Reason #3: There Are a Lot of Risks and Consequences for Being Wrong

This is a particularly sensitive case for the pediatrician, who depends on precise lab measurements, because his estimate could affect the life-or-death situation of a child. "Everybody that's been in medicine very long has made errors—grievous errors—by believing lab data that are wrong, and they've made errors by not believing lab data that are right. A lot of times, the only way to solve something is to do it [measure it] all over again." Although he may still make estimates about critical measures, he does not take medical action until he has the real numbers to work with when attending to a patient.

Reason #4: There May Be Unfamiliarity or Uncertainty with the Task at Hand

Over the years, the automotive technician or mechanic has seen a lot of changes in the way vehicles are made and repaired. For him, familiarity with a task determines whether he will estimate or not. "I'll find the appropriate reference material and appropriate tool and do it the way it's supposed to be done the first few times until I am comfortable enough to make an estimate."

SUGGESTIONS FOR INSTRUCTION TO ENHANCE ESTIMATION EXPERIENCES

The information in the interviews of professionals supports several suggestions for the inclusion of estimation in the school curriculum. These suggestions cut across various levels of school mathematics.

Build the Skill

Many of the professionals declared that they have built up their estimation skills over time. When the restaurant kitchen manager or cook first began his profession, he made most of his dishes by precise measurement. Even now the automotive technician or mechanic says that when challenged with a new repair task, he follows the manual the first few times before he begins to estimate attributes like pressure and calibrations. Students also need time to build the skill of estimation. A simple task like estimating how many jelly beans are in a jar is not a wasted task as a beginning estimation experience.

The ice-cream "scooper" is trained to feel the weight of a scoop of ice cream. This "feeling" does not come about after scooping ice cream only once. It takes many days of practice of scooping and checking the weight of the scoop before the scooper estimates a proper portion of ice cream. Students need many opportunities to "try and try again" with estimation experiences in order to build their skills and confidence in their estimates.

Multiple Senses

Plan for students to engage in estimation for which a variety of senses are used (e.g., hearing, seeing, touching) to help broaden students' perspective of ways to develop estimates. Do not limit estimation exercises to what students can just touch (e.g., weight) but encourage them to "eye" substances and objects as well in order to develop perceptions.

One activity in which students can be encouraged to use hearing as a means of estimating length or distance is to take the class into the school gymnasium or auditorium. Have a student stand with her back to the class. Another student will make some planned noise from a given distance. The student with her back to the class will estimate how far away the student making the noise is standing. The only sense the student can use for making the estimate is hearing.

External Sources and Prior Knowledge

Encourage students to use external sources or prior knowledge to develop opinions and make judgments about measurements. This will help students test and reinforce estimates, particularly those that are useful for long-term experiences (e.g., estimates for a pound or meter) and help them build references and benchmarks for estimating. One strategy that basketball referees use to monitor the three-second time limit is to say "one one thousand, two one thousand, three one thousand." On average, this phrase takes three seconds to repeat. For estimating small increments of time, like seconds, students can employ this strategy.

Present students with situations where the students see the results of overestimating or underestimating to support their future estimation experiences. Use a cooking session where one-half of the class completes the recipe with exact measurements and the other half of the class completes the recipe with estimates only. Perhaps the teacher can modify a recipe on paper so that parts of it have been accidentally erased or stained. Students using estimation will need to decide estimations for some or all ingredients. A recipe for a simple food like gelatin will work well for this exercise. Perhaps the students know everything except how much water the recipe requires. An overestimation of water will result in a mixture that will not mold. An underestimation of water will result in a dish that may mold but that has a very strong, perhaps unpalatable taste.

Multiple Steps

Plan learning experiences so that students can perform estimates that require several steps before a suitable estimate can be obtained. For example, before making final judgments regarding estimates of things like initial body weight, the pediatrician may review a child's medical chart if it is available, make inquiries to the child's caregiver, and perform a physical examination of the child. This is a good example to model for students so that they gain understanding about different ways estimates are obtained. One simple exercise to do is to ask the students to make estimates regarding the cost of a consumer product. Give them the opportunity to ask questions about the product that might help in developing the estimated cost. Explain to them that people like landscapers, contractors, and carpenters have to engage in this kind of exercise to furnish estimates for their customers.

Also consider the restaurant kitchen manager or cook, who must make estimates for meat portions based on the raw condition of the meat as well as the cooked condition of the meat. Create word problems that require students to estimate how much meat (or other type of food) is needed to feed a certain number of people on the basis of this paradigm. Other foods shrink during cooking (e.g., cabbage). For an exploration, students (and teachers) create a display of foods, in raw form and cooked form, so that students can make estimates about shrinkage that occurs during cooking. Students can then assess their estimates by comparing the measurement (e.g., weight) of a particular food in both forms. This can also form the basis for a valuable home-school connection.

Decision Making

Students can engage in discourse about when it is an appropriate time to estimate and when estimation is not an appropriate or desired method. This will empower them to make decisions that are based on a critical analysis of the situation, instead of developing the idea that estimates should and can always accompany or take the place of precise measurements. At the ice-cream parlor, the scooper uses estimation when a customer orders a scoop of ice cream, but the scooper measures with a kitchen scale when a customer requests a pint or quart of ice cream. Students can explore other professions that have to make similar decisions about when to estimate and when to measure.

Connecting with Professionals

Seek cooperation from professionals in your community and plan "job shadowing" days for students to accompany professionals to work to explore how the professionals use estimation to complete job tasks. (Plan this according to any guidelines provided by your school or district.) Professionals such as cosmetologists and seamstresses often learn skills from observing experts in their fields. Cosmetologists go to and participate in hair shows. Seamstresses go to and participate in clothing fairs and fashion shows. These professions are conducive to observation and learning by doing. Seek these and similar professionals who are willing to allow students to observe them at work. A good time for students to engage in this experience is on teacher work days when students are not required to attend school.

Job-shadowing experiences will help answer students' questions about why they should develop estimation skills. It will also connect school mathematics to real-life situations, reinforcing the idea that mathematics is a tool for living. Students can develop a profile of the professional shadowed, write a report, and share both with the whole class. These professionals may be parents of students or even teachers who are engaged in other professional endeavors outside school time (e.g., teachers who are referees for night and weekend sports games).

Develop lessons in which students act out real-life estimation scenarios for whole-class exploration and discussion. For example, students can act as land surveyors and survey school or community property by using information from interviewing a land surveyor and how the professional uses measurement to survey land. In addition, a land surveyor can be invited to participate in the activity by presenting a demonstration of estimation used in a surveying project.

Interviews with professionals can serve to engage students in discourse about the similarities and differences between the ways different professionals use estimation. One of the first professions students can begin with is the teaching profession. For example, what is the most common attribute estimated by the mathematics teacher? Does the science teacher use estimation more than the physical education teacher? For what tasks does the art teacher estimate? There are many avenues for students to explore when interviewing teachers as well as other professionals that will help the students develop positive conceptions and skills for estimation.

Conclusion

One important aspect of school mathematics is measurement, and estimation is a vital component of measuring. Estimation helps students recognize when a measurement is reasonable (National Council of Teachers of Mathematics 2000). According to the interviews of professionals, estimation can be the primary tool for completing tasks in real-life situations. Not only should students be given opportunities to learn how to estimate for the purpose of judging the precision and accuracy of measurements, but they should also have opportunities to learn how to estimate for the purpose of developing a lifelong skill that can be used in the workplace and at home. Estimation is also very useful in recreation environments. For example, we considered only the referees from the basketball game, but basketball players must also estimate the various time limits during a game to prevent forfeiting the ball to the opposing team.

These interviews have shown us that estimation is at work all around us. As employees, customers, consumers, and participants in recreation, we often engage in estimation or benefit from the estimates of professionals who provide goods and services to us. Exploring the workplace tasks of a variety of professionals is a lucrative source for showing students that estimation is really at work!

REFERENCE

National Council of Teachers of Mathematics (NCTM). *Principles and Standards for School Mathematics.* Reston, Va.: NCTM, 2000.

18

Exploring Measurement Concepts through Literature: Natural Links across Disciplines

Richard A. Austin

Denisse R. Thompson

Charlene E. Beckmann

LITERATURE provides a natural opportunity to motivate middle school students in their mathematics study. Stories are not only a springboard into measurement activities but are also a bridge between the mathematics of measurement and other curricular areas. As middle school teachers increasingly team with teachers of other disciplines, literature offers an opportunity to coordinate lesson content across a variety of fields. The authors' experiences in middle school classrooms have convinced them that middle school students respond positively to stories and to mathematical investigations that are based on the story presentation (Austin and Thompson 1997; Thompson, Austin, and Beckmann 2002).

Many investigations that focus on measurement deal with the conversion of units as well as with ratios and geometry. This article provides detailed ideas from three books to illustrate how measurement ideas can connect several mathematics topics as well as connect with other curricular areas. There is also an annotated list of other literature references, each with a brief idea for a possible lesson.

Connecting Measurement, Science, and Social Studies

The Librarian Who Measured the Earth (Lasky 1994) is a story that ties measurement with a geographical area and a historical time period as it connects mathematics with science and social studies. This story encourages students' interest in the mathematical aspects of computing the circumference of the Earth on the basis of measuring the distance between two ancient Egyptian cities.

Eratosthenes is best remembered as the first person to compute the circumference of the Earth. *The Librarian Who Measured the Earth* provides an overview of the life and education of Eratosthenes. He was born to Greek parents living in North Africa and completed his formal education in Athens. Later, he was employed by the Egyptian ruler Ptolemy III. The story describes the Alexandrian library and museum at which Eratosthenes worked and focuses on inventions and discoveries made at the museum.

The ties to social studies certainly involve an investigation of Egypt or of the influence of the Greek culture in northern Africa and other parts of the ancient world. The question of how the Ptolemy line of pharaohs was established and when they ruled Egypt is a possible area of investigation that clearly comes from the story. Students can trace the name of the city, Alexandria, and investigate when and why the capital of Egypt was moved from Alexandria. The other major Egyptian city referred to in the story is Syene. When students try to find the two cities on a modern-day map of Egypt, they will find that the ancient city of Syene has had its name changed to Aswan.

Geometry arises in the story through setting up the ratio to compute the Earth's circumference. There is a bit of science here, too; the light arriving from the Sun is considered to be hitting the Earth as a set of parallel lines. Setting up the ratio provides students an opportunity to apply the concept of alternate interior angles formed by parallel lines being cut by a transversal.

Eratosthenes knew that at noon on a certain day, the sun cast no shadow on the walls of a well at Syene. He also knew that there was a shadow cast in Alexandria at noon on that same day. At noon on this day, Eratosthenes erected a pole of a known height in Alexandria and measured the length of the shadow it produced. Using this information he was able to determine the angle of the sun's rays at Alexandria. On the basis of these measures and his investigations at the library, he was able to determine the angle at the center of the Earth determined by the sector connecting Alexandria and Syene. Hence, he could determine the number of same-sized sectors required to circumnavigate the Earth. The really difficult problem then facing Eratosthenes was determining the distance between the two cities of Alexandria and Syene. Once this distance was obtained, he simply had to apply the ratios already found and multiply the distance by the number of sectors needed to complete the sphere. His overall method is outlined in figure 18.1.

The story indicates that linear distances were measured by bematists, surveyors trained to walk in equal-sized steps. These bematists determined the distance from Alexandria to Syene to be 5040 stades, with a stade being the length of a Greek *stadium*, or about 515 feet. This information, together with the number of sectors determined in figure 18.1, enabled Eratosthenes to

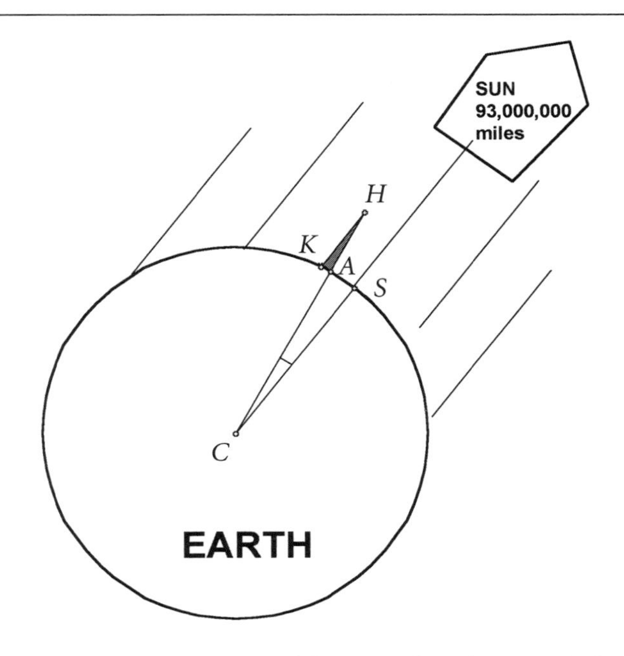

1. To determine the measure of the central angle, $\angle ACS$, formed by the center of the Earth, point *C*, and the cities Alexandria and Syene, points *A* and *S* respectively in the diagram:
   a. assume that light from the sun arrives at Earth in parallel rays;
   b. erect a pole (exaggerated) in Alexandria, line segment AH in the diagram;
   c. calculate or directly measure $\angle KHA$;
   d. note that $\angle KHA \cong \angle ACS$ and both have measure 7.2 degrees.
2. To find the Earth's circumference, divide 360 degrees by 7.2 degrees and note that the circumference is 50 times the distance between Alexandria and Syene.

Fig. 18.1. Eratosthenes' model for his project

approximate the circumference of the Earth as 252,000 stades, which converts to about 24,662 miles.

The story provides an opportunity to examine the concept of estimation and the degree of accuracy needed in different cases. In every class in which this story was used, at least one student asked, "What is the real distance around the Earth?" Finding the distance was left for the student to investigate with some possible resources mentioned. Students' further investigation was encouraged by asking them, "Was Eratosthenes really off by only about 200 miles, as the author stated?"

A natural extension of Eratosthenes' discovery is to have students "pace off" a certain known distance so that they can determine how big their average step is. Then students can measure large distances by pacing them off. This method is still used by geologists in the field to get quick estimates of distances. For example, how large is the perimeter of the school property? What is the perimeter of the parking lot? If students also time themselves taking 100 paces, they could determine how long it would take to walk at that pace across a major bridge, to a neighboring city, across their state, or the length of the Appalachian trail.

CONNECTING MEASUREMENT, GEOGRAPHY, AND ECONOMICS

Another book that can be used to focus students' attention on distance measurement and the subjects of geography and economics is *The Silk Route: 7,000 Miles of History* (Major 1995). This story explores the route that a load of silk might have followed as caravans set out from the center of China to the Middle East during the eighth century A.D. Here the measurement of distance and time offers a clear mathematical focus. Students can estimate the distance on the first leg of the Silk Route from Chang'an to Dunhuang using the map included in the book. In addition, they can estimate the distance across the Trans-Oxiana, the lawless part of the Silk Route, or between any of the identified trading centers.

There are many possible areas of investigation concerning measurement and economics topics. One can ask how value is placed on goods from far away. For example, is silk from China more valuable in Baghdad than a Persian rug? The opposite question might be asked as well. Suppose a bolt of silk costs $10 in Chang'an and doubles in value at each of the major trading centers along the route. What if, instead, a merchant adds $20 to the value of the silk at each trading center? How much would that same bolt of silk be worth in Baghdad, the fifth major trading center? Which of these models seems more likely? The book sets up the notion of the value of goods, which today is usually measured in money but might have been measured in other ways in the past. Other social studies connections can be made as students investigate the various cultures of the peoples along the silk route and the impacts of trade across borders. There are also links to the geography of Asia and the Middle East regions of the world.

For many students, it is interesting to compare how long it takes to transport goods by truck compared to caravans of old. Even today, donkey caravans or small boats are used to transport food or other supplies in parts of the world where automobile or truck traffic is difficult or impossible. How much can an animal or small boat carry? How many animals or small boats are needed to carry a load of a given weight? How long would it take a caravan of animals or boats to travel from a distribution center to the remote area?

CONNECTING MEASUREMENT AND THE BIOLOGICAL SCIENCES

One book that links measurement and the biological sciences is *Ride the Wind: Airborne Journeys of Animals and Plants* (Simon 1997). This book highlights stories of the migration of various insects, birds, and seeds. One story highlights the locust, one of the most dreaded air travelers because a swarm eats everything in its path, causing ruin and starvation to many communities. A large swarm can often tower to a mile in height and cover more than a hundred square miles. In addition, locusts are powerful fliers, often flying at up to ten miles an hour for twenty hours at a time.

Simon tells the reader that a large swarm of desert locusts could easily contain fifty billion locusts capable of eating three thousand tons of crops daily. Middle grades students can be asked, "How much, on average, would each locust eat?" This question allows students to grapple with measurement issues using very large and very small numbers and simplifying computations using scientific notation. In addition, this is a good problem to emphasize the limitations of technology because many calculators will display only up to 50 million. However, writing the division in fraction form helps students recognize that they can simplify the fraction by dividing both the numerator and the denominator by 1000.

$$\frac{3000\text{ tons}}{50,000,000,000\text{ locusts}} = \frac{3\text{ tons}}{50,000,000\text{ locusts}}$$

To provide meaning to this result, students need a more common unit of measure. Hence, there is a natural context for converting tons to pounds and ultimately pounds to ounces.

$$\frac{3000\text{ tons}}{50,000,000,000\text{ locusts}} \bullet \frac{2000\text{ pounds}}{\text{ton}} = \frac{6\text{ pounds}}{50,000\text{ locusts}}$$

$$\frac{6\text{ pounds}}{50,000\text{ locusts}} \bullet \frac{16\text{ ounces}}{\text{pound}} = \frac{.002\text{ ounce}}{\text{locust}}$$

Although .002 ounce per locust is a very small amount, students recognize that there are an enormous number of locusts.

This work naturally suggests further investigations with weight. How many blades of grass it would take to weigh an ounce? How many leaves weigh an ounce? Students can convert ounces to grams and compare the 0.002 ounce to an aspirin tablet or to other small common items to get an idea of the amount that a single locust eats. Students can be encouraged to locate other sources in which they can find facts about the desert locust, such as their size and the amount they typically eat.

Middle grades students might be asked to investigate how closely together locusts fly when they swarm by examining the size of the swarm. According to the story a swarm of 50 billion locusts would be contained in about 100 cubic miles. On average, how many locusts would be contained in a space of 1 cubic foot? This activity challenges students to consider the difference between converting linear versus cubic measures.

Elementary school teachers have commonly used literature to explore mathematics. Our experience tells us that literature is a powerful tool for middle school students as well. Middle grades teachers are encouraged to explore the use of literature to study mathematics and connect mathematics topics with other curricular areas. Although the reading levels vary from book to book, the mathematical levels of the suggested activities are all appropriate for students at the middle grades level. Teachers can adjust the complexity of the activities to make them appropriate in grades 3 through 8.

REFERENCES

Austin, Richard A., and Denisse R. Thompson. "Exploring Algebraic Patterns through Literature." *Mathematics Teaching in the Middle School* 2 (February 1997): 274–81.

Lasky, Kathryn. *The Librarian Who Measured the Earth.* Boston: Little, Brown & Co., 1994.

Major, John S. *The Silk Route: 7,000 Miles of History.* New York: HarperCollins Publishers, 1995.

Simon, Seymour. *Ride the Wind: Airborne Journeys of Animals and Plants.* San Diego: Harcourt Brace & Co., 1997.

Thompson, Denisse R., Richard A. Austin, and Charlene E. Beckmann. "Using Literature as a Vehicle to Explore Proportional Reasoning." In *Making Sense of Fractions, Ratios, and Proportions,* 2002 Yearbook of the National Council of Teachers of Mathematics (NCTM), edited by Bonnie Litwiller and George Bright, pp. 130–37. Reston, Va.: NCTM, 2002.

ANNOTATED LIST OF OTHER MEASUREMENT RESOURCES

The books in this list are children's literature, appropriate for reading in a wide range of grades from upper elementary through middle school. The mathematics lessons based on the stories can be adapted to be appropriate to the mathematics level of the students in the class. Because we generally read these books in mathematics class, we choose books that can be read in 10–15 minutes in order to have sufficient time in class to engage in a meaningful mathematics lesson.

Systems of Measurement

1. Adler, David A. *How Tall, How Short, How Far Away?* New York: Holiday House, 1999.

   This book introduces students to a number of different measuring units that have been used over time, from cubits and spans to current customary and metric units.

   - Have students measure an object in cubits, using the guide in the book, as well as in current customary or metric measures. Use the differences as a starting point for discussing the need for standard measures.

2. Axelrod, Amy. *Pigs on a Blanket.* New York: Simon & Schuster Books, 1996.

   The pigs want a change of pace, so they decide to go to the beach. Do they run out of time?

   - Have students keep track of the time it takes them to do various activities at home and at school. Classify activities of like type; if appropriate, determine statistics (mean, median, or mode) to describe the amount of time that students engage in the activity. Also, gather timetables from airport, bus, or train schedules, and have students determine the total traveling time from one location to another.

3. Hightower, Susan. *Twelve Snails to One Lizard: A Tale of Mischief and Measurement.* New York: Simon & Schuster Books, 1997.

   Many measurement activities take place as a beaver tries to measure the distance across a creek so it can build a dam.

   - Have students measure a set of given objects in inches, feet, and yards. Record the values in a table, and have students look for patterns. This is one way to have students develop the relations among various units.

4. Myller, Rolf. *How Big Is a Foot?* New York: Dell Publishing, 1990.

   An apprentice tries to make a bed for the queen. Until he measures with the same foot size as the king, he is not able to build a bed that is the proper size.

- Have students measure the dimensions of the classroom using their own feet. Use the different measures as a way to discuss the need for standard units. Have students consider how the size of the unit impacts the number of units needed.

Measuring Length, Weight, Capacity, and Speed

5. Allen, Pamela. *Who Sank the Boat?* New York: Coward-McCann, 1982.

   A number of animals try to get in a boat. Which one causes the boat to sink?

   - Have students fill a cup or lid floating in water with objects of the same size. How many can the lid hold before it sinks?

6. Axelrod, Amy. *Pigs in the Pantry: Fun with Math and Cooking.* New York: Simon & Schuster Books, 1997.

   The pigs follow a recipe as they make Firehouse Chili.

   - Have students bring in a favorite family recipe. How would the ingredients need to be modified to make enough of the recipe for the entire class? Have students consider which measuring tools would be used to measure various ingredients. For instance, if they need four 1/3-cup measures, help students realize that they could use the 1/3-cup measure four times or they could use a 1-cup and a 1/3-cup measure.

7. Barner, Bob. *How to Weigh an Elephant.* New York: Bantam Doubleday Dell, 1995.

   Different animals compare their weight to that of an elephant.

   - Have students write ratios or word problems comparing their weight with the weight of an animal. What percent of an animal's body weight does it eat in food each day? What percent of their body weight do students eat in food each day?

8. Brenner, Barbara. *Wagon Wheels.* New York: HarperTrophy, 1978.

   A family travels from Kentucky to Kansas in the late 1800s seeking free land.

   - Have students plan a trip from their home to a destination of their choice. How long does it take to travel there by car? By plane?

9. Hoban, Tana. *Is It Larger? Is It Smaller?* New York: Mulberry Paperbacks, 1985.

   Relative sizes are explored through photographs of real-world objects.

   - Have students estimate sizes of objects and then actually measure the objects. Also have students make estimates for real-life objects in their environment, such as the height of their school, their classroom, or a famous local building. Compare their estimates to the real measures to help students improve their visual measuring skills with estimation.

10. Jenkins, Steve. *Biggest, Strongest, Fastest.* New York: Ticknor & Fields Books, 1995.

    Seventeen animals are highlighted who represent the strongest, or biggest, or fastest, and so on, in some area.

    - Similar comparisons to those made with other books can be done with this book as well.

11. Ling, Bettina. *The Fattest, Tallest, Biggest Snowman Ever.* New York: Scholastic, 1997.

    A child makes a snowman and measures it with a paper-clip chain.

    - Have students build a chain, recording the number of paper clips and the length of the chain in a table. Write a rule to describe the pattern observed in the table.

12. Malam, John. *Highest, Longest, Deepest: A Fold-Out Guide to the World's Record Breakers.* London: Simon & Schuster, 1996.

    This book provides many natural examples that are record breakers for being the longest, highest, or deepest in their category.

    - The various comparisons in this book provide a natural resource for creating word problems of all types. In addition, have students compare two objects of their own choosing, writing ratios and proportions that describe the comparisons. The context of record breakers in the natural environment provides an opportunity to blend math with geography.

13. Medearis, Angela Shelf. *Poppa's New Pants.* New York: Holiday House, 1995.

    During the night, several relatives shorten the father's new pair of pants. In the morning, they are just the right length for the son.

    - Have students measure the length of the pants, shorts, or other outfits of class members. Find the mean, median, and mode of these values.

14. Nathan, Cheryl, and Lisa McCourt. *The Long and Short of It.* Bridgewater Books, 1998.

    Long and short features of several animals are illustrated, with comparisons made to common objects. For example, the tail of a ring-tailed lemur is longer than a skateboard.

    - Have students research animals other than those listed in the book and make their own book of similar comparisons.

15. Neuschwander, Cindy. *Sir Cumference and the Dragon of Pi.* Watertown, Mass.: Charlesbridge, 1999.

    Sir Cumference has to determine the ratio of circumference to diameter in order to save his father's life.

- Have students measure the circumference and diameter for a large number of circular objects of their own choosing. Record the values in a table, together with the ratio of circumference to diameter. Look for patterns to develop the concept of p.

16. Pinczes, Elinor J. *Inchworm and a Half.* Boston: Houghton Mifflin Co., 2001.

    Inchworms of different lengths fall from the sky in order to measure with smaller and smaller measures.

    - Have students measure objects to the nearest inch, nearest half-inch, and nearest quarter inch.

17. Schwartz, David M. *If You Hopped Like a Frog.* New York: Scholastic Press, 1999.

    Comparisons are made to a variety of animals. For instance, if a person had the strength of an ant, they would be able to lift a car.

    - Have students research facts about animals so they can write similar comparison statements of their own.

18. Silverstein, Shel. "One Inch Tall." In *Where the Sidewalk Ends: The Poems and Drawings of Shel Silverstein*, p. 55. New York: HarperCollins Publishers, 1974.

    How would you function if you were only one inch tall?

    - Have students write ratios comparing common objects to their height. For instance, what is the ratio of the length of their toothbrush to their height? At the same ratio, what would be the length of a toothbrush for a one-inch person?

19. Stevens, Janet, and Susan Stevens Crummel. *Cook-a-Doodle-Doo!* San Diego: Harcourt Brace, 1999.

    A rooster makes a family strawberry shortcake recipe.

    - Have students modify recipes to make enough for an entire class. This activity provides lots of opportunities to work with various measures and fractions.

20. Tompert, Ann. *Just a Little Bit.* Boston: Houghton Mifflin Co., 1993.

    Which animal will be big enough to allow the elephant to go up and down on the see-saw?

    - Have students try a weight experiment similar to that with *Who Sank the Boat?* If there is a see-saw on the playground, have the children actually try the experiment. Does the distance one sits from the pivot affect the number of children needed to balance the see-saw?

21. Wells, Robert E. *How Do You Lift a Lion?* Morton Grove, Ill.: Albert Whitman & Co., 1996.

Numerous comparisons are made to ways of lifting a wide range of objects and weights.

- Work together with science classes to design pulleys, and determine the amount of weight that can be lifted with the pulleys.

22. Wells, Robert E. *Is a Blue Whale the Biggest Thing There Is?* Morton Grove, Ill.: Albert Whitman & Co., 1993.

    Sizes of objects on earth and throughout the galaxy are compared using the size of a blue whale.

    - Have students compare the sizes of the different planets in the solar system. How long does light take to travel from the sun to the different planets? How long would it take the shuttle to travel from earth to the various planets and return home?

23. Wells, Robert E. *What's Faster than a Speeding Cheetah?* Morton Grove, Ill.: Albert Whitman & Co., 1997.

    Speeds of different objects are compared through wonderful pictures and illustrations.

    - Have students determine the distance they can walk or run in a set amount of time. Then determine their speed.

24. Wells, Robert E. *What's Smaller than a Pygmy Shrew?* Morton Grove, Ill.: Albert Whitman & Co., 1995.

    Sizes of objects are compared to the size of a pygmy shrew.

    - Have students compare the sizes of very small objects. Students can make natural connections to science, discussing sizes and weights of molecules and atoms.

Area, Perimeter, and Volume

25. Caselli, Giovanni. *Wonders of the World.* New York: Dorling Kindersley, 1992.

    This book describes ancient and modern wonders of the world.

    - Have students compare the sizes of ancient wonders to the sizes of many modern wonders. This book provides a natural opportunity to teach mathematics in a social studies context.

26. Grifalconi, Ann. *The Village of Round and Square Houses.* Boston: Little, Brown & Co., 1986.

    In the African village of Tos, the men live in square houses and the women live in round houses. The story explains how this practice came to be.

    - Have students explore the maximum area that can be enclosed with a given perimeter or the minimum perimeter that is used with a given area.

19

Using the Geoboard to Enhance Measurement Instruction in the Secondary School Mathematics Classroom

Barbara A. Burns

Gail A. Brade

MANIPULATIVES are teaching tools commonly used to facilitate mathematics instruction and understanding in elementary school classrooms. It is less likely to find them being implemented in secondary school classes. One such tool, the geoboard, lends itself to enhancing the study of many topics in measurement that are covered or reviewed in these later grades. For example, areas of polygons and ratios of similar figures can be explored in the middle grades, and the geometry of the circle can be explored in high school.

By definition measurement is "the assignment of a numerical value to an attribute of an object, such as the length of a pencil" (National Council of Teachers of Mathematics 2000, p. 44). The term itself most likely conjures up images of rulers being used to ascertain lengths. Measurement in the secondary school classroom, however, goes far beyond the use of this basic tool to include measurement of angles, areas of figures, and lengths of arcs.

Geometry exploration, using the geoboard, can be guided so that students recognize relationships, formulate conjectures, develop and deepen their understanding of conjectures, justify results and construct arguments that validate relationships. This is in contrast to simply being told by the teacher the facts that she or he desires the students to learn. "Sociocultural and constructivist theorists both highlight the crucial role that activity plays in mathematical learning and development" (Cobb 1994, p. 14). It is assumed that the constructed knowledge will be synthesized more thoroughly and therefore applied more easily (readily) in future situations. Therefore, we believe that the following approaches to measurement, although possibly requiring a greater time commitment, will enhance students' relational understandings of these topics.

One such topic, seen primarily in the middle grades, is the area of polygons. Formulas for areas of squares, rectangles, and triangles can be reviewed using the geoboard. Since this may be too time-intensive as a classroom activity, it may be more valuable for those students who are having difficulty conceptualizing the formulas. In brief, the area formulas for squares and rectangles might be more fully internalized by counting the areas of many different squares (and rectangles) formed on the geoboard and comparing these counts to the side lengths of the figures. In this way, the relationship of the figure's area to its formula is more apparent.

Triangles can be explored by first investigating right triangles (with two sides parallel to sides of the geoboard), whose area relationship to an appropriate rectangle is easily recognizable. Placing a rectangle on top of this right triangle (another color rubber band will help distinguish between the two figures) will produce two right triangles that are congruent (see fig. 19.1). Students should recognize that the length and width of the rectangle are the same segments making up the base and height of the right triangle. The formula $A = (1/2)lw$ (or more commonly $A = (1/2)bh$) can easily be seen. Students should be encouraged to check this relationship using right triangles of different dimensions.

For triangles that have base and height parallel to orthogonal sides of the geoboard, a convenient triangle for students to start with is an isosceles triangle enclosed by a 4×4 rectangle. Students can use a rubber band to mark the height of the isosceles triangle. Using the grid lines and the rubber bands

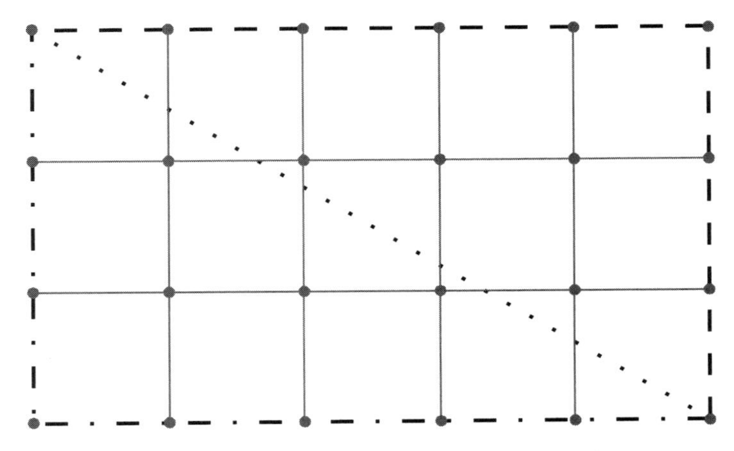

Fig. 19.1. Right triangle inscribed in a rectangle

as guides, students find that there are four congruent right triangles (verified by the side-angle-side [SAS] postulate) enclosed by the rectangle (two of which form the triangle in question; see fig. 19.2). The area of the triangle is therefore one-half the area of the rectangle. Looking at the dimensions of the rectangle and again observing that the measures of the length and width of the rectangle are the same as the measures of the base and height of the triangle, the students should arithmetically verify the formula $A = (1/2)bh$.

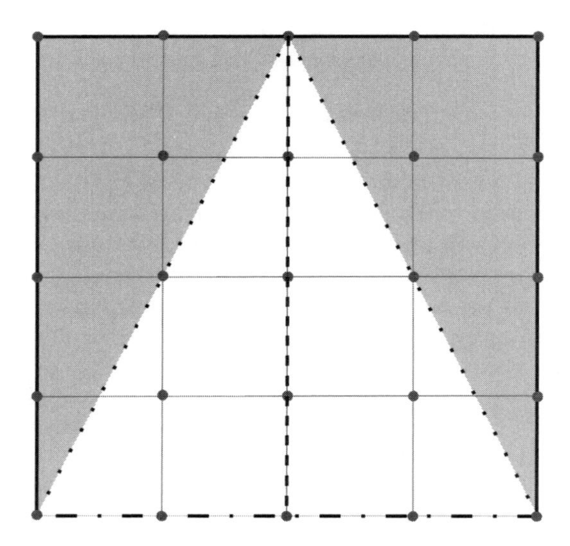

Fig. 19.2. Isosceles triangle inscribed in a 4 × 4 rectangle

If an obtuse triangle formed with the longest side parallel to one side of the geoboard is enclosed in a rectangle, the altitude drawn to that base will divide the triangle in a similar manner as the previous example. The altitude forms two pairs of congruent right triangles (see fig. 19.3). One triangle of each pair is inside the obtuse triangle, whereas the other is inside the rectangle but not the obtuse triangle. Continuing with additional examples should enhance students' conceptual understanding of this area formula.

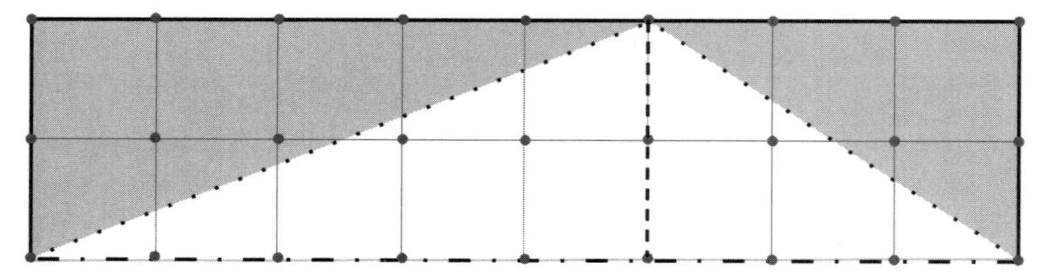

Fig. 19.3. Obtuse triangle inscribed in an 8 × 2 rectangle

To develop the area formula for parallelograms, a rectangle having the same base and height can be placed on top of a parallelogram on the geoboard (see fig. 19.4). Students can then be directed to notice (by SAS) that right triangle 1 is congruent to right triangle 2, therefore their areas are equal in measure. Because of this equivalence, triangle 1 can be translated onto triangle 2, illustrating that the area of the parallelogram and that of the rectangle must be the same. Adjusting the labels used transforms $A = lw$ into $A = bh$. Students can explore several other parallelograms to determine that the relationship holds.

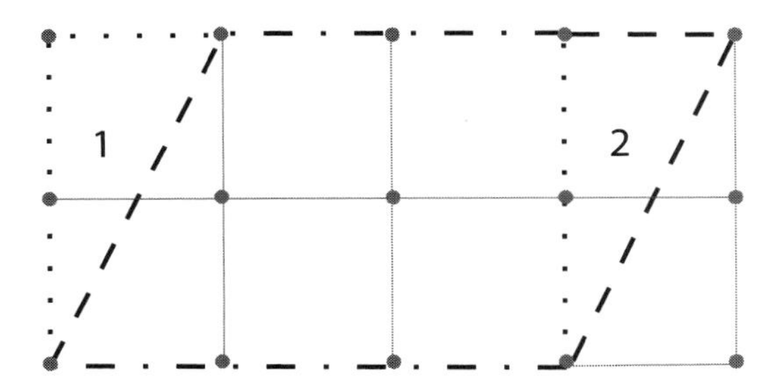

Fig. 19.4. Parallelogram with a rectangle overlay

Students can use their knowledge of the area formulas already described to explore another common area formula. The following vignette illustrates the use of the geoboard in developing the formula for the area of a trapezoid.

> A teacher announces to the class, "Our goal today is to find a formula that we can use to calculate the area of any trapezoid. Recall that trapezoids are quadrilaterals with exactly one pair of parallel sides. Enclose a trapezoid on your geoboard using one rubber band." As the teacher moves about the room, various types of trapezoids are observed. Attention is first called to those that are right trapezoids. Jose reconstructs his right trapezoid on the overhead geoboard. The teacher questions the class as to how its area might be found. Katie suggests that the trapezoid be divided into a rectangle and a triangle. She places a rubber band on Jose's trapezoid to illustrate her idea [see fig. 19.5]. James observes that it is possible to add the rectangle's area and the triangle's area to find the trapezoid's area. The teacher writes the formula $A = lw + (1/2)bh$ on the board. Amanda notices that the height of the triangle and the length of the rectangle are the same. The teacher then modifies the formula to produce $A = hw + (1/2)bh$.
>
> The class is then asked if this formula can be applied to Michael's trapezoid (he reproduces his isosceles trapezoid on the overhead geoboard). Krista observes that this trapezoid needs to be divided into three figures instead of two [see fig. 19.6]. She places rubber bands that divide the trapezoid into a rectangle and two right triangles. The teacher writes the area formula for this trapezoid using these three figures: $A = (1/2)bh + lw + (1/2)Bh$. Again, the length of the rectangle is the same

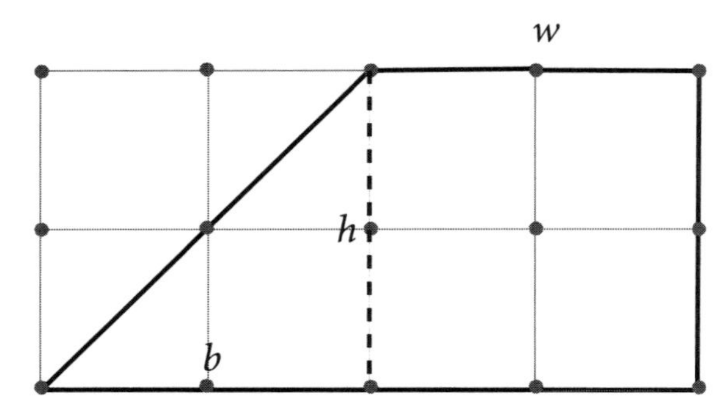

Fig. 19.5. Right trapezoid

as the height of both triangles so the formula is modified: $A = (1/2)bh + hw + (1/2)Bh$. Several students agree that this formula can also be applied to their trapezoids, even though they are not isosceles. Amanda complains that there are now two formulas for the area of a trapezoid and they do not appear easy to remember. The teacher agrees that they are rather cumbersome and reminds the class that the goal is to find one formula that applies to all trapezoids. Other students have trapezoids that are still different from the cases already examined and are concerned because they cannot apply either of the formulas already developed.

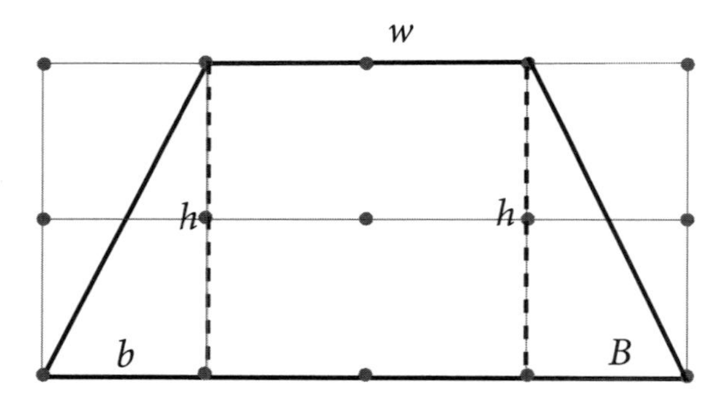

Fig. 19.6. Isosceles trapezoid

Jason reproduces his trapezoid on the overhead geoboard [see fig. 19.7]. The teacher asks the class if this trapezoid can be divided into figures whose area formulas are known. The class is encouraged to work on their own geoboards to find a way to do this. The teacher checks their progress while circulating about the room. After a few minutes, the teacher asks for a volunteer to show how the trapezoid might be divided. Maria connects two opposite vertices of the trapezoid on the overhead figure, creating two triangles [see fig. 19.8]. The teacher suggests labeling the top of the trapezoid b_1, the bottom b_2. Sam becomes confused about how to find the height of the triangles. Theresa reminds Sam that the altitude must be drawn outside the triangles. The teacher writes the trapezoid area formula using the sum of the areas of the two triangles. $A = (1/2)b_1h + (1/2)b_2h$. Jordan suggests that the "$(1/2)h$" can be factored out to produce $A = (1/2)h(b_1 + b_2)$.

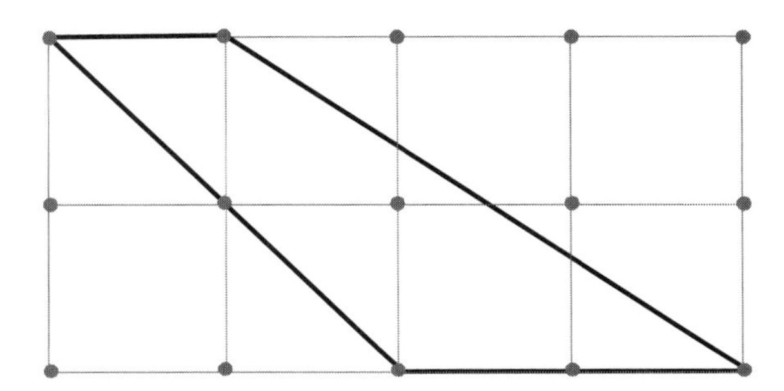

Fig. 19.7. Obtuse trapezoid

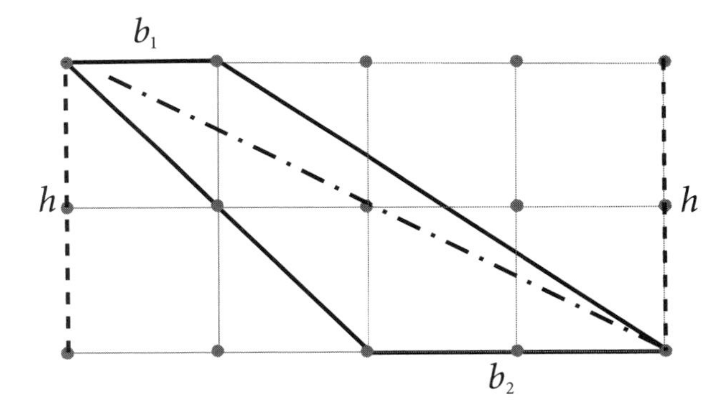

Fig. 19.8. Obtuse trapezoid with diagonal

The teacher asks the class if this formula can be applied to any trapezoid. She questions, "Can any trapezoid be broken into two triangles by connecting opposite vertices as was done here? Will the height of both triangles always be the same and equal to the height of the trapezoid?" Each student checks his or her original trapezoid and the class confirms that this formula can be applied to the trapezoids for which other formulas had previously been found.

The class further observes that the heights of the triangles formed will always be equal because the bases of any trapezoid are parallel. Therefore this formula can be used to calculate the area of any trapezoid, the goal of this activity.

After these basic formulas have been discussed, attention can be turned to any polygon with sides (or altitudes) whose length cannot be easily determined (because they are not parallel to a side of the geoboard). These can be enclosed by an appropriate rectangle. Students can be directed to think about how this rectangle can assist them in finding the area of the original figure. (Colored rubber bands will be helpful for this activity). Teachers can guide the students in determining that the sum of the areas of the figures enclosed by the rectangle and the given polygon needs to be subtracted from the area of the rectangle. Consider this example in figure 19.9. Each student should find the area of the rectangle (9 square units). The area of each triangle enclosed by the rectangle and the polygon is 1 square unit. The area of the square is also 1 square unit. Therefore, the area of the polygon is 9 – (1 + 1 + 1 + 1) = 5 square units.

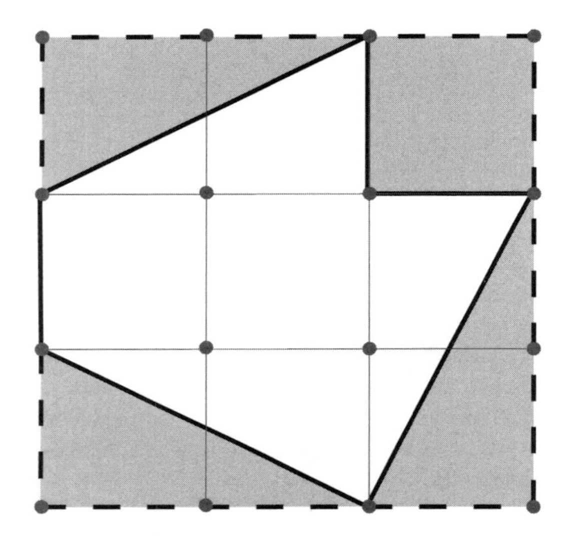

Fig. 19.9. Polygon enclosed by a rectanglee (square)

Another middle school–level measurement investigation involves ratios of similar figures. An overview is provided here; more detailed activity sheets are included in the Classroom Companion. Have students enclose a square of side length six units. Using a corner of this square and another rubber band, students should enclose a "similar" square of side length two units. Investigating the ratio between the sides of the squares and the ratio between the perimeters of the squares will show a ratio of 1:3 for each.

When this activity is repeated for pairs of other-sized squares, the side ratio will again match the perimeter ratio. This activity can be repeated for similar polygons such as triangles and rectangles.

A discussion about the areas of these polygons may then be conducted. Students may hypothesize that the area ratio will be the same as the perimeter ratio and the side ratio. The versatility of the geoboard will allow students to see that this is not true. However, they will then need to formulate the true relationship. Beginning with squares, a square of side length 5 will have an area of 25 square units, whereas a square with side 2 will have an area of 4 square units. Students will see that the ratio of sides (5:2) does not equal the ratio of areas (25:4).

Several carefully chosen examples using various pairs of similar polygons will help students discover and confirm the fact that the area ratio will always be the square of the side (perimeter) ratio.

These middle school activities are only a sample of what can be done with geoboards within the topic of measurement. Other topics can also be explored, and the use of the geoboard is not limited to middle school. High school students may use the other side of the geoboard (or an individual circular geoboard) to explore the geometry of the circle. Circular geoboards come in many sizes. There are models having different sized radii as well as those with a different number of pegs forming the circle. Some of these circular geoboards also have concentric circles.

A limitation to make students aware of when working with a circular geoboard is that a rubber band stretched to form the "circle" is truly not circular. Rather, the "circle" is formed by many segments that actually form an n-gon (depending on the number of pegs). A geoboard containing 24 pegs versus 12 pegs will more closely approximate a circle (see fig. 19.10). Therefore, the terms *circle* and *arc* are used loosely in the following discussion. Despite this limitation, these circular geoboards can still be used to investigate many relationships including circumference, angles formed by segments of a circle, and the relationships among those segments.

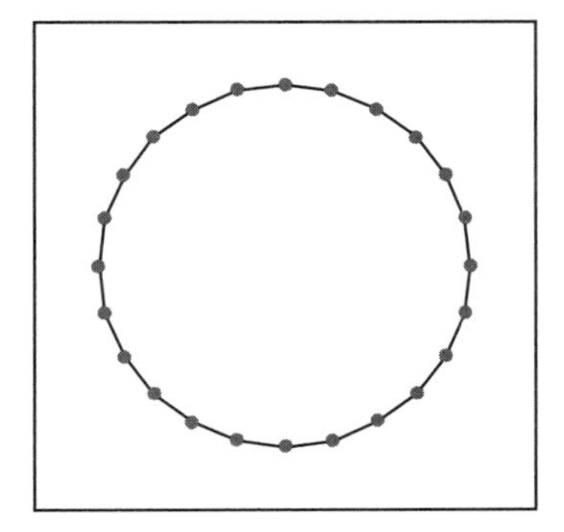

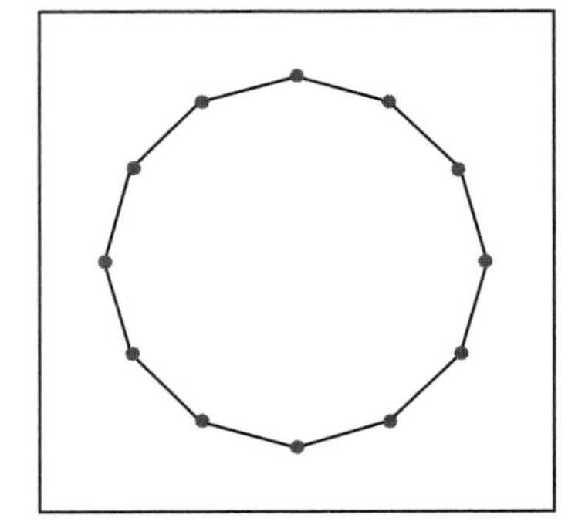

Fig. 19.10. Different "circular" geoboards

DEGREE MEASURES OF ARCS

Regardless of the size, the first investigation is the relatively simple task of finding the degree measure of each arc formed by two consecutive pegs. Reiterating that there are 360 degrees in every circle, have students count the total number of spaces between pegs on the circle and divide 360 by that result. Knowing the degree measure of the individual arcs will help us in later investigations. One immediate use is to emphasize the fact that there are 180 degrees in the measure of a semicircle. This can be illustrated by having the students stretch a rubber band across the diameter of the circle, count the number of arcs forming the semicircle, and multiply by the individual arc measure.

CIRCUMFERENCE

Begin first with an exploration of the measure of the distance around a circle. Have the students attach a piece of string to one peg. They wrap the string around the circle, attach it to the same peg, and cut off all loose edges. Then they cut the string and unwind it, leaving one long segment. This segment can now be measured with a ruler and classified as the distance around the circle (called the *circumference*). An interesting discovery can be made using this segment. Have the students place the segment across the diameter as many times as possible. They will notice that this piece of string fits along the diameter three times with a little left over. Since ≠ is approximately 3.14, this activity provides a physical model emphasizing the circumference formula $C = \neq d$.

RADIAN MEASURE

A piece of string cut to the size of the radius formed on the geoboard is useful for this next activity. When this segment is placed along the circumference of the circle, we can speak about another unit of measure called a *radian*. To facilitate this activity, press a piece of plain paper over the pegs of the circular geoboard. Students can place this string (of radius length) along the circumference of the circle marking the endpoints. Have them use a straightedge and a pencil to connect each endpoint to the center of the circle. By definition, a radian is the measure of the central angle of a circle that intercepts an arc equal in length to the radius of the circle. Therefore, they have just created a central angle measuring one radian. By also measuring this central angle with a protractor, students can notice the equivalence 1 radian ⊕57.3 degrees. (These numbers will vary but should be close to 180/≠ degrees.) With a little guidance, students can establish the relationship 1 radian = 180/≠ degrees and use it in future conversions.

ARC LENGTH

Further connections between arc length and degree measure can also be made. Students can observe that the measure of an arc's length is simply the fraction of the circle cut off by the arc multiplied by the circumference ($L = n/360 \times \neq d$ where n is the number of degrees in the arc or $L = m/2\neq \times \neq d$ where m is the number of radians in the arc). To connect this fact with the definition of a radian, students can draw central angles measuring one radian using the concentric circles. They can then observe that although both angles measure one radian, their arc lengths are different because the measures of the radii differ. In each of these circles, slightly more than six of these radian lengths (radii) form the circumference (see fig. 19.11). Since $2\neq \oplus 6$, students can connect this new information to the circumference formula $C = 2\neq r$.

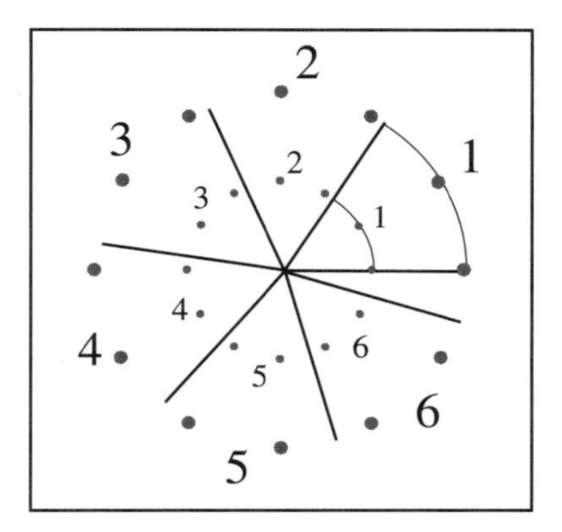

Fig. 19.11. Geoboard with concentric circles showing radian measure

When students are familiar with conversions between degrees and radians, another arc-length relationship can be investigated. Students can enclose any sector of a circle with string. They will also need to find the length of the radius of their circle (using a ruler) as well as the central angle's measure (converted to radians). (The angle measure can be found using a protractor if the fact that the measure of a central angle of a circle is equal to the measure of the intercepted arc has not been established.) Have them mark the string at the endpoints of the arc with a pen, cut the string at these marks, and measure this length with a ruler. The teacher can record each student's measurements on the board and ask the class to come up with a conjecture concerning the relationship among these three values. The relationship $S = \theta r$ can be suggested and checked using a different sector.

ANGLES FORMED BY SEGMENTS OF A CIRCLE

Using the degree measure of the arcs of the circle, students can now find relationships among arcs and angles formed by segments of the circle. Have students enclose any sector with a rubber band. The angle formed with the circle's center is called a *central angle.* Have students measure this angle with a protractor and record this measure. Next, have students find the degree measure of the arc. Several repetitions using different-sized sectors should be performed and measurements of arcs and central angles listed in a table. A comparison between the measures will illustrate that the measure of a central angle is the same as the measure of the intercepted arc.

Students can then be instructed to move the vertex of the central angle to a peg on the circle forming an inscribed angle. Students may construct a table listing the measure of the inscribed angle (found by using a protractor) and the degree measure of the intercepted arc. Several repetitions will lead the students to discover that inscribed angles have measures equal to one-half the measure of the intercepted arc. One special case reveals the fact that a triangle inscribed in a semicircle is a right triangle.

Developing the formulas for measures of angles formed by two intersecting chords, a tangent and a chord, two tangents, two secants, or a secant and a tangent all build on similar procedures.

Have students place two intersecting chords (not diameters) on their circle. Students should measure one angle formed by these chords (using a protractor), as well as observe the measures of the arcs that both the angle and its vertical angle intercept. The teacher can record each student's measurements on the board, and the class can look for a connection among these measures. The proposed conjecture can be tested using other pairs of chords (including diameters). The formula $m\angle A = (1/2)(\text{arc}_1 + \text{arc}_2)$ should be established. Similar investigations involving angles formed by other pairs of segments can be performed in order to establish related formulas. However, only a few examples can be constructed using rubber bands, depending on the size of the geoboard. This limitation may cause false conjectures that apply only in special cases. Therefore, a teacher's guidance is strongly recommended.

RELATIONSHIPS AMONG THE SEGMENTS OF A CIRCLE

Begin by establishing that the diameter is the longest chord in the circle. Students can proceed to discover that congruent chords intercept congruent arcs. This fact can best be developed by first observing that any two diameters intercept arcs whose measures are 180 degrees. Moving on to any other pair of congruent chords, students can observe that their intercepted arcs have the same degree measure.

The formula involving the lengths of the segments determined by two intersecting chords can be developed in the following manner. (For the next activity, replacing the rubber bands with string will enable students to measure the segments more conveniently.) Have the students place string to form two intersecting chords. Next, using a small-sized centimeter ruler, they can measure the pair of segments associated with each chord. The teacher can then direct students to examine the product of each pair of measurements. These products, although allowing for some measurement error, should be close enough to one another for students to realize the desired connection (the product of the lengths of the segments of one chord is equal to the product of the lengths of the segments of the other chord).

Other formulas involving lengths of segments of a circle (combinations of tangents, secants, and chords) can be similarly developed.

ALTERNATE INTERIOR ANGLES

A connection between the inscribed angles of a circle and angles formed by two parallel lines cut by a transversal can be made using the circular side of the geoboard. Have students form two parallel segments using pegs on the circle. Next have them connect the right endpoint of one to the left endpoint of the other, forming a transversal (see fig. 19.12). The measures of the alternate interior angles formed can be easily determined because they are both inscribed angles of the circle. After observing that these measures are equal, students can investigate other sets of parallel lines and transversals to conclude that the alternate interior angles formed by them will always be congruent. This conclusion may be extended to transversals other than ones

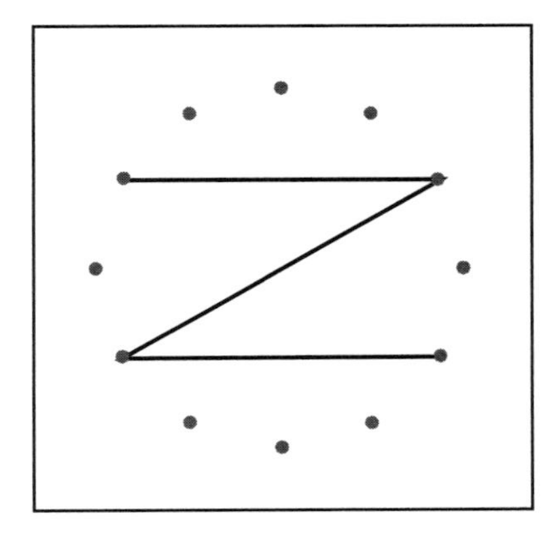

Fig. 19.12. Alternate interior angles on a "circular" geoboard

connecting the endpoints of the parallel segments. The measures of the angles in these instances, however, will need to be found using a protractor.

SUM OF THE INTERIOR ANGLES OF POLYGONS

(Note: the development of this formula requires knowledge that the measure of an inscribed angle in a circle is one-half the measure of the intercepted arc.)

Have students use the circular side of the geoboard, to inscribe any triangle using the pegs on the circle. The measure of each angle of the triangle can be found, since each is an inscribed angle. Each student will determine that the sum of the angles of their triangle is 180 degrees. Proceed to any quadrilateral, and determine the measure of each angle in the same manner. Have each student find the angle sum. The teacher can begin a master list on the blackboard relating the number of sides in each polygon to the sum of the measures of the interior angles. Repeat this procedure for pentagons and hexagons (e.g., see fig. 19.13). After the results of each figure are recorded, have the students examine the list to see whether any conjectures can be proposed. At any point when the class agrees on a possible formula, the next n-gon can be investigated to see whether that relationship holds. The formula $180(n - 2)$ degrees should eventually be discovered and checked using polygons with different numbers of sides.

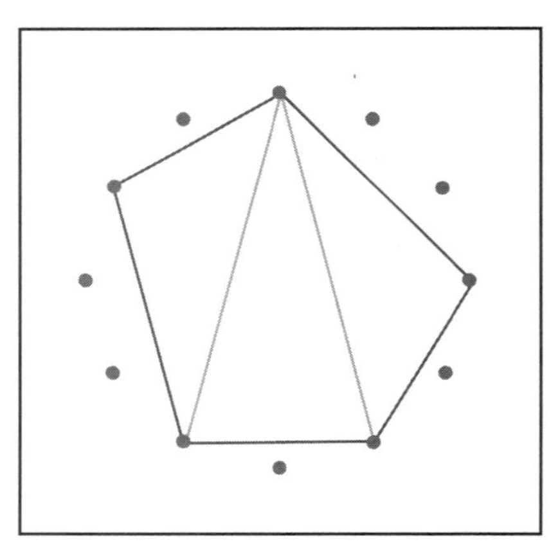

Fig. 19.13. Pentagon inscribed in a circle

This concept can be illustrated on the geoboard in another way. Start with any nontriangular polygon inscribed in a circle. Have students connect one vertex to each nonadjacent vertex forming nonoverlapping triangles. Recalling that every triangle contains 180 degrees, the students can then multiply

180 by the number of triangles found in their polygon to find the total number of degrees in this polygon. By examining polygons with different numbers of sides, students can conclude that the number of triangles formed in this manner is always two less than the number of sides in the polygon. Thus the formula already established, $180(n-2)$ degrees is reinforced conceptually.

SUM OF THE EXTERIOR ANGLES OF POLYGONS

Using the circular side of the geoboard, again begin with a triangle enclosed using pegs on the circle. The measure of each interior angle can again be determined using the inscribed angle relationship. By imagining that each side of the polygon is extended, students should observe that the measure of each exterior angle is the supplement of the measure of the adjacent interior angle. Have students keep a record of the measure of each exterior angle (one at each vertex), and have them find their sum. Repeat this procedure with quadrilaterals, pentagons, and hexagons until a conjecture is reached. Have them check this conjecture using polygons with more sides. They should conclude that the sum of the measures of exterior angles (one at each vertex) in any polygon is 360 degrees.

Several applications of geoboard use for measurement topics have been discussed. Other sources containing geoboard activities are included in the reference list (Arvold, Turner, and Cooney 1996; Hutcheson 1975; Jones 1976; Kennedy and McDowell 1998; Lichtenberg 1975; Schmidt 1975; Smith 1990). It is our hope that this article might provide a basis for teachers to realize other secondary school topics that might be taught effectively through the use of manipulatives. Our intent was not to provide an exhaustive list of measurement topics to which the geoboard can be applied but simply to encourage the use of hands-on teaching tools in the secondary school classroom.

REFERENCES

Arvold, Bridget, Pamela Turner, and Thomas J. Cooney. "Analyzing Teaching and Learning: The Art of Listening." *Mathematics Teacher* 89 (April 1996): 326–29.

Cobb, Paul. "Where Is the Mind? Constructivist and Sociocultural Perspectives on Mathematical Development." *Educational Researcher* 23 (October 1994): 13–20.

Hutcheson, James. "The Circular Geoboard: A Promising Teaching Device." *Mathematics Teacher* 68 (May 1975): 395–98.

Jones, Robert L. "The Nine-Point Circle on a Geoboard." *Mathematics Teacher* 69 (February 1976): 141–42.

Kennedy, Joe, and Eric McDowell. "Geoboard Quadrilaterals." *Mathematics Teacher* 91 (April 1998): 288–90.

Lichtenberg, Donovan R. "From the Geoboard to Number Theory to Complex

Numbers." *Mathematics Teacher* 68 (May 1975): 370–75.

National Council of Teachers of Mathematics (NCTM). *Principles and Standards for School Mathematics.* Reston, Va.: NCTM, 2000.

Schmidt, Philip A. "A Nonsimply Connected Geoboard—Based on the 'What If Not' Idea." *Mathematics Teacher* 68 (May 1975): 384–88.

Smith, Lyle R. "Areas and Perimeters of Geoboard Polygons." *Mathematics Teacher* 83 (May 1990): 392–98.

20

Using Measurement to Develop Mathematical Reasoning at the Middle and High School Levels

Mary Enderson

WHAT does it mean to develop mathematical reasoning? How does one prepare students to reason mathematically? What kinds of situations should be used to foster mathematical reasoning in the classroom? All these questions are important and can be addressed by using measurement as a bridge to develop mathematical reasoning.

When students study geometry, they often focus on measurement and proof. Measurement is quite familiar to many students because they usually have had many experiences in measurement since the early grade levels. Proof, however, is not as well known to students and is often a part of geometry that presents them with new challenges. Pierre and Dina van Hiele, whose work focused on five levels of geometric thought and how students progress through these stages, classified work with proof as the fourth level of geometric thinking (Fuys, Geddes, and Tischler 1988). It is at this level that students reason deductively within a given mathematical system. Although this is indeed important, all too often students are not developmentally ready to arrange the necessary components for a proof and instead memorize steps without any regard for the meaning or reasoning involved in organizing a proof. When students are presented with exploratory measurement environments rather than proof-based environments, they can build their thinking and reasoning skills, which in turn can help in the development of mathematical reasoning.

As noted in *Principles and Standards for School Mathematics* (National Council of Teachers of Mathematics [NCTM] 2000), the process of reasoning is essential to understanding mathematics: "By developing ideas, exploring phenomena, justifying results, and using mathematical conjectures in all content areas and … at all grade levels, students should see and expect that mathematics makes sense" (p. 56). It can also be said that "mathematical reasoning requires analytical thinking, but also requires creative and practical thinking" (Sternberg 1999, p. 43). Providing students with situations in which they must make decisions about how to approach or represent problems, as well as how to organize strategies to solve them, will help them develop "creative and practical thinking." This type of thinking will prepare students to apply some of the techniques and skills they have learned to new and different situations.

FOSTERING MATHEMATICAL REASONING

Measurement situations that involve students in "making sense" of the tasks at hand can be very rich learning environments. Teachers should strive to place students in mathematical situations where they can build conjectures, test their conjectures by measuring certain attributes, and validate them by using specific data sets. Students can then use the information to generate claims related to the concepts. Such activity is essential in attempting to develop students' mathematical reasoning.

These actions can be accomplished through the support of a dynamic interactive computer environment. For the situation presented here, Geometer's Sketchpad (Jackiw 1995) is used as a vehicle to foster students' development of mathematical reasoning through measurement situations. Other types of dynamic software, such as Cabri (from Texas Instruments), could be used in a similar fashion.

Dynamic geometric software allows users to construct geometric objects and measure their attributes. Figures can be transformed by dragging vertices or component parts in the workspace. Students gain immediate feedback from the technology, which can guide their investigation and experimentation. All these aspects are important in helping students' growth and movement to a more advanced level of reasoning.

For measurement situations explored in the classroom environment, a protocol similar to that presented in Thinking Mathematically (Mason, Burton, and Stacey 1982) is used quite regularly: ENTRY, ATTACK, and REFLECT.

- ENTRY—How can I set the problem up? What can I use to solve the problem?
- ATTACK—Follow through with the plan from the entry stage.
- REFLECT—How did the attack go? Did I solve the problem? If I solved the problem, does it seem reasonable? If I didn't solve the problem, do I need to consider other options? If so, then go back to the entry stage for other plans of actions.

For students to make use of this process, they are presented with a situation to organize, investigate, discuss, and make attempts to solve. They use the computer environment to set up the given situation (ENTRY) and find measurements related to it (ATTACK). These initial phases are often areas of specialization where students find specific examples that fit the given conditions and organize the measurement data into a table format. After specific examples have been identified, students use the measurement data to initiate small-group dialogues where they discuss any assertions or conjectures and try to validate them. If possible, groups may progress toward making generalizations based on their specific situations as well as observing reported measurement data collected from other groups. It is through the REFLECTION process—revisiting approaches taken to investigate the situations, how data sets were created, and whether any findings were similar—that students use measurement situations generated and supported by the software to help build mathematical arguments.

It should also be noted that nurturing the development of students' mathematical reasoning depends a great deal on the teacher and his or her selection of tasks to investigate. *Principles and Standards for School Mathematics* (NCTM 2000) describes the importance of fostering reasoning by having students formulate plausible conjectures, test the conjectures, and share assertions for evaluation by others. Although this process may not be as rigorous as proof, experiencing such actions will certainly promote one's progress toward the development of proof.

MEASUREMENT SITUATIONS

Two specific measurement situations have been quite fruitful in expanding the mathematical reasoning of students: exploring midpoints, and exploring perimeters and areas. These situations not only deal with content that students usually cover in a typical geometry unit or course but also furnish instances where the ideas and goals of *Principles and Standards* (NCTM 2000) can be put into practice. Specifically, these measurement situations have students—

- develop and use measurement concepts and skills throughout the school year rather than exclusively in a separate unit of study;
- measure objects to help them gain a better understanding of "sensible" measurements;
- model situations to explore mathematical relationships;
- represent, analyze, and generalize data organized in table format;
- generate or draw geometric figures with specific properties;
- use collected data (from their measurements) to make conjectures to test and validate.

One might observe that the activities presented are not grouped in the Measurement sections of *Principles and Standards* but instead are found in the Number and Operations, Algebra, Geometry, and Data Analysis and Probability sections. This certainly supports the notion that measurement should not be treated as a separate unit of study.

Exploring Midpoints

The first measurement situation focuses on exploring the midpoints of the sides of a quadrilateral. Students are given some general information about what is needed to begin the investigation in the geometric computer environment (ENTRY). They are to record their findings for the class handout (see fig. 20.1) to share with group partners when they join their team. How they decide to explore the situation (ATTACK) is completely up to individual students as long as they follow the guidelines of the handout.

One strength of this exploration is that students may begin with any quadrilateral—no restrictions are placed on the quadrilateral. Thus, as students come together in their working groups to share measurement data and responses to questions, they will see that their classmates have different quadrilaterals with some similar and different measurements (ATTACK). This again will help build arguments for making generalizations based on specific measurement data (REFLECTION). Two different examples of student-generated computer files for the midpoint exploration are shown in figures 20.2 and 20.3.

Groups should present their findings related to the investigation and support them with the measurement data from the table on their handout. Students should observe that the figure generated by the midpoints will have an area that is one-half the area of the original figure. Students may also build a convincing argument that the interior figure is a parallelogram regardless of the initial quadrilateral. If a teacher is interested in moving beyond the reasoning and building of arguments, then she or he can certainly help students organize their ideas into a proof. One perspective that supports the development of a proof for an exploration of this kind can be found in *Rethinking Proof with The Geometer's Sketchpad* (de Villiers 1999).

Another advantage of this software is the variety of mathematical concepts that can be addressed by continuing this measurement exploration. Areas of focus may include properties of convex and concave quadrilaterals, slopes, and parallel lines. After students have generated an initial quadrilateral, they can carry out investigations where they find the sums of interior and exterior angles and compare them as well as observe any similarities or differences between convex and concave quadrilaterals. Students can also use the slope-measurement feature of the software to observe the relationships of slopes of parallel and perpendicular segments or the sides of a quadrilateral.

Exploring Perimeter and Areas of Similar Figures

The concept of similar figures permeates the geometry curriculum. Unfortunately, students often have little understanding of the relationships of scale factors—ratios—to perimeters and areas of similar figures. Using dynamic geometric software allows students to explore these relationships without

Name: ______________________ Date: ______________

Midpoint Exploration with Quadrilaterals

1. Begin the exploration with any quadrilateral. Label the vertices so that you have Quadrilateral *ABCD* (upper left vertex is *A*; move counterclockwise for other labels). Make a simple sketch below.

2. Construct the midpoints of the sides of the quadrilateral. Label the midpoints, *E*, *F*, *G*, and *H* (beginning with side $\overline{AD}$ and moving in a counterclockwise direction).

3. Join or connect points *E*, *F*, *G*, and *H* with segments. Make a simple sketch below.

4. Construct the interiors of *ABCD* (Poly1) and *EFGH* (Poly2) and measure their areas.

5. Record the initial sketch and the area measurements in the table below. Collect data for seven new quadrilaterals. (Remember that you can drag and drop any point to get a "new" quadrilateral.)

Simple Sketch of Quadrilaterals	AREA of Poly1 (*ABCD*)	AREA of Poly2 (*EFGH*)

6. What do you notice about the relationship of the area of Poly1 compared to the area of Poly2? Is the relationship consistent, or the same, for all examples?

7. Get together with your group members to share your measurement data. Are your findings similar to theirs?

8. Using the software, try to investigate properties that the interior quadrilateral, Poly2, possesses. Can Poly2 be identified as a specific type of quadrilateral? Will it change depending on what the initial quadrilateral was?

Fig 20.1. Midpoint exploration

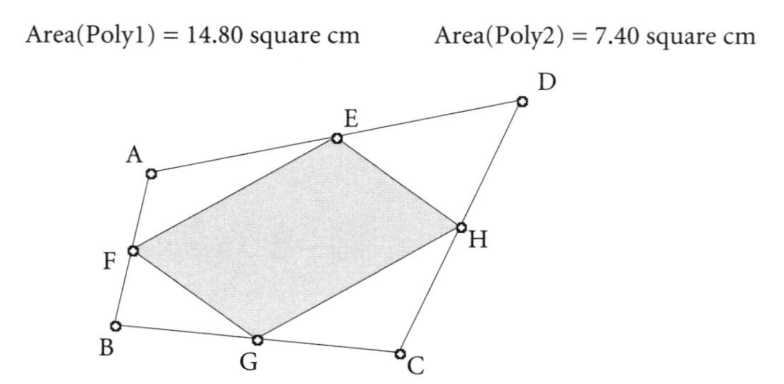

Fig. 20.2. Convex quadrilaterals *ABCD* (Poly1) and *EFGH* (Poly2) and area measures

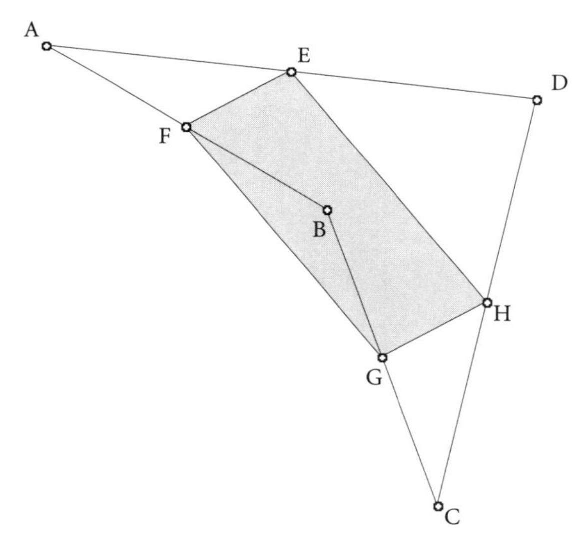

Fig. 20.3. Concave quadrilaterals *ABCD* (Poly1) and *EFGH* (Poly2) and area measures

getting bogged down with "nice numbers." They can use the measurement feature and organize data into a table to help develop conjectures for the relationships. As previously described in the midpoints exploration, students are supplied with some general information about what they need to set up in the computer workspace (ENTRY) and are instructed to record responses to a class handout that gives some direction to the exploration (see fig. 20.4).

Name: ____________________________ Date: __________

Perimeter and Area of Similar Figures

1. Begin the exploration with any convex polygon. Label the vertices. Make a simple sketch below.

2. Use the dilate tool with a scale factor of 2 to create a figure similar to, and twice the size of, your convex polygon (original: dilation). Make a simple sketch of this figure.

3. Verify the properties of similar polygons by comparing corresponding angle measurements and corresponding side lengths as ratios. Identify the original polygon as Poly1 and its dilation as Poly2. Record the measurements in the table below. If your polygon and its dilation have more than six sides, then select any six corresponding angles and any six corresponding sides to compare.

Corresponding Angle Measures		**Corresponding Side-Length Ratios (Scale Factor)**		
Poly1	**Poly2**	**Poly1**	**Poly2**	**Ratio of sides**
$\angle A =$	$\angle A' =$	side 1 =	side 1' =	s1/s1' =
$\angle B =$	$\angle B' =$	side 2 =	side 2' =	s2/s2' =
$\angle C =$	$\angle C' =$	side 3 =	side 3' =	s3/s3' =
$\angle D =$	$\angle D' =$	side 4 =	side 4' =	s4/s4' =
$\angle E =$	$\angle E' =$	side 5 =	side 5' =	s5/s5' =
$\angle F =$	$\angle F' =$	side 6 =	side 6' =	s6/s6' =

4. Measure and record the perimeters of the original polygon (Poly1) and its dilation (Poly2). Using these values, record the ratio of the perimeters. Measure and record the areas of Poly1 and Poly2. Using these values, record the ratio of the areas. Compare each ratio to the scale factor of the similar figures (refer to step 3).

	Poly1	**Poly2**	**Ratio**
Perimeter			
Area			

5a. What do you notice about the ratio of perimeters and the scale factor?

b. What do you notice about the ratio of areas and the scale factor?

6. When you have completed these activities for the initial figure and its dilation, click and drag any vertex of the original polygon to change the side lengths or the angle measurements. Drop the point you have been dragging, and record the "new information" you have on the computer screen. Collect measurements for at least three different adjustments. Record your data in the following tables:

Fig 20.4. Similar figures exploration (continued on next page)

	Poly1	Poly2	Ratio
Perimeter			
Area			

	Poly1	Poly2	Ratio
Perimeter			
Area			

	Poly1	Poly2	Ratio
Perimeter			
Area			

7. How do the ratios of the perimeters compare for the three instances above? How do the ratios of the areas compare with one another? Are the findings any different from what you found for step 5? Why or why not?

8. Repeat the investigation (steps 1–5) with a different initial convex polygon. Change the scale factor of the dilation to a value other than 2. How do the ratios of the perimeters compare with the previous findings? How do the ratios of the areas compare? Record your data below to support your response.

SCALE FACTOR: _________

	Poly1	Poly2	Ratio
Perimeter			
Area			

9a. Write a statement about the ratios of perimeters of similar figures.

b. Write a statement about the ratios of areas of similar figures.

Fig 20.4. Similar figures exploration (continued from previous page)

Prior to carrying out this particular investigation, students have had instruction on similar and congruent figures. They are familiar with scale factors, so little or no time is needed to discuss dilations and how to construct them in the geometric computer environment. The focus here is on relationships of similar figures—more specifically, the perimeters and areas of similar figures. Students are free to begin this exploration with any convex polygon. Examples of student-generated similar figures with measurement data can be seen in figures 20.5 and 20.6.

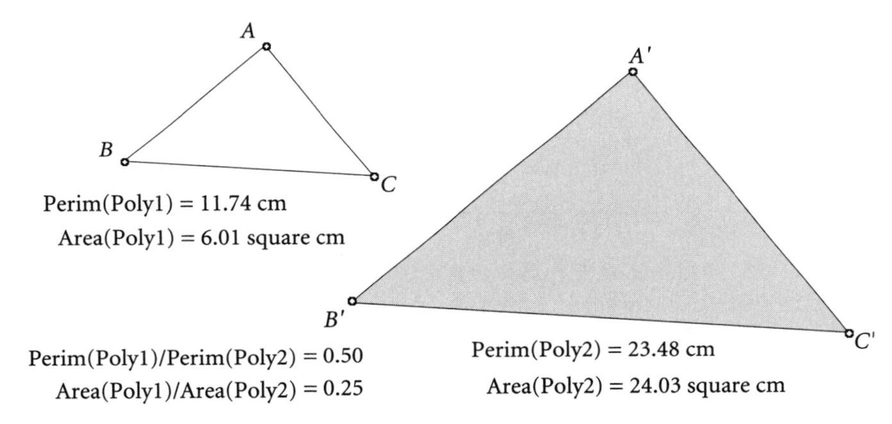

Fig. 20.5. Perimeter and area measurements for triangles (ratio of 1:2)

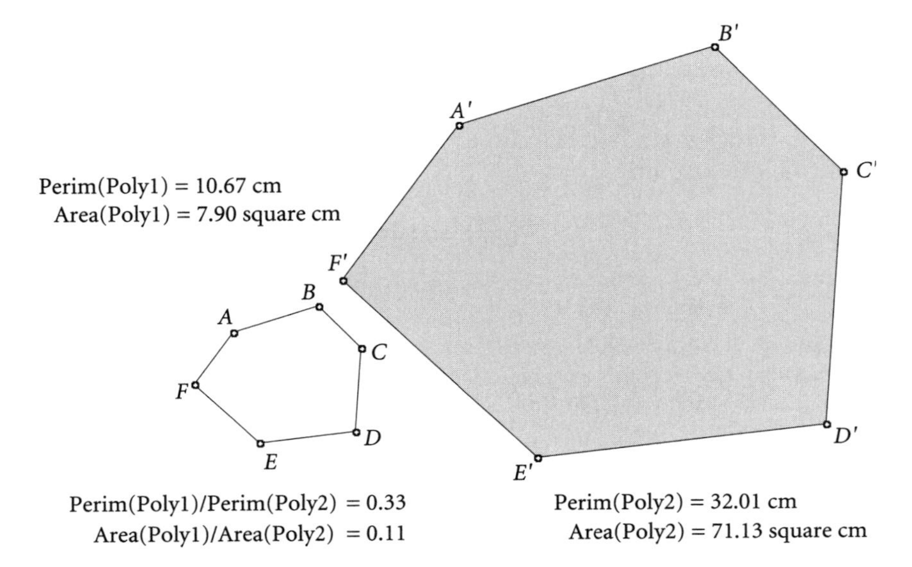

Fig. 20.6. Perimeter and area measurements for hexagons (ratio of 1:3)

In the similar-figures exploration, students are placed in a situation in which they can use specific data to help them make conjectures and in turn build arguments that will help develop their mathematical reasoning. Students are also actively involved in discussing the mathematics as well as presenting it to their peers. They should observe and use the data to conclude that the scale factor and perimeter ratios are the same and that the area ratio is the scale factor squared. These findings are quite powerful, and possibly the discovery by these students was due to the nature of the dynamic workspace, which provides students with a number of different examples to share with one another. Such mathematical concepts have traditionally been overlooked in discussions on similar figures. This type of environment is very student-centered, allowing them to use the technology to address many of their own questions and actually to see that certain properties exist in mathematics.

COMMENTS ABOUT THE EXPLORATIONS

Furnishing students with rich learning environments is important but not always as simple as it sounds. Planning and organizing tasks that will engage students in "making sense" of mathematics requires the teacher to have a level of understanding about the content that is not always presented in a textbook format. It also requires that the teacher take the time to rethink how to ask questions that will help students develop their thinking and reasoning strategies. Using dynamic interactive software to present, explore, question, test, and validate is a powerful vehicle to aid students' learning.

Measurement is a strand of mathematics that is an integral part of our daily lives and is a part of many mathematics topics in the school curricula. Dynamic software, along with exploratory mathematical measurement situations, can engage students in "doing mathematics" while developing their mathematical reasoning.

REFERENCES

de Villiers, Michael D. *Rethinking Proof with The Geometer's Sketchpad.* Emeryville, Calif.: Key Curriculum Press, 1999.

Fuys, David, Dorothy Geddes, and Rosamond Tischler. *The Van Hiele Model of Thinking in Geometry among Adolescents.* Journal for Research in Mathematics Education Monograph 3. Reston, Va.: National Council of Teachers of Mathematics, 1988.

Jackiw, Nicholas. The Geometer's Sketchpad. Software. Emeryville, Calif.: Key Curriculum Press, 1995.

Mason, John, Leone Burton, and Kaye Stacey. *Thinking Mathematically.* Menlo Park, Calif.: Addison-Wesley Publishing Co., 1982.

National Council of Teachers of Mathematics (NCTM). *Principles and Standards for School Mathematics.* Reston, Va.: NCTM, 2000.

Sternberg, Robert J. "The Nature of Mathematical Reasoning." In *Developing Mathematical Reasoning in Grades K–12*, 1999 Yearbook of the National Council of Teachers of Mathematics (NCTM), edited by Lee V. Stiff, pp. 37–44. Reston, Va.: NCTM, 1999.

21

Is Our Teaching Measuring Up?

Race-, SES-, and Gender-Related Gaps in Measurement Achievement

Sarah Theule Lubienski

THIS article reports on recent national data indicating that measurement is the mathematical strand with the largest race-, socioeconomic-status– (SES), and gender-related achievement gaps. It is generally known that U.S. mathematics achievement tends to correlate with race, SES, and gender. That is, white students tend to outperform Latino, Latina, and African American students; wealthy students tend to outperform poorer students; and males often outperform females on standardized mathematics tests. This article goes beyond previous discussions by exploring how these disparities vary by mathematical strand.

In this article, U.S. students' achievement is examined in relation to the five mathematical strands assessed by the National Assessment of Educational Progress (NAEP): (1) number and operations, (2) data analysis/statistics, (3) algebra/functions, (4) geometry, and (5) measurement. The data to be discussed indicate that measurement is the mathematics strand with the most marked race-, SES-, and gender-related disparities.

BACKGROUND: UNDERSTANDING NAEP DATA

The NAEP is the only ongoing national assessment of academic achievement in the United States. Using a representative sample of thousands of U.S. students, the NAEP measures achievement at fourth, eighth, and twelfth

---

The author wishes to thank Kayonna Camara, Matthew Bettis, and Rachel Randall for their valuable research assistance.

grades in mathematics and other subject areas. In the most recent main[1] NAEP mathematics assessment (2000), the sample included approximately 14,000 fourth graders, 16,000 eighth graders, and 13,000 twelfth graders in both public and private schools across the country.

In 1990, the framework guiding the main NAEP mathematics assessment was revised in accordance with the *Curriculum and Evaluation Standards for School Mathematics* published by the National Council of Teachers of Mathematics (NCTM) (1989). Since that time, the NAEP was administered in 1990, 1992, 1996, and 2000. Although several other documents summarize various aspects of recent NAEP results (e.g., Braswell et al. 2001; Silver and Kenney 2000; Mitchell et al. 1999), this article uniquely focuses on ways in which equity-related disparities vary by mathematical strand.

The NAEP mathematics composite score is computed as a weighted average of the five mathematical strands, with the weights varying slightly by grade level. In order to help the reader interpret both the strand and composite scores reported here, some information about NAEP scores is necessary. NAEP uses a consistent 500-point scale for both the strands and composites for all three grade levels. The consistency of the scale enables score comparisons across grade levels. In 2000, fourth graders scored an average of 228; eighth graders, an average of 275; and twelfth graders, an average of 301. Hence, in rough terms, a 9- or 10-point achievement gap can be thought of as approximately a one-year difference.

Students' Performance on NAEP Mathematics Strands

The findings reported in this article tend to point toward areas of concern. However, it is important to note that average NAEP scores increased between 1990 and 2000 for every mathematics strand in every grade.

Table 21.1 reports overall students' performance in 2000 for each of the five mathematical strands measured by NAEP. The overall means varied only slightly by strand, and achievement in measurement was not particularly high or low in comparison with other strands.

---

1. Although the focus of this article is the main mathematics assessment, it is worth noting that there are actually three different NAEP mathematics assessments: the main assessment, the long-term trend assessment, and the state assessment. The framework that determines the content of the main assessment is responsive to national education reforms, such as the National Council of Teachers of Mathematics (NCTM) *Standards*. The long-term trend assessment was created in 1973, and the content and format of its questions have remained constant over time. The more recent "State NAEP" is a separate assessment that measures students' achievement within each participating state. More information about NAEP mathematics achievement is available at the NAEP mathematics Web site, nces.ed.gov/nationsreportcard/mathematics.

TABLE 21.1

NAEP Achievement by Grade and Mathematical Strand, 2000

	Number	Data	Algebra	Geometry	Measurement	Mathematics Composite
Grade 4	225	230	232	227	228	228
Grade 8	276	278	277	272	273	275
Grade 12	296	301	303	304	300	301

Despite the relative consistency in overall achievement across the five strands, examinations of race-related differences in mathematics achievement, both overall and within strands, reveal severe disparities.

As figure 21.1 reveals, the average mathematics composite score for white students was 31, 39, and 34 points higher than that of black students at the fourth, eighth, and twelfth grades, respectively.[2] White-Hispanic gaps were slightly smaller, with white students scoring an average of 24, 33, and 25 points higher at the fourth, eighth, and twelfth grades, respectively.

One way to interpret the severity of these differences is to note that the average white eighth-grade score (286) was actually higher than the average black (274) and Hispanic (283) twelfth-grade score. However, an analysis by mathematical strand reveals that disparities in measurement achievement are even larger, and that they are consistently the greatest of all the strands.

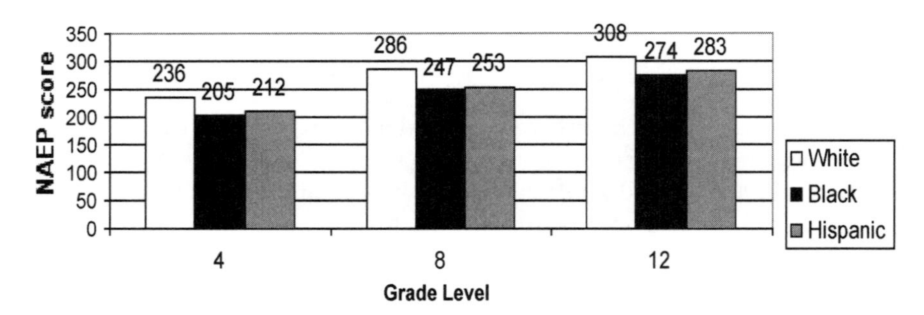

Fig. 21.1. Average NAEP mathematics composite scores by race, 2000

Table 21.2 shows the average number of points (on NAEP's 500-point scale) that white students outscored black and Hispanic students. The table reveals that, across every strand, the average score for white students was sig-

2. Although differences of opinion exist about the appropriateness of the terms "black" and "Hispanic," the terms are used here to be consistent with NAEP data.

nificantly higher than the average for black or Hispanic students. The strand with the largest disparities in fourth, eighth, and twelfth grades was measurement, with differences particularly marked at eighth grade, where the black-white gap is 58 points and the Hispanic-white gap is 44 points. Since 1990, these eighth-grade gaps have increased a significant 18 and 14 points—more than twice as much as any other strand. (At the fourth- and twelfth-grade levels, the gaps increased only slightly.)

Table 21.2

NAEP Achievement Gaps by Race, Grade, and Mathematical Strand, 2000

	Number	Data	Algebra	Geometry	**Measurement**	Mathematics Composite
Grade 4						
White/black gap	29	32	26	29	**37**	31
White/Hispanic gap	23	24	20	23	**29**	24
Grade 8						
White/black gap	34	43	33	34	**58**	39
White/Hispanic gap	31	39	30	26	**44**	33
Grade 12						
White/black gap	32	38	29	34	**42**	34
White/Hispanic gap	26	27	24	24	**28**	25

The data analysis strand tended to place second in gap size at all three grade levels, with gaps ranging between 24 and 43 points. In contrast, algebra tended to be the strand with the smallest disparities, with race-related gaps ranging from 20 to 33 points.

Figure 21.2 shows the actual measurement scores for white, black, and Hispanic students. Again, comparisons across grade levels help clarify the severity of these gaps. Note that the average measurement achievement of white fourth graders (238) was higher than that of black eighth graders (230) and just slightly lower than that of Hispanic eighth graders (244). Similarly, white eighth graders (288) scored 21 points higher than black twelfth graders (267) and 7 points higher than Hispanic twelfth graders (281).

NAEP data are also reported by three indicators relevant to students' SES: parent education level, free or reduced lunch eligibility, and literacy resources in the home. In brief, an analysis of achievement gaps relating to these factors revealed patterns similar to the race-related gaps noted above, with measurement consistently being the strand with the largest disparities. For example, the mathematics composite scores of fourth, eighth, and twelfth graders who did not qualify for free or reduced lunch were 26, 30, and 24 points higher than the scores of their less advantaged peers. The gaps in measurement achievement between these two groups of students were 29, 40, and 29 points at the fourth, eighth, and twelfth grades.

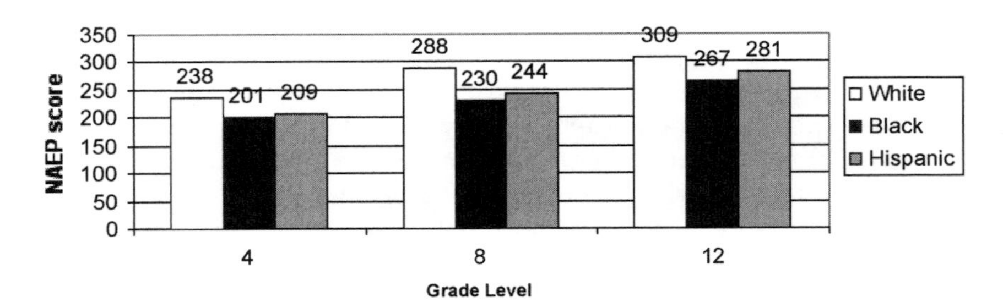

Fig. 21.2. Average NAEP measurement scores by race, 2000

Although the race- and SES-related patterns were similar, one should not conclude that race-related gaps are simply due to SES-related factors. An examination of race and SES together revealed that SES differences account for some, but not all, race-related gaps. In an analysis of 1996 NAEP data, the highest-SES quartile of black eighth graders scored a significant 13 points lower than the lowest-SES white eighth graders (252 versus 265). (See Lubienski 2001 for more information.) The implication of these NAEP findings is that lower-SES students, as well as minority students from all SES levels, need more opportunities to develop understanding in all strands of mathematics, but especially in the area of measurement.

Relative to race and SES, education researchers have given significant attention to the gender gap in mathematics education over the past three decades (Lubienski and Bowen 2000). Although some gender gaps remain and must continue to be addressed, the small size of these gaps give perspective to the severity of the race- and SES-related disparities discussed above. In 2000, the boys outscored girls on the overall NAEP mathematics assessment by 3 points at the fourth- and eighth- grade levels and 4 points at twelfth grade. A closer look at gender differences in performance in each of the five mathematical strands revealed that across all three grade levels in every NAEP assessment since 1990, measurement is the strand with the largest gender disparities. In 2000, these disparities were 6 points in fourth and eighth grades and 8 points in twelfth grade. Concerned about the gender gap in measurement at the fourth-grade level, Ansell and Doerr (2000) analyzed the disparities by item and found that males tended to outperform females on measurement items requiring the reading or use of a measurement instrument (e.g., a ruler or speedometer).

Some Specific Examples

Mathematics questions from the 2000 NAEP are not available. However, analyses of questions from previous assessments can shed light on the types of measurement problems that seem particularly problematic regarding equity.

Race-related disparities were significant but relatively small on some multiple-choice items, such as those asking students to identify the best unit of measurement in a particular situation or to calculate an area or volume with a given formula. For example, when students were asked whether a centimeter, meter, kilometer, foot, or yard would be the best unit for measuring plant growth, the percent of eighth-grade students who correctly chose centimeter was 84, 63, and 66 for white, black, and Hispanic students. The same question was asked at the twelfth-grade level, where 90, 77, and 83 percent of white, black, and Hispanic students answered correctly. When twelfth graders were asked to identify the volume of a right circular cylinder (with formula and five answer choices provided), 73 percent of white students, 54 percent of black students, and 57 percent of Hispanic chose the correct answer.

Questions involving multistep problems or extended responses tended to have larger disparities. NAEP conducted a special "theme assessment" in 1996 that involved having students solve mathematics problems set in real-world contexts. Students' overall performance on these relatively complex problems was poor. Problems involving measurement had particularly large equity-related disparities. Here I draw from the detailed examples provided by Mitchell et al. (1999) in order to exemplify some of the trends shown in the data above. Similar to other NAEP data, the trends in this special assessment indicate the existence of major inequities in students' opportunities to learn both basic measurement skills (such as using a ruler to measure), as well as fundamental measurement-related concepts.

A block of fourth-grade questions on the theme assessment focused on the creation of a "butterfly booth" at a school science fair. One question in this block asked students to measure the wingspan of butterflies using the centimeter ruler provided (see fig. 21.3). As Mitchell et. al. (1999) report, 47 percent of the white students tested made the measurements correctly, in comparison with 20 percent of black students and 26 percent of Hispanic students. Additionally, 2 percent of white students, 8 percent of black students, and 12 percent of Hispanic students omitted the question. One possibility is that the term *wingspan* was more familiar to some groups of students than others (although an example was provided to help ensure that students understood the term).

An eighth-grade problem also showed a disparity in knowledge about using a ruler. This problem asked students to measure, in inches, three aspects of a model doghouse (to be assembled out of pieces provided). Fifty-four percent of white students tested made the three measurements correctly, compared with only 17 percent of black students and 36 percent of Hispanic students. Also, 6 percent of white students, compared with 17 percent of black students and 8 percent of Hispanic students, omitted the question. (It is possible that the difficulties students encountered related not just to measuring with a ruler but also to constructing the doghouse from the pieces provided.)

One pivotal mathematical idea linked with measurement is ratio. Eighth-grade students were asked how a scale of 1 inch = 1.5 feet should be used to convert measurements of the model doghouse to real measurements. Students did not need to actually convert any measurements but needed to

On the Butterfly Information Sheet the wingspan of the Monarch butterfly is shown. Use your ruler to measure the wingspans of the other two butterflies on the sheet, the Black Swallowtail butterfly and the Common Blue butterfly, to the nearest centimeter.

Black Swallowtail Wingspan:__________ centimeters

Common Blue Wingspan:__________ centimeters

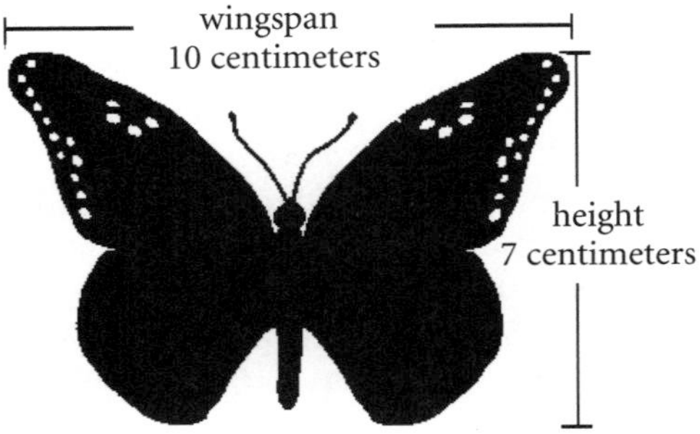

wingspan
10 centimeters

height
7 centimeters

Monarch Butterfly

Black Swallowtail Butterfly

Common Blue Butterfly

Source: National Center for Education Statistics, National Assessment of Educational Progress 1996 Mathematics Assessment

Fig. 21.3. Measuring the wingspan of butterflies

explain that the measurements in inches would be multiplied by 1.5 to get the number of feet. Whereas 38 percent of white students answered the question at least partially correctly, only 9 percent of black students and 18 percent of Hispanic students gave at least a partially correct response. (A response was deemed partially correct if it contained converted measurements but no general explanation of how to make the conversion.) Additionally, 16 percent of white students, versus 36 percent of black students and 31 percent of Hispanic students, omitted the question.

Another measurement-related problem involved designing a rectangular dog pen requiring 36 feet of fence. This problem asked students to find the dimensions of the pen that would produce the maximum area. The correct response involved writing an explanation that would convince the dog owner that a 9×9 pen would produce the maximum area for the given perimeter. Thirty-four percent of white eighth graders, compared with 10 percent of black eighth graders and 13 percent of Hispanic eighth graders, answered the problem at least partially correctly. Twenty-one percent of white students, compared with 45 percent of black students and 44 percent of Hispanic students, omitted it.

Finally, on the main 1996 NAEP assessment, an item involving area comparisons was used on the fourth-, eighth-, and twelfth-grade assessments (see fig. 21.4). The question asked students to find a way to compare the areas of a triangle and a square. Students were expected to compare the two equal areas through measuring and calculating the areas, or through tracing, folding, or covering the shapes in some way. An example of a correct response is shown in figure 21.5. In fourth grade, only 7 percent of white students, and 0 percent of black and Hispanic students answered the question correctly. In eighth grade, 32 percent of white students, compared with 8 percent of black and 18 percent of Hispanic students, responded correctly. In twelfth grade, 40 percent of white students, compared with 12 percent of black and 25 percent of Hispanic students, responded correctly. Very few students omitted the problem at any grade level, presumably because the question was essentially a multiple-choice question, albeit with some explanation required.

Discussion

According to data on NAEP achievement, measurement is the mathematical strand with the largest race-, SES-, and even gender-related disparities. As Ansell and Doerr (2000) discuss, the gender-related disparities seem consistent with spatial-related differences previously found by mathematics education researchers. Scholars tend to attribute such differences to environmental factors, such as which toys are viewed as appropriate for boys and girls (e.g., tool sets versus dolls). It is less clear why race- and SES-related differences would be large on measurement items. The NAEP disparities

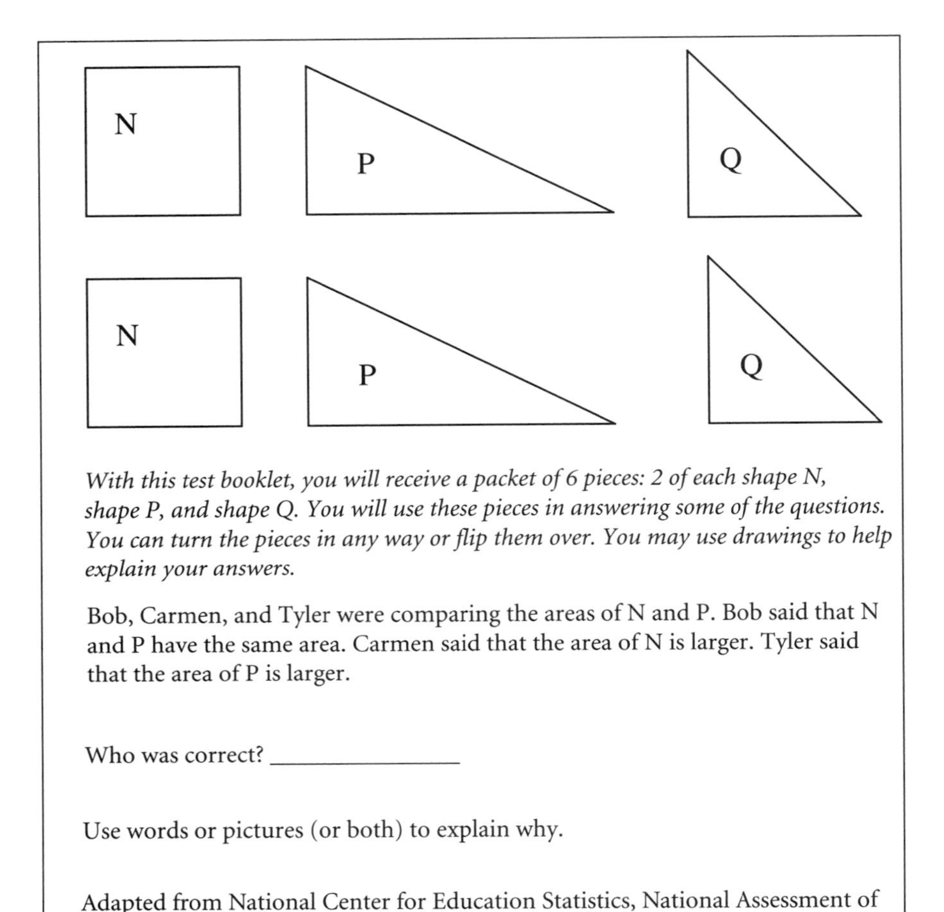

With this test booklet, you will receive a packet of 6 pieces: 2 of each shape N, shape P, and shape Q. You will use these pieces in answering some of the questions. You can turn the pieces in any way or flip them over. You may use drawings to help explain your answers.

Bob, Carmen, and Tyler were comparing the areas of N and P. Bob said that N and P have the same area. Carmen said that the area of N is larger. Tyler said that the area of P is larger.

Who was correct? ________________

Use words or pictures (or both) to explain why.

Adapted from National Center for Education Statistics, National Assessment of Educational Progress Web site nces.ed.gov/nationsreportcard/itmrls/qtab.asp

Fig. 21.4. Comparing the areas of a square and triangle

discussed above point toward differences in students' opportunities both to solve multistep problems and to learn fundamental concepts and skills relating to measurement. Regardless of whether such disparities are due more to differences in experiences within or outside the mathematics classroom, mathematics teachers are in a position to help equalize students' exposure to complex problems involving measurement-related ideas.

Traditional textbooks have tended to emphasize arithmetic at the expense of other topics, such as measurement. Teachers of black, Hispanic, and low-SES students need to be particularly careful when planning their curricula each school year. Measurement must not wait until the last few weeks of the year. It is a topic that can be introduced early each school year, serving as a

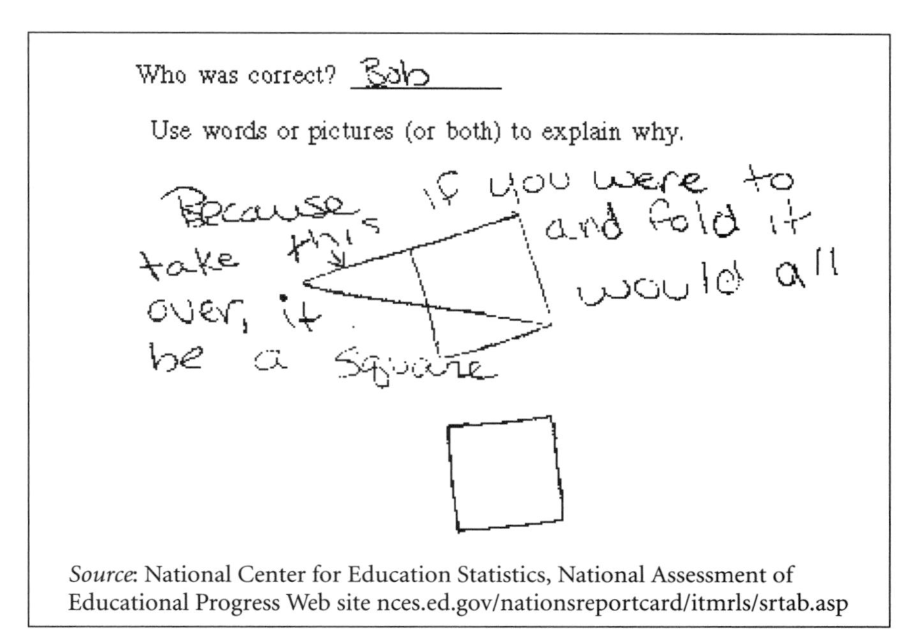

Source: National Center for Education Statistics, National Assessment of Educational Progress Web site nces.ed.gov/nationsreportcard/itmrls/srtab.asp

Fig. 21.5. One correct response

theme through which many other mathematical topics can be taught throughout the year. For example, measurement problems have natural links with algebra, geometry, data analysis, and number sense.

However, care must be taken to ensure that students develop conceptual understanding of the measurement process. Simply giving students rulers and other manipulatives to use is not enough. NAEP data on mathematics classroom practices indicate that black and low-SES fourth and eighth graders get at least as much exposure to rulers and other physical materials as white and high-SES students (Lubienski 2001). However, current achievement disparities clearly indicate that such exposure is not sufficient. Students need to understand the attributes they are measuring (e.g., length, area, weight, and so on), as well as what it means to measure.

One way to promote these understandings is through the use of fanciful, engaging problems, such as guessing how many pencils long the hallway is, how many candy bars it would take to cover the classroom floor, or how many pieces of popcorn it would take to fill the classroom. Such problems give students a sense of what it means to measure length, area, or volume in nonstandard units. These experiences can help students understand the need for, and meaning of, standard units and measuring instruments. (See Van de Walle 2001 for a helpful description of this approach.) The knowledge students gain from such an approach can help students determine and interpret measurements with greater ease and understanding.

The measurement problems mentioned above (i.e., the pencils, candy bar, and popcorn problems) have the additional advantage of being accessible yet requiring students to persist through several steps before reaching a solution. Teachers of underserved students must help students learn how to both approach and write clear solutions to multistep problems, because these were the problems with the largest achievement disparities (with many black and Hispanic children omitting them completely).

Overall, the NAEP findings presented in this article reveal alarming race- and SES-related disparities in students' measurement knowledge. This article highlights the need for teachers of underserved students, in particular, to strengthen their teaching of measurement. Teacher educators, administrators, and policymakers must provide the support necessary for such improvements to be made.

REFERENCES

Ansell, Ellen, and Helen M. Doerr. "NAEP Findings Regarding Gender: Achievement, Affect, and Instructional Experiences." In *Results from the Seventh Mathematics Assessment of the National Assessment of Educational Progress*, edited by Edward Silver and Patricia A. Kenney, pp. 73–106. Reston, Va.: National Council of Teachers of Mathematics, 2000.

Braswell, James S., Anthony D. Lutkis, Wendy S. Grigg, Shari L. Santapau, Brenda Tay-Lim, and Matthew Johnson. *The Nation's Report Card: Mathematics 2000.* Washington, D.C.: U.S. Department of Education, National Center for Education Statistics, 2001.

Lubienski, Sarah T. "Are the NCTM *Standards* Reaching All Students? An Examination of Race, Class, and Instructional Practices." Paper presented at the American Educational Research Association Annual Meeting, Seattle, April, 2001.

Lubienski, Sarah T., and Andrew Bowen. "Who's Counting? A Survey of Mathematics Education Research 1982–1998." *Journal for Research in Mathematics Education* 31 (December 2000): 626–33.

Mitchell, Julia H., Evelyn F. Hawkins, Frances B. Stancavage, and John A. Dossey. *Estimation Skills, Mathematics-in-Context, and Advanced Skills in Mathematics.* Washington, D.C.: U.S. Department of Education, National Center for Education Statistics, 1999.

National Council of Teachers of Mathematics (NCTM). *Curriculum and Evaluation Standards for School Mathematics.* Reston, Va.: NCTM, 1989.

Silver, Edward A., and Patricia A. Kenney, eds. *Results from the Seventh Mathematics Assessment of the National Assessment of Educational Progress.* Reston, Va.: National Council of Teachers of Mathematics, 2000.

Van de Walle, John A. *Elementary and Middle School Mathematics: Teaching Developmentally.* 4th ed. New York: Longman, 2001.

22

Measurement, Representation, and Computer Models of Motion

Christopher Hartmann

Jeffrey Choppin

THE authors of the National Council of Teachers of Mathematics *Principles and Standards for School Mathematics* call for a greater emphasis on connections and an increased emphasis on mathematical modeling and data analysis. In support of this goal we discuss two ideas for studying the motion of falling objects in an algebra classroom. The first approach employs Boxer, a programming language designed for educational use that is distributed freely on the Web (www.soe.berkeley.edu/boxer). The second employs the Texas Instruments TI-83 graphing calculator and Calculator Based Ranger (CBR). We compare these two approaches to illustrate the benefits of using multiple modeling tools to help students elaborate their understanding of the physics of motion through the exploration of mathematical concepts of measurement and representation. Our goal is to demonstrate how an instructional sequence combining these technologies can increase students' mathematical power to reason about measurement, graphing, and modeling.

The two activities we describe in this article focus on students' conceptualizations of the motion of a ball dropped from a given height. In the first, the students use a Boxer microworld to generate, revise, and test models. In the second, the students use a CBR to measure and model the motion of the dropping ball. In both instances the students are describing the same phenomenon. However, the representations look quite different. We describe how these differences can strengthen students' conceptions of motion and, in the process, their understanding of representations and measurement.

For algebra teachers the ball-drop situation has several interesting dimensions. First, a well-known scientific model exists ($h = -0.5gt^2 + h_0$, where h is height of ball at time t, g is acceleration due to gravity, and h_0 is initial height), and current technologies allow students to explore the ball-drop phenomenon and construct their own models for comparison to the standard model. Second, issues of the mathematics of measurement arise in relation to collecting data about observed phenomena. Just like Galileo in his original explorations of simple motion, students need to construct an understanding of how and why to collect data about distance and time in order to model motion. Third, translating observations of the ball drop into the Cartesian plane provides an opportunity to discuss the power of this abstract representation with students. This power is not obvious to the developing mathematical minds of beginning algebra students.

EXAMINING THE PROBLEM OF ABSTRACT REPRESENTATIONS

Imagine dropping a ball straight to the ground and using an electronic motion detector (e.g., a CBR) and graphing calculator to record its vertical motion beginning with its first bounce and continuing until it returns to the ground (see fig. 22.1). With appropriate programming a graphing calculator can produce a distance-versus-time graph similar to figure 22.2a. Although this graph provides a conventional representation, we also note that it physically resembles a sideways view of the motion of a bouncing ball traveling from right to left across the screen. As a result it is common for students to interpret the curvature of the graph not as depicting the height of the ball over time, but instead as a picture (see fig. 22.3) of the path of a bouncing ball. The students' confusion reflects the similarity of the graph to a phenomenon that they observe in their daily lives (i.e., a bouncing ball moves horizontally as well as vertically).

In a typical classroom activity sequence, students use the initial graph and data to produce an algebraic model as shown in figure 22.2b. The students can generate a model through a guess-and-check procedure, an analytic approach, or the use of the calculator's statistical functions. Regardless of the method, the students' model is an abstraction from the graphical representation. The func-

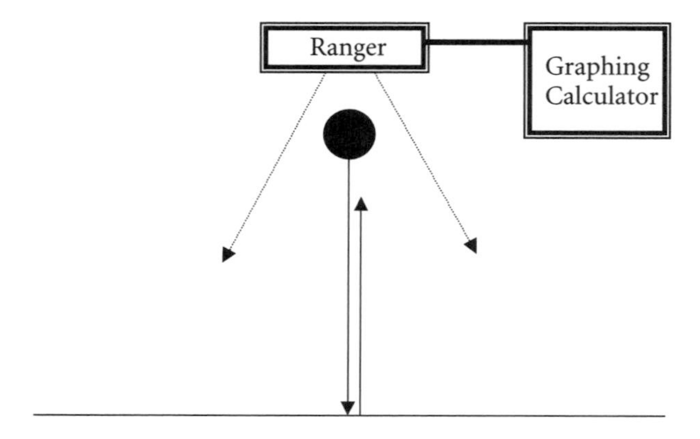

Fig. 22.1. Ball-drop experiment

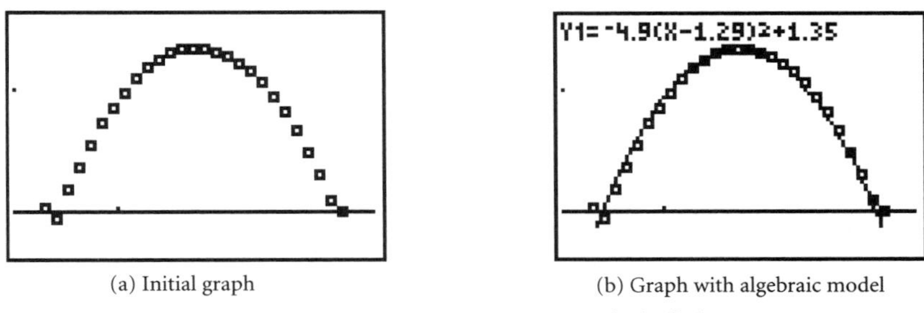

(a) Initial graph (b) Graph with algebraic model

Fig. 22.2. TI-CBR representations of a ball drop

tion does not necessarily connect with either the students' interpretation of the graphical representation or their understanding of the physics of a ball drop. In the students' minds the model can exist as an isolated, symbolic model of the observed motion in the form of a quadratic function.

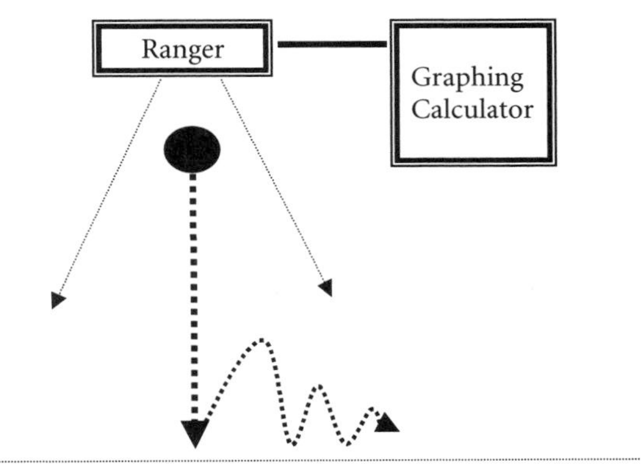

Fig. 22.3. Literal observation of the path of a ball drop

We believe that it is common in a situation such as this for students to understand the relationship between the equation and the graph (i.e., "a parabola equals a quadratic function") while simultaneously finding it difficult to connect the meaning of either representation to their own observations of the ball drop. This problem is further compounded if the students misinterpret the original graph as a literal depiction of the path of a bouncing ball. It is our experience as classroom teachers that this sequence of disconnected representations is common in the presence of technology. However, we suggest that this example illustrates students' attempts to use their existing knowledge to try to solve a problem, a habit that we wish to promote. Consequently, we recommend trying to build on students' existing knowledge by making connections between their own observations of the event and the representations produced by advanced technologies. This will enhance students' abilities to observe and analyze motion.

USING TECHNOLOGY TO BUILD ON STUDENTS' EXISTING KNOWLEDGE

Several recent journal articles for classroom teachers promote classroom activities similar to the graphing-calculator modeling exercise described above (e.g., Beckmann and Rozanski 1999; Doerr, Rieff, and Tabor 1999; Grant 2000; Hale 2000; Lapp and Cyrus 2000). Authors often either assume the transparency of the graphical representation provided by the calculator (i.e., they assume that students see a direct relationship between the graph and the phenomenon), or they argue that effective classroom leadership by the teacher can overcome this problem. We offer an alternative—increasing the presence of technology in the classroom to approach this problem from two perspectives. We think that employing an additional technology, Boxer, maintains important dynamic features of technologies in general while providing students with experiences that promote representational understanding and allow for greater connection between their observations and mathematical models.

We support the use of Boxer because this software allows students to design models that explicitly simulate their own observations of the ball drop. Thus, students can begin their explorations with models that graphically simulate a ball falling to the ground. DiSessa et al. (1991) described how students are able to use Boxer to explore concepts in depth by beginning with dynamic simulations of their own observations. In the example of the ball-drop simulation, students are able to approach concepts such as the measurement of distance and time, velocity, and acceleration through a process of designing increasingly accurate models in the form of computer programs.

USING BOXER TO MODEL MOTION

Figure 22.4 displays a Boxer microworld created by the Boxer Group at the University of California at Berkeley. (To download this microworld and the other classroom resources described in this article, visit the Web site dewey.soe.berkeley.edu/ boxer/projects.html on the Internet and follow the links for Modeling and Motion.) The microworld provides a starting point for an investigation of one-vector motion in the case of a falling object. The screen-shot in figure 22.4 contains a question, a graphics box, a menu box, and a one-line computer program that provides a simulation of a ball being dropped to the floor. In contrast to the CBR graph in figure 22.2a, the graph in figure 22.4 has the quality of resemblance with students' observations of a ball being dropped. The "go" program is a model that animates the ball, instructing it to make twenty consecutive five-unit movements producing a dot after each movement to trace its "steps." The model is dynamic, and it simulates the effect of multiple exposure photography in tracking the move-

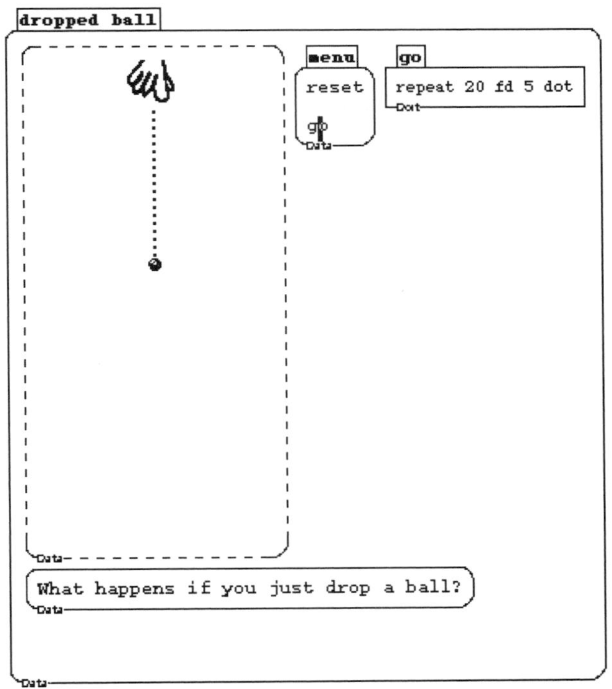

Fig. 22.4. Initial Boxer model of a ball drop

ment of the ball. In practice, we found it effective to provide the students with this initial program to scaffold their explorations using Boxer. Our experience is that with very basic guidance students are able to make the connection between their observations of a ball being dropped and the representation produced by the "go" model.

After a brief introduction to the microworld, students quickly engage with the problem of critiquing and revising the "go" program. The revisions occur as students rewrite and test new versions of "go." Most students are unhappy with "go" as a model because neither the movement of the ball during the simulation nor the resulting picture of the motion is consistent with the observed acceleration of a falling object. From these critiques the students progress to the design of simple experiments that allow them to generate data that can be used to revise the model presented in figure 22.4. For example, students can use a stopwatch to time a ball falling from various heights and compare time ratios. These observations allow students to conjecture about the ball's velocity as it falls. Through successive cycles of experimentation, observation, and revision of the "go" program, the students can use the microworld to construct simulations that are consistent with their observations of the physics of a ball drop.

Figure 22.5 presents an example of a revised program that represents an intermediate stage of students' model development. This program is typical of the way that the students' models progress in practice. In this intermediate stage the students often use the basic structures of the "go" model in the revised models. For example, in "go2" four repeat statements are employed to demonstrate increasing velocity during the fall. The phenomenon of acceleration is demonstrated by increasing the distance between the dots. This change in step size increases the distance between the dots and makes the ball appear to move faster as it drops. An important point is that, as the model incorporates the students' ideas about the ball drop, it maintains the quality of resemblance to the actual phenomenon. Obviously, this intermediate model has flaws. The ball in this model accelerates in stages as if shifting gears.

One pitfall of the "go" model is the potential for confusion between two functions—"Drop Distance = Number of Steps × Length of Each Step" and "Distance = Rate × Time." In practice, students generally consider this relationship but are quickly able to understand the difference and make appropriate use of the microworld to construct simulations that fit their observations and knowledge about an object falling to the ground. We believe that students overcome this pitfall easily because the model has face validity for the students, due to its resemblance to their observations of the phenomena.

Typically, students' initial revisions vary the rate of acceleration of the ball

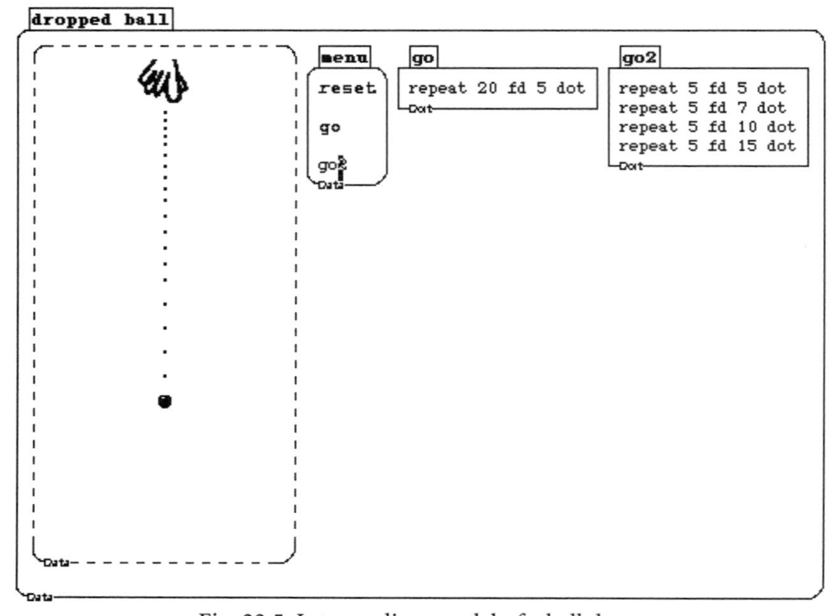

Fig. 22.5. Intermediate model of a ball drop

as occurs in the "go2" program. Through their efforts to simulate the changes in speed "precisely," students wrestle with both the concepts of gravitational force and the problems of measurement. During discussions of their simulations the students revise their thinking and eventually arrive at a model similar to the one shown in figure 22.6. This model resembles not only students' observations but also the theories of physics. The "go3" program simulates the ball drop by increasing the distance between the dots with each step. We believe that this simulation of constant acceleration builds an important conceptual foundation for students, enabling increased understanding during the mathematization of motion in a standard algebraic model. With further experimentation and redesign, the students can continue to improve the accuracy of their model. However, we do not think that this is a necessary step. Instead, we advocate switching to the CBR at this point, in order to build on students' grounded understanding of representing and analyzing motion.

BALL-DROP MODELING WITH A CBR

In this activity, students use a motion detector and a graphing calculator to record the motion of a ball dropped from a height of about two meters. A convenient program for this is the BALL BOUNCE program found in the APPLICATIONS menu of the RANGER program that can be downloaded

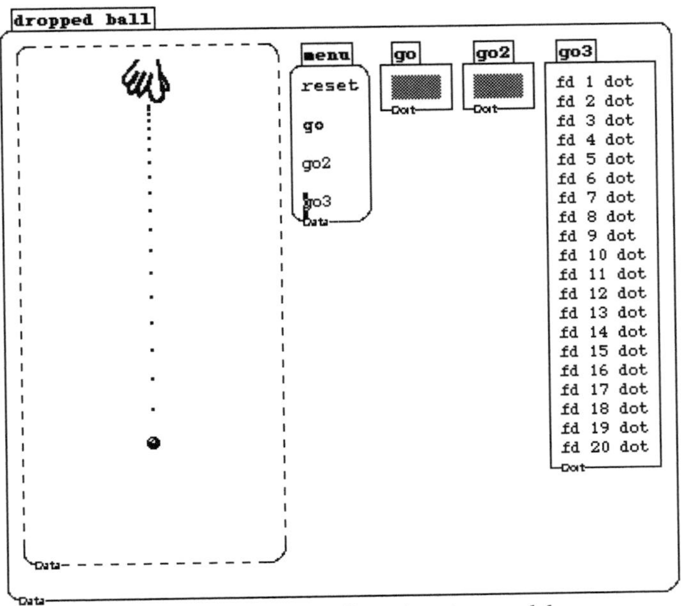

Fig. 22.6. A "constant" acceleration model

from the CBR into the calculator. After recording several bounces, students can use the SELECT DOMAIN option of the PLOT TOOLS in the RANGER program to select the portion of the bounce that represents the ball falling from its initial height to the ground. That is, they should select from the peak of a parabola to the inflection point where the next parabola begins. This portion of the graph represents a ball falling to the ground from rest.

Figure 22.7a presents a sample graph produced by the BALL BOUNCE program. We note for the reader that the BALL BOUNCE program manipulates the data so that it looks like the ball was bouncing on the motion detector instead of under it. Figure 22.7b gives a clearer representation of the actual measurements recorded by the CBR. In figure 22.7b, the *x*-scale tick marks represent 0.1 seconds and the *y*-scale tick marks represent 0.25 meters. In order to produce a graph like the one in figure 22.7b, it is necessary to quit the RANGER program, go to the STAT PLOT menu, choose PLOT 1 (which should already be turned on), and choose the discrete graph TYPE (first choice) and the box MARK.

What Do the Students See in This Graphical Representation?

This representation differs from the Boxer representation in several aspects. First, it is no longer a single-dimensional graph, and second, the graph curves even though it represents a one-vector motion. In order to determine why this happens, it will be important for students to consider what the CBR is measuring and how it transforms those measurements into the graph. Strictly speaking, the CBR is only measuring distance, but it is doing so at regular time intervals. Teachers must assess how students are viewing both the time and distance dimensions in the CBR view. It is important for the students to understand how the CBR explicitly represents aspects of the ball drop that are implicit in the Boxer model.

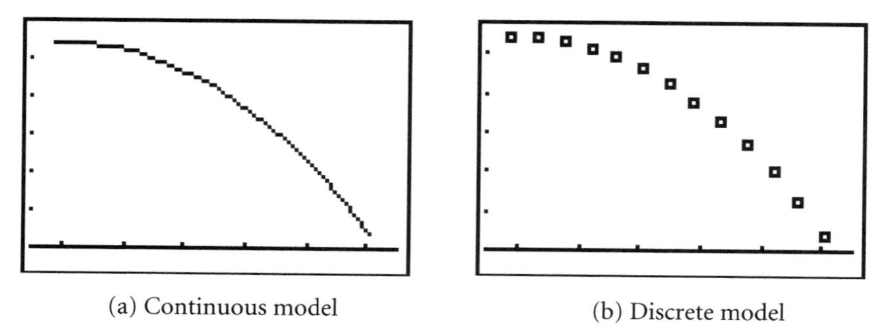

(a) Continuous model (b) Discrete model

Fig. 22.7. TI-CBR models of a ball drop

What Is Being Measured?

The students have already constructed a model of the ball drop in Boxer that incorporates time, distance, velocity, and acceleration. Although they were not measuring and representing these quantities in their model, they were producing a qualitatively descriptive model. In contrast, using the CBR, time and distance have already been measured, but the qualities of the model need to be inferred from the data. In order to make these qualities explicit to the students, it is helpful to discuss what has been measured, how it has been measured, and how those measurements are represented in the graph. For example, the teacher can lead a discussion of how the CBR collects data about the falling ball. Using the trace feature of the calculator on a graph such as figure 22.7b, the students can observe that the CBR collects data about height at constant intervals of time and consider how this influences the shape of the graph.

Representing Time

In the Boxer model, time was an implicit feature; in the CBR view, it is explicitly represented as a dimension. Although the CBR does not measure time, it does record and graph it. The students need to recognize that time is recorded in constant intervals. It is helpful to calculate how often the CBR collects data about the height of the ball (about twenty-four times a second in this experiment) and how these intervals are represented on the graph.

Representing Distance

The students incorporate distance in their Boxer programs by specifying how far the ball would drop in each step. This is similar to the CBR view, but in order for the students to see this, it will be important for them to analyze the graph. Class discussion should center on the vertical gaps between points as a way to discuss how the distance is being measured and how it varies. The two-dimensional aspect of the graph demonstrates the covariation between time and distance. The manner in which the two dimensions relate leads to a discussion of the curvature of the graph.

Velocity and Acceleration in the Cartesian Representation

The biggest anomaly between students' observations of the ball drop and the Cartesian representation of distance and time produced by the CBR is the curvature of the graph. In the Boxer model, students increased the distance traveled by the ball at each moment in its free-fall to reflect constant acceleration. In doing so, the students demonstrate an awareness of acceleration that provides a foundation for discussing the curvature of the graph. The Cartesian representation differs from the Boxer model in figure 22.4 primarily in its explicit representation of two variables, time and distance, effectively translating the dots in the Boxer model horizontally by moving each successive point a constant horizontal distance. This distance reflects the use of constant time intervals between the data points. In order to help students understand this relationship, it is helpful to focus on pairs of consecutive points on the graph and to calculate differences, by height and time, from one point to the next. As this analysis progresses, students' observations will focus on the constant horizontal differences and the increasing vertical distances between adjacent data points. These increasing vertical distances should connect with their representation of acceleration in the Boxer models. For example, some students might suggest shifting the points in figure 22.7b to line them up vertically, thereby recreating the picture in the Boxer model.

In some instances, recognizing the increasing vertical distances might not be sufficient for students to understand the meaning of the curvature in the graph. In these situations it may be helpful to have the students draw a graph, similar to figure 22.8, where both the horizontal and vertical components change at constant rates, and discuss this graph as akin, in terms of physics, to the original Boxer model in figure 22.4. This graph portrays a ball that does not accelerate. This comparison will help students to understand curvature as an indication of acceleration in the CBR view.

TOWARD A MULTIPLE-MODELING-TOOLS CURRICULUM

The two modeling tools described in this article allow students to represent the phenomenon of a falling object in quite different ways. Although each approach has its proponents, we argue for using a combination in order to take full advantage of existing technologies. The model produced using Boxer, although not conventional, furnishes an opportunity for students to build representational competence and conceptual understanding of motion. Important physical concepts such as acceleration must be built into the model; they cannot be inferred from an existing representation. Furthermore, the Boxer model has the quality of "resemblance to observed phenomenon" that promotes students' initial efforts at the design and revision of models. The Cartesian representation produced by the CBR and motion detector is more conventional, accurately documents changes in both time and distance, and lends itself to algebraic modeling. However, the tool converts the measurements into this representation without input or observation from the user. Students need to learn how to interpret Cartesian representations if they are to understand a coordinate graph as an accurate representation of the data. It is our experience that issues of measurement,

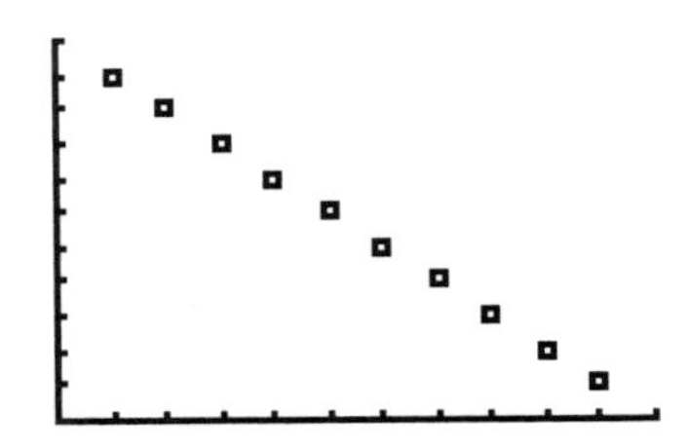

Fig. 22.8. Graph of a falling object that does not accelerate

representation, and modeling need to be discussed in the classroom if the benefits of technologies are to be fully realized there.

The instructional approach that we advocate builds conceptual understanding of the relationships among concepts of motion, issues of measurement and modeling, and the qualities of representation by employing both computer modeling and graphical calculator technologies. In combination these two approaches to algebraic modeling of motion provide students with experiences that allow them to uncover the assumptions behind the Cartesian model and to build an understanding of representation and measurement prior to taking advantage of the powerful functions of handheld graphing technologies.

REFERENCES

Beckmann, Charlene E., and Kara Rozanski. "Graphs in Real Time." *Mathematics Teaching in the Middle School* 5 (October 1999): 92–99.

diSessa, Andrea A., David Hammer, Bruce Sherin, and Tina Kolpakowski. "Inventing Graphing: Metarepresentational Expertise in Children." *Journal of Mathematical Behavior* 10 (1991): 117–60.

Doerr, Helen M., Cathie Ann Rieff, and Jason Tabor. "Putting Math in Motion with Calculator-Based Labs." *Mathematics Teaching in the Middle School.* 4 (March 1999): 364–67.

Grant, Cathy Miles. "Beyond Just Doing It: Making Discerning Decisions about Using Electronic Graphing Tools." *Learning and Leading with Technology* 27 (May 2000): 14–17.

Hale, Patricia. "Kinematics and Graphs: Students' Difficulties and CBLs." *Mathematics Teacher* 93 (May 2000): 414–17.

Lapp, Douglas A., and Vivian Flora Cyrus. "Using Data Collection Devices to Enhance Students' Understanding." *Mathematics Teacher* 93 (September 2000): 504–10.

23

Measuring the Unmeasurable: Using Technology to Study the Irrational

Maurice Burke

WITH the advent of graphing calculators, students can pry open many dusty passageways in the history of mathematics and shed some light on old topics. One such topic is the use of rational numbers to measure irrational quantities. It is most appropriate that the calculator that replaced the massive tables of values for irrational quantities—so important in the measurement formulas of scientists, engineers, and others in the past—should be used by students to study the mathematics behind those tables and not simply to reproduce them. Down this passageway you will find the footprints of many great mathematicians. By focusing on attempts to measure the irrational quantity π, this article hopes to stir up some dust and illustrate the potential for using graphing calculators to study precision and convergence as these ideas relate to measuring (approximating) irrational numbers with rational numbers.

HISTORICAL PRELUDE

To the early Greeks, *number* meant "natural numbers." Zero was not a number (Gundlach 1969, p. 29); it was a limiting concept for magnitude, much like "∞," which today is not commonly treated as a number but as a limiting concept. Early Greeks believed all geometric magnitudes (lengths, areas, volumes) could, in theory, be exactly measured by natural numbers or by ratios of natural numbers. They believed magnitudes were "commensurable" in the sense that for any two magnitudes of the same type, for example, two lengths, one could find a unit that divided each of them exactly an integer number of times. This implied that the ratio of any two magnitudes of the same type could be exactly represented as a ratio of natural numbers. With the discovery that the ratio of the diagonal of a square to the length of the square's side cannot be represented by the ratio of two natural numbers, the early Greeks realized there were geometric magnitudes that they could not measure exactly within their system of numbers and ratios of numbers. Thus, in Greek mathematics the notions of geometric magnitude and number were kept distinct and "irrational magnitudes" were not thought of as numbers (Edwards 1979, pp. 10–12).

The Greeks did not know if π, the ratio of the circumference of a circle to its diameter, was an irrational magnitude. Either way, it was not considered a number; it was a ratio. Not until Diophantus (ca. fourth century A.D.) did the Greeks consider fractions to be numbers and not simply ratios of numbers (Gow 1968, p. 112). However, the Greeks (Eudoxus, ca. 408–355 B.C.) developed a theory of proportions allowing them to compare ratios of magnitudes, whether rational or irrational, to ratios of numbers. This enabled them to approximate ratios like π with ratios of natural numbers. Archimedes (ca. 287–212 B.C.) was the first to give a method (and prove it worked) for calculating π to any degree of accuracy (Beckman 1971, pp. 62–64). He proved $223/71 < \pi < 22/7$. It should be noted, though, the approximation $\pi \oplus 22/7$ was in use even at the time of Euclid (ca. 300 B.C.) and before Archimedes (Gow 1968, p. 235 n).

With this brief historical account we realize that the real number system taught in our schools today represents a sophisticated measurement system. Within the real number system π is a number—an irrational number. (This was first proved in 1766 by Lambert.) The concept of irrational number, however, is not trivial, and students need to be given the opportunity to explore the relationship between the rational numbers and the irrationals. With technology today, we have the opportunity to help students realize that irrationals can be approximated by rationals to any degree of precision. We can help students to make sense of some different methods for approximating irrationals and to realize that some of these methods converge faster than others.

A GRAPHING-CALCULATOR APPROACH TO PRECISION

One common measure of π used in schools is 22/7. Calculators can help students understand why this is used. Since π is irrational, its decimal expansion is infinite and nonrepeating. I have a TI-92 calculator. It calculates and retains all decimal results with up to fourteen significant digits. Therefore, my calculator uses 3.1415926535898 for π. This is "precise to thirteen decimal places," since the calculator rounds π at the thirteenth decimal place with the resulting error of approximation less than 5.0×10^{-14}. Likewise, 22/7 is rounded to 3.1428571428571. An estimate of the "absolute error" made when approximating π with 22/7 is $|3.1415925535898 - 3.1428571428571|$, or 0.0012644892673. Thus, the approximation of π by 22/7 is precise to two decimal places, since the true value of π is within 0.005 of 22/7.

To appreciate 22/7 as an approximation, students need to consider other fractions that might be used to approximate π. Using number lines marked off in fractional units such as thirteenths, students can be challenged to find the location of π. Students might begin by rolling a piece of paper into a cylinder with a diameter of length 1 as measured on their number line. The circumference of their cylinder is, therefore, approximately π. By marking the paper appropriately, students can unroll it and use their mark(s) to determine the length of paper equal to the circumference of the cylinder. Placing that length of paper on their number line, students can estimate the location of π (Burke and Taggart 2002; Coffey 2001).

Such number line investigations lead immediately to interesting questions. For example, what fraction with denominator 13 best approximates π? That is, what integer n makes $n/13$ closest to π? Students might note that $n/13 \oplus \pi \Rightarrow n \oplus 13\pi$. This tells us that n is the integer closest to 13π. Rounding 13π to the nearest integer—on the calculator, Round(13π,0)—yields $n = 41$. Thus, 41/13 is the fraction with denominator 13 closest to π. Its absolute error is about 0.012, indicating that 41/13 is precise to only one decimal place.

The power of technology is realized in the repeated calculation and graphing of the errors for different denominators. The error of approximation for the best fraction with denominator d is given by

$$\text{Error}(d) = \left| \pi - \frac{\text{Round}(d \times \pi, 0)}{d} \right|.$$

Using sequence mode on the calculator, we get the following graph for the errors associated with denominators from 1 to 50. Note that the relative minimums in figure 23.1 correspond to the fractions 22/7, 44/14, 66/21, … 154/49. These are all equivalent to 22/7, suggesting that 22/7 is the best fraction (with small denominator) to use for approximating π.

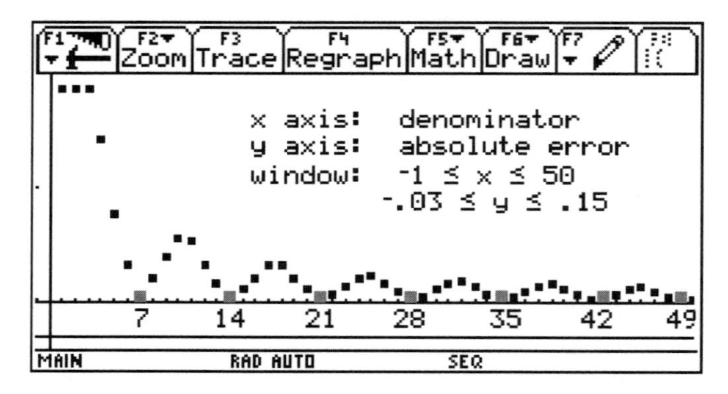

Fig. 23.1. Graph of Error(d) function for π

Exploring the graph of Error(d) for larger denominators (fig. 23.2), students come across many historical benchmarks and can discover several significant theorems from number theory. For example, such graphs powerfully illustrate the pivotal approximation theorems found in Ivan Niven's book *Numbers: Rational and Irrational* (1964), written for high school students before the advent of graphing calculators.

In figure 23.2, the point labeled A corresponds to the error for the fraction 176/56. The error for 179/57, labeled B, is smaller, thus showing that the minimums in the graph do not always occur at multiples of 7. Point C corresponds to the fraction 355/113 discovered in China by Tsu Ch'ung-chih (ca. A.D. 429–500) and in the late sixteenth century by Adriean Anthoniszoon and others of the Netherlands. This fraction is precise to six decimal places! (The next denominator yielding a smaller error than 355/113 is 16604.) Point D corresponds to 333/106. It was also discovered in the sixteenth century by Dutch mathematicians using continued fractions to estimate π. Point E corresponds to Archimedes' lower bound (223/71), and F corresponds to 377/120, the fraction used by the astronomer Ptolemy in his calculations.

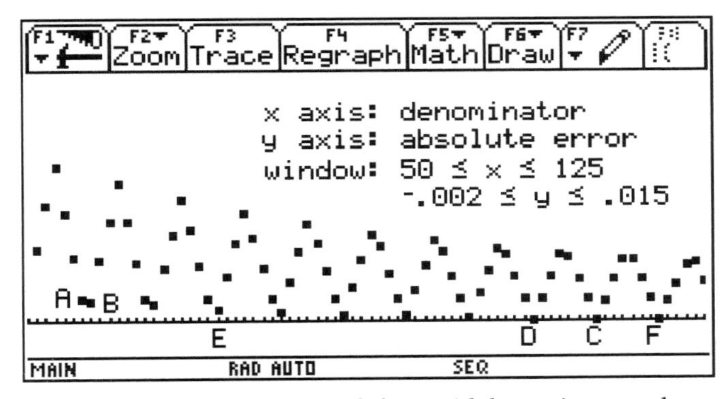

Fig. 23.2. Graph of Error(d) for π with key points noted

It is worth mentioning that students with a calculator and a good grasp of fractions can investigate continued fraction approaches leading to several of the approximations above for π. This is nicely illustrated in the NCTM *Student Math Notes* issue "To Be Continued ..." (Masunaga and Findell 1993). For example, $\pi = 3 + 0.14159....$ However,

$$0.14159... = \frac{1}{\frac{1}{0.14159...}} = \frac{1}{7.06251...}.$$

Therefore,

$$\pi = 3 + \frac{1}{7.06251...} = 3 + \frac{1}{7 + 0.06251...}.$$

Repeating the process, we find that

$$0.06251... = \frac{1}{\frac{1}{0.06251...}} = \frac{1}{15.99659...}.$$

Thus,

$$\pi = 3 + \frac{1}{7 + \frac{1}{15 + 0.99659...}}.$$

Continuing, we get the following approximations:

$$\pi \approx 3 + \frac{1}{7} = \frac{22}{7}, \quad \pi \approx 3 + \frac{1}{7 + \frac{1}{15}} = \frac{333}{106}, \quad \pi \approx 3 + \frac{1}{7 + \frac{1}{15 + \frac{1}{1}}} = \frac{355}{113}.$$

The next fraction in the sequence is 103993/33102, which differs from π by less than 6×10^{-10}.

ARCHIMEDES' METHOD

The calculator makes searching for rational approximations to irrational numbers look easy—perhaps too easy. The Error(d) function and the continued-fractions calculations described above depend on already knowing the value of π to many decimal places. What if you didn't know anything about π except that it was the ratio of the circumference of a circle to its diameter and it was the ratio of the area of a circle to the square of its radius? Archimedes was the first to provide a rigorous proof that these two ratios were in fact the same (Edwards 1979, p. 31)! Archimedes was also the first to find a method for approximating π and establishing a bound on the error of the approximation (Beckmann 1971, p. 64). From his time to the invention of the calculus in the seventeenth century, Archimedes' method was the primary tool used to approximate π. The calculator enables students to study this method from several points of view.

Archimedes inscribed and circumscribed a circle with regular polygons (see fig. 23.3). If S_n and T_n represent the length of the sides of the inscribed and circumscribed polygons, respectively, then their perimeters are $P_{insc,n} = n \times S_n$ and $P_{crcm,n} = n \times T_n$. Archimedes knew that $P_{insc,n} < C < P_{crcm,n}$, where C is the circumference of the circle, and that both perimeters got arbitrarily close to C as the number of sides increased. Therefore, $(P_{insc,n}/2r) < (2\pi r/2r) = \pi < (P_{crcm,n}/2r)$ and, as n gets larger, the difference $(P_{crcm,n}/2r) - (P_{insc,n}/2r)$ goes to 0. Thus, $P_{crcm,n}/2r$ and $P_{insc,n}/2r$ both converge to π as the number of sides increases. Furthermore, the absolute error of the approximation when using either $P_{insc,n}/2r$ or $P_{crcm,n}/2r$ to estimate π is less than $(P_{crcm,n}/2r) - (P_{insc,n}/2r)$.

Suppose the circle has radius 1. If one knows the values of S_n and T_n, one can double the number of sides of the inscribed and circumscribed polygons and find the values of S_{2n} and T_{2n} by taking appropriate square roots. Consider the case for S_{2n} shown in figure 23.3. From the Pythagorean theorem, one has

$$h_n = \sqrt{1 - \frac{S_n^{\ 2}}{4}}$$

and

$$S_{2n} = \sqrt{\frac{S_n^{\ 2}}{4} + (1 - h_n)^2}\,.$$

Replacing h_n in the second equation and simplifying, one gets

$$S_{2n} = \sqrt{2 - \sqrt{4 - S_n^{\ 2}}}\,.$$

In a similar fashion, one can show that

$$T_{2n} = \frac{2\sqrt{4 + T_n^{\ 2}} - 4}{T_n}\,.$$

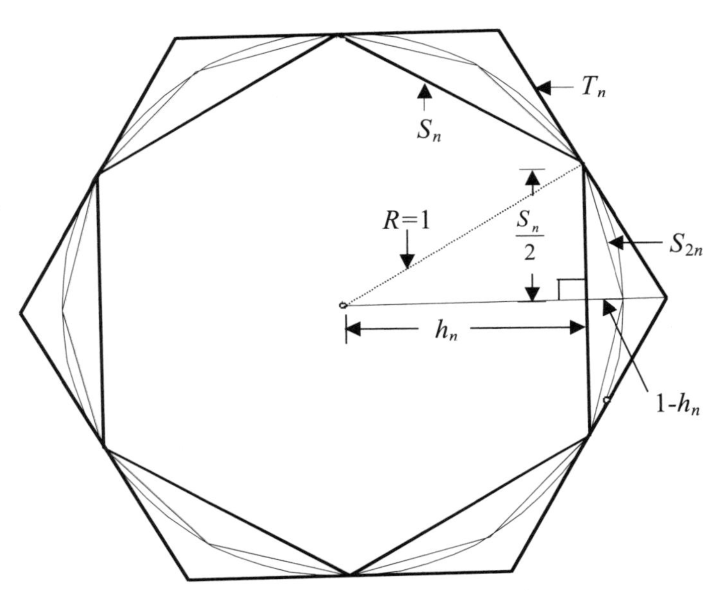

Fig. 23.3. Inscribed and circumscribed polygons

By starting with regular hexagons and a circle with radius 1, Archimedes successively doubled the number of sides and obtained estimates for the perimeters for polygons with 12, 24, 48, and 96 sides. To start the calculations, Archimedes first noted that with hexagons and a circle of radius 1, $S_6 = 1$ and $T_6 = 2/\sqrt{3}$. His method for taking square roots is not known. To avoid error in the inequality he sought, Archimedes rounded up the values for T_{2n} and rounded down the values for S_{2n} (Edwards 1979, p. 34). He thereby proved

$$3\frac{10}{71} < \pi < 3\frac{1}{7}.$$

A CURVE-FITTING APPROACH TO PRECISION

The results of imitating Archimedes with the aid of a calculator are shown in table 23.1. The table shows the first six stages and gives the value of $(P_{crcm,n}/2) - (P_{insc,n}/2)$ as an upper bound of the absolute error of the approximation. By analyzing the trend in this upper bound, students can estimate how rapidly Archimedes' method converges to π.

TABLE 23.1

First Six Stages of Archimedes' Process

Stage	Sides	S_n	Estimate for π	T_n	Estimate for π	Upper Bound Error
1	6	1	3	1.1547005	3.464102	0.464102
2	12	0.5176381	3.105829	0.5358984	3.215390	0.109561
3	24	0.2610524	3.132629	0.2633050	3.159660	0.027031
4	48	0.1308063	3.139351	0.1310869	3.146086	0.006735
5	96	0.0654382	3.141034	0.0654732	3.142715	0.001681
6	192	0.0327235	3.141452	0.0327278	3.141873	0.000421

For example, except perhaps for the first stage, the error seems to drop by a factor of one-fourth at each successive stage. Thus, a possible model for the error at stage m is Error(m+1) ⊕(1/4)Error(m) with Error(1) = 0.464102. Noting the pattern in the model's successive terms— Error(1) ⊕0.464102, Error(2) = (1/4)×Error(1) = (1/4)(0.464102), Error(3) = (1/4)×Error(2) = $(1/4)^2$(/0.464102) —students can discover an explicit form for the model: Error(m) = $0.464102(1/4)^{m-1}$. Thus, rounding to two significant digits, we find that Error(m) = $0.464102 \times 4 \times (1/4) \times (1/4)^{m-1} = 1.9 \times (1/4)^m$. Graphically, Error($m$) appears to be a very good fit for the data in table 23.1 (see fig. 23.4). Using the exponential regression option on their calculators and

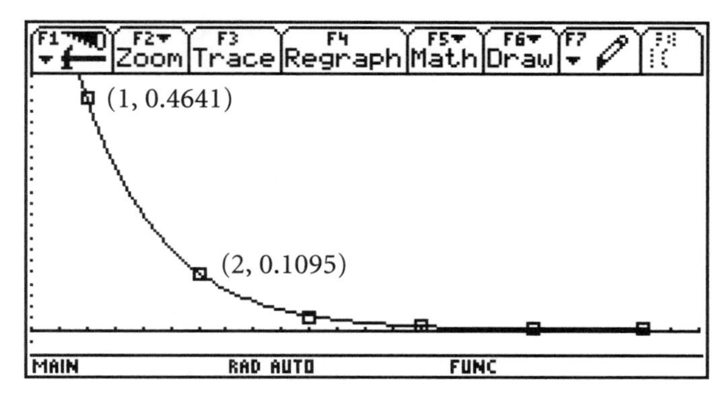

Fig. 23.4. Scatterplot of Error(m) function

excluding the first stage, which seems to be an exception, students find a similar error estimate: Regression(m) = $1.756944(0.249055)^m \oplus 1.8 \times (1/4)^m$.

Although the error function appears to work well for the data in table 23.1, students should test their empirical model beyond that domain. Using the sequence mode on the calculator, students can quickly generate the first twenty stages. At the tenth stage, one gets the approximation $3.1415921061353 < \pi < 3.141593784027$ (Error $\oplus$ 0.0000017); at the seventeenth stage, $3.1414242559735 < \pi < 3.1415835712738$ (Error $\oplus$0.00016). Something has gone wrong! The estimate for π is getting worse! This offers a wonderful chance to help students realize the impact of roundoff error in a calculator.

Although seemingly insignificant, the roundoff error of the TI-92 and most other calculators can accumulate quickly when using iterative processes. Archimedes' process is particularly vulnerable because as the roundoff error in computing the length of the sides gradually increases, the multiplication by the rapidly increasing number of sides, which doubles at each stage, quickly magnifies the error. To see this, one can use software packages or Web sites that allow high precision calculations. (For example, the "calculator" at wwwisis.ufg.edu.sv/labvirtual/math/math2/21/BigCalculator.html does computations with up to 10,000 significant digits.)

Comparing the values of the lengths of the sides of the inscribed polygons generated on the TI-92 to the lengths of those sides accurately computed with fourteen significant digits, one sees the roundoff error grow. Table 23.2 shows these comparisons with the differences underlined. Using the more precise results for the length of the sides from table 23.2, we obtain, at stage 10, $3.1415921059927 < \pi < 3.1415937477135$ with an error of 1.6×10^{-6}, and at stage 17, $3.1415926535564 < \pi < 3.1415926536566$ with an error of 1.0×10^{-10}. Thus, at stage 10 the estimates from the inscribed and the circumscribed polygons would both be precise to at least five decimal places, whereas at stage 17 the estimates are precise to at least nine decimal places. It is interesting to note that the error function Error(m) = $1.9 \times (1/4)^m$ predicts

TABLE 23.2.

Comparisons Showing How the Roundoff Error Increases

Stage	Length of side calculated using TI-92 (inscribed)	Length of side precise to 14 significant digits (inscribed)
3	0.26105238444017	0.26105238444010
10	0.0020453073607651	0.0020453073606766
17	0.000015978110025907	0.000015978966540305

an error of 1.8×10^{-6} at the tenth stage and 1.1×10^{-10} at the seventeenth stage, very close to the actual errors. (The regression equation predicts the errors even better.)

VALIDATING THE EMPIRICAL MODEL FOR PRECISION

Although the calculator appears to be of limited use in carrying out Archimedes' process due to the multiplication of very large with very small numbers and the roundoff error in the square root operation, our prediction of the error at any stage seems remarkably good. However, our error analysis is only an empirical model. To validate the model, one must look more closely at the upper bound on the error in Archimedes' process. A student equipped with trigonometry and a calculator can show that the error function we generated is in fact reasonable.

By using trigonometry (see fig. 23.5), we have $S_n = 2 \times \sin(\pi/n)$ and $T_n = 2 \times \tan(\pi/n)$, since the central angle subtending a side of the regular polygon of n sides has measure $2\pi/n$ radians. Thus $(P_{insc,n}/2) = (n \times S_n)/2 = n \times \sin(\pi/n)$ and $(P_{crcm,n}/2) = (n \times T_n)/2 = n \times \tan(\pi/n)$. From these formulas we have the upper bound $(P_{crcm,n}/2) - (P_{insc,n}/2) = n[\tan(\pi/n) - \sin(\pi/n)]$. Call this upper bound on the error E_n.

The first thing to note is that the trigonometric representation of the estimates of π do not involve the iterative rounding off that caused problems before. In fact, for the tenth and seventeenth stages, the TI-92 provides the same results, precise to fourteen significant digits, that we reported above. By comparing E_n to E_{2n}, we can determine how fast the upper bound goes to 0. $E_{2n} = 2n[\tan(\pi/2n) - \sin(\pi/2n)]$. Therefore,

$$\frac{E_{2n}}{E_n} = \frac{2\left(\tan\left(\frac{\pi}{2n}\right) - \sin\left(\frac{\pi}{2n}\right)\right)}{\left(\tan\left(\frac{\pi}{n}\right) - \sin\left(\frac{\pi}{n}\right)\right)}.$$

This expression can be

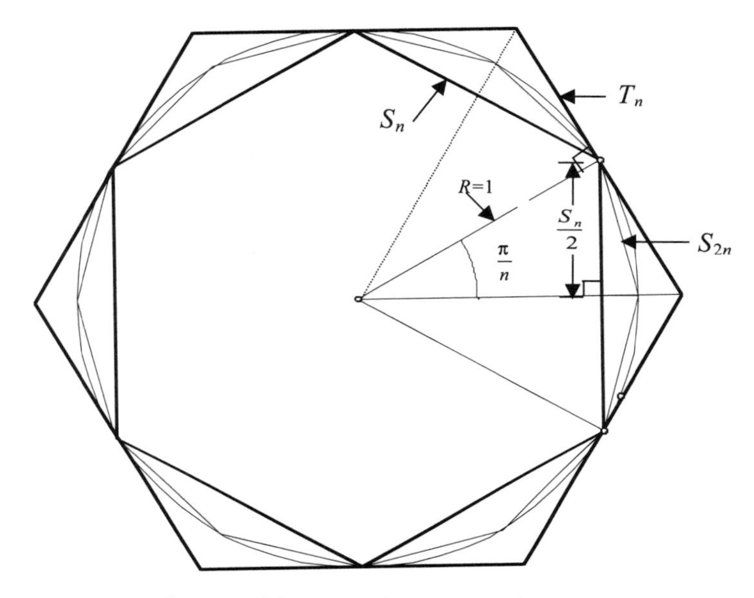

Fig. 23.5. Trigonometric representation

easily explored using a calculator. Students can generate a table of values for this ratio revealing that when $n \geq 6$, the ratio converges quickly to 1/4, in a monotonically increasing fashion, as the number of sides keeps doubling. In fact, the TI-92 calculator can take limits, and one finds that the limit of this ratio, as n increases, is 1/4. Thus, though not proved here, $E_{2n}/E_n < 1/4$ or $E_{2n} < 1/4 \times E_n$. This implies that from the first stage (when $n = 6$) on, the upper bound shrinks by a factor of at least 1/4 each time the number of sides is doubled. Our empirical error function Error(m) = $1.9 \times (1/4)^m$, therefore, provides a valid upper bound on the error for all stages m.

Historical Note

The trigonometric representations used above involve π. This raises the question, Was our error analysis really free of knowing π to many decimal places? The answer is yes. We could have used degrees instead of radians in the representations. Students might raise the question, even using degrees, do we need to know π to many decimal places in order to find good rational approximations to sine, cosine, and tangent values, another widely used class of generally irrational numbers? The answer is no and points back to Ptolemy and how he calculated trigonometric tables. In fact, Ptolemy used his computation of the sin(1°) to generate his estimate of π (Boyer 1968, p. 186). Once again, we find another passageway worth exploring.

ALTERNATIVES TO ARCHIMEDES

Many students do not have the background needed for making sense of the mathematics in Archimedes' method for generating estimates of π. However, by slightly modifying the approach and focusing on area instead of perimeter, students with a basic algebra and geometry background can accomplish similar results. Students can use the Web to find simulations of Archimedes' process. A good example is the "Pi Experiment" at www.hypercomplex.org/pi.htm. Here, students enter the number of sides of the polygons. The applet draws the inscribed and circumscribed polygons and reports their perimeter and area. Students can use these data to investigate the error of approximation pattern using the curve-fitting method described above.

Another, more concrete approach is to have students construct with rulers and compasses (or a geometry software package) the first two stages of Archimedes' process shown in figures 23.6 and 23.7. In the unit circle, the area of the circle is π. Therefore, students can quickly arrive at some appropriate inequalities: $A_{insc,n} < A_{circle} = \pi < A_{crcm,n}$ or $0 < \pi - A_{insc,n} < A_{crcm,n} - A_{insc,n}$. Students can observe that the amount of error in the approximation of π by either the inscribed or the circumscribed polygons is bounded above by $A_{crcm,n} - A_{insc,n}$. They also can see that this upper bound on the error, $A_{crcm,n} - A_{insc,n}$, consists of the area of n congruent triangles, which, for ease of reference, we will call the "error triangles" for polygons with n sides. (See the shaded area in figs. 23.6 and 23.7.)

There are many ways for students to estimate how quickly the upper bound on the error of approximation decreases as the number of sides doubles. For example, students can take measurements with rulers or geometric software packages and then compute the sum of the areas of the error triangles. Students can even physically cut out the error triangles from their 6-gon and 12-gon constructions and compute the ratio of the "weights" of the error for the 12-gon compared to the 6-gon. Although discussion is needed to justify why the weight of the paper is related to the needed areas, students get a concrete sense of the idea of convergence and, specifically, how rapidly Archimedes' method converges to π.

A more rigorous analysis can be accomplished with some basic geometry. The process of constructing the 12-gons in figure 23.7 from the 6-gons in figure 23.6 is the same process used at each stage in the Archimedean method. One finds the points of intersection of the perpendicular bisectors of the sides of the inscribed n-gon with the circle. These points, combined with the vertices of the inscribed n-gon, form the vertices of the inscribed $2n$-gon. Finally, by constructing the tangent lines at these new vertices and marking their intersections with the sides of the circumscribing n-gon, one gets the vertices of the circumscribing $2n$-gon (see fig. 23.8).

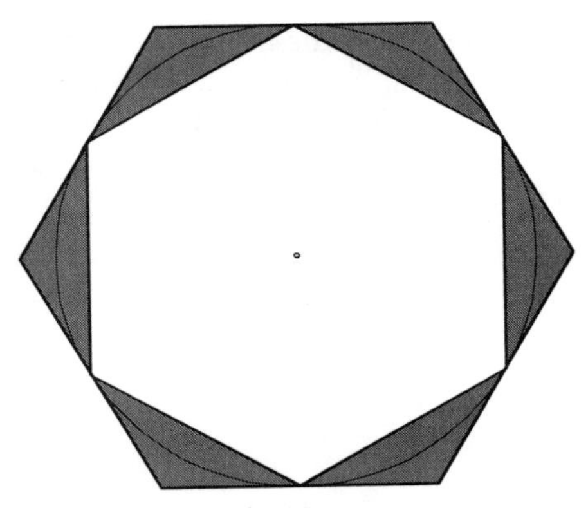

Fig. 23.6. Inscribed and circumscribed hexagons with "error triangles" shaded

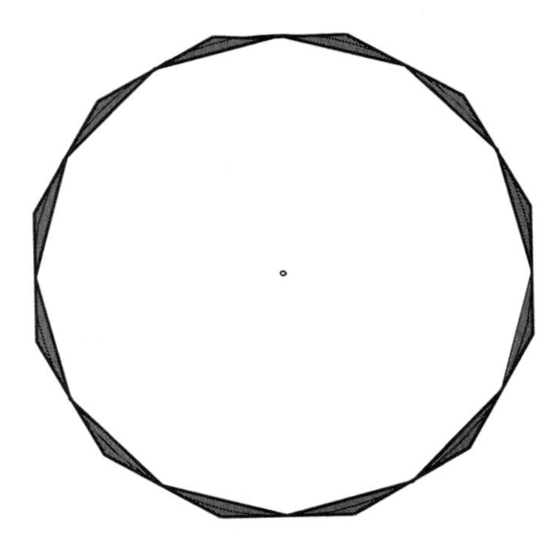

Fig. 23.7. Inscribed and circumscribed dodecagons with "error triangles" shaded

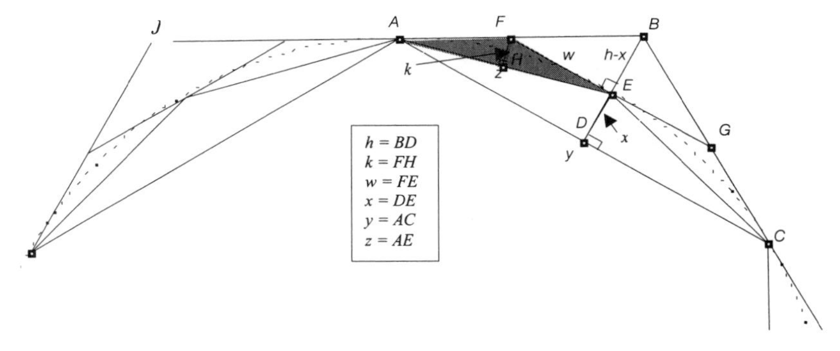

Fig. 23.8. A comparison of areas of error triangles for n-gon and $2n$-gon

In figure 23.8, $\overline{AC}$ is one side of the inscribed n-gon, $\overline{JB}$ is a side of the circumscribing n-gon, $\overline{AE}$ is a side of the inscribed $2n$-gon and $\overline{FG}$ is a side of the circumscribing $2n$-gon. Therefore, ΔABC = error triangle for n-gons and ΔAFE = error triangle for $2n$-gons. Since $\overline{FG} \parallel \overline{JB}$, $\Delta FEB \sim \Delta ADB$ and $\Delta FEH \sim$ EAD. Using ratios of corresponding sides yields:

$$w = \frac{y(h-x)}{2h}, k = \frac{wx}{z};$$

thus Area(ΔAFE) = $kz/2$ = $wxz/2$ and Area(ΔABC) = $yh/2$. Thus,

$$A_{crcm,n} - A_{insc,n} = n \times \frac{yh}{2}$$

and

$$A_{crcm,2n} - A_{insc,2n} = 2n \times \frac{wx}{2} = 2n \times \frac{y(h-x)x}{4h},$$

The ratio of these upper bounds on the error is

$$\frac{2n \times \frac{y(h-x)x}{4h}}{n \times \frac{yh}{2}}.$$

or simply $(hx - x^2)/h^2$. By thinking of x as a variable and h as fixed, we see that this quantity is a quadratic whose vertex is a maximum. From algebra, students know that the vertex will occur at $x = h/2$, at which value the ratio is 1/4. Thus, as with perimeters, the upper bound for the error of approximation from stage to stage decreases by a factor of at least 1/4 when using areas to estimate π. Students can also discover this fact by constructing figure 23.8 in a geometry editor and dragging point E along $\overline{DB}$, keeping $\overline{FG} \parallel \overline{AC}$ while measuring the relevant areas. They can even vary ΔABC to generalize the result.

CONCLUSION

The efforts of mathematicians to "measure" and make sense out of the "unmeasurable" led ultimately to the development of our most important measuring tool, the real number system. This article makes use of the calculator to explore a passage from the history of approximating π. Other topics that can be studied in this manner are square roots, the values of trigonometric functions or logarithmic functions, and many other crucially important "irrational" tools used in measurement formulas today.

Whatever the passageway we choose to enter, our students can gain deep insights about the real number system by exploring methods for approximating irrational numbers by rational numbers and the error patterns in the approximations. Technological tools, even with their limitations as noted in this article, can be used by students to investigate visually, analytically, or numerically the rates at which the approximations converge. These investigations can often be carried out with only a few simple tools such as graph paper, rulers, scissors, and protractors. Whatever the teaching method, students can learn to appreciate the richness of the real number system by attempting to measure the unmeasurable.

REFERENCES

Beckmann, Petr. *A History of Pi.* New York: St. Martins Press, 1971.

Boyer, Carl. *A History of Mathematics.* New York: John Wiley & Sons, 1968.

Burke, Maurice, and Diana Taggart. "So That's Why 22/7 Is Used for Pi!" *Mathematics Teacher* 95 (March 2002): 164–69.

Coffey, Margaret E. "Irrational Numbers on the Number Line: Perfectly Placed." *Mathematics Teacher* 94 (September 2001): 453–55.

Edwards, Charles Henry, Jr. *The Historical Development of the Calculus.* New York: Springer-Verlag, 1979.

Gow, James. *A Short History of Greek Mathematics.* Rev. ed. New York: Chelsea Publishing Co., 1968.

Gundlach, Bernard H. "The History of Numbers and Numerals." In *Historical Topics for the Mathematics Classroom,* Thirty-first Yearbook of the National Council of Teachers of Mathematics (NCTM), edited by John K. Baumgart et al., pp. 18–36. Washington, D.C.: NCTM, 1969.

Masunaga, David, and Carol Findell. "To Be Continued..." *Student Math Notes* (September 1993): 1–4.

Niven, Ivan. *Numbers: Rational and Irrational.* New York: Mathematical Association of America, New Mathematical Library, 1964.

24

Measurement in Adult Education: Starting with Students' Understandings

Myriam Steinback

Mary Jane Schmitt

Martha Merson

Esther Leonelli

WHAT measurement concepts do adults truly understand? Experiences with measurement build understanding, making adults (and children) more cognizant of the dimensions of our world and more comfortable manipulating them. Meaningful mathematical experiences occur in students' everyday lives, but they can also happen in the classroom. In this article, we include vignettes from two teachers' classes where such experiences occurred, allowing us a glimpse into students' understanding of measurement. We use these vignettes to spin out the implications of this kind of instruction. As bookends to the descriptions of actual classes, we place on one side, a vision of an ideal classroom for adult education, and on the other, an analysis of several standards and policy documents that outline what students should know about measurement, followed by a description of the EMPower curriculum, which seeks to support the classroom vision we articulate below.

A Classroom Vision

Our vision of a mathematics classroom of adult learners resembles reform mathematics classrooms of younger learners. Students tackle problems that stem from realistic contexts. They work in small groups, puzzling over problems, strategizing about solutions, sharing these solutions with others, and listening to others' reasoning. As they collaborate, students "talk math." Through opportunities for discussion, students refine their mathematical vocabulary and their logic. The teacher acts as a facilitator, guiding the learning and making the mathematics explicit.

Underlying this classroom vision is a set of core beliefs about teaching and learning. When we listen, there is much for us to learn; therefore, we provide our students with opportunities to articulate their mathematical thinking. We not only encourage students to air their ideas but also pose relevant problems in order to connect students' experiences and intuitions to the mathematics at hand. When we develop communities of mathematical investigation, we build on the foundation of students' understandings.

THE ADULT BASIC EDUCATION CONTEXT

Few adult education teachers are certified as mathematics teachers or specialists (Ward 2001). Similar to elementary school teachers, most have expertise in the field of literacy or language, yet find themselves teaching mathematics as well. In the United States approximately four million adults are enrolled in adult basic education (ABE), an umbrella term for instruction in literacy, numeracy, English language, and adult secondary education (ASE), which includes preparation for high school equivalency or general education development (GED). Each year, participants in federal- and state-funded programs are a tiny percentage compared with estimates of 20 to 40 percent of the nation's adult population in need of educational services (Sum 1999).

It is hard to find a teacher in an adult learning center who does not enjoy the work. When asked why, the reply frequently is, "I love the students!" The varied backgrounds, ages, and experiences of the students make ABE classes vibrant places to teach. There are women who have worked in markets in Haiti, truck drivers from Maine, farm workers from the Appalachians and Mexico, health care workers, and determined parents of all ethnic and racial groups. Adult learners often overcome tremendous obstacles to attend class. Motivated by tangible goals, adults strive to obtain a high school diploma or the GED, help their children with homework, prepare for a job, change careers, or gain entrance into college or training programs. The diversity of backgrounds and goals presents a feast of possibilities for teachers who want the mathematics curriculum to relate to their students' lives.

In spite of many teachers' expressed desire to do just that, making those connections is difficult. Why? For years adult educators have relied on workbooks that intend to enhance adults' conceptual understanding. Yet in most workbooks, rules take precedence over concepts and rich contexts are sparse. For example, workbooks typically present measurement as a set of rules for converting between units. Even when these materials attempt to teach basic measurement skills, such as how to read a ruler or a measuring cup, pictures take the place of real tools. The instructional materials seem to distance learners from previous and future experiences with measurement. In addition, they lend themselves to an individualized learning environment with little opportunity for students to communicate with one another. As Schmitt (2000) says when describing the workbooks, "success is defined as the ability to follow successfully a sequence of rule-based instructions that can be matched to one-step or two-step word problems" (p. 3). This hardly seems adequate given adults' aspirations and needs.

Workbooks are not the only obstacle to connecting school mathematics to learners' lives. Researchers have documented a gap between street or work mathematics and school mathematics—a gap that interferes with learning (Lave 1988; Nunes, Schliemann, and Carraher 1993; Coben 2000; Beach 2001). Coben refers to a similar phenomenon describing "invisible mathematics" as the math people do but do not recognize as mathematics. Some, like the carpenters studied by Lesley Millroy (1992) in South Africa, have a deep understanding of measurement but do not see it as mathematics. When adults equate what they do as common sense rather than mathematics, they often view what they cannot do as mathematics. In making this distinction, they undermine their success at school mathematics. First, they cut themselves off from their experientially based knowledge. Second, they make assumptions or self-fulfilling prophecies about what they can and cannot learn.

Our own experiences confirm this debilitating separation. We have met adults who do construction for a living but are paralyzed by fraction computations in class; others are successful at finding common denominators but are stumped by a tape measure. Clearly the biggest challenge is how to incorporate what learners know about measurement—and about mathematics in general—into the classroom.

It is possible to bridge the gap between street math and school math by creating instructional environments that invite learners to draw on the personal resources they bring from their experiences at work, at home, in the community, and in other informal educational settings. Below we introduce two vignettes, from adult education classrooms, that illustrate the complexities, challenges, and successes that teachers experience when they make changes to their class setup. Both teachers facilitated lessons where students measured real things using common measuring tools, where they speculated, sharing strategies and findings. The instructional environment provided an opening for learners to use what they already knew about measurement. Creating such an environment accomplishes another purpose—it makes adults' understandings about measurement visible and audible. Following the vignettes, we analyze the teachers' choices and the evidence of learners' understandings on the basis of the classroom discourse.

AN ENVIRONMENT FOR LEARNING ABOUT MEASUREMENT

The Case for Real Tools

Debra Richard (1994) taught a GED class at a church three times a week. The classroom was crowded. "We do not have tables, which really is a problem when working cooperatively with manipulatives. We have desks, which makes pairing up difficult, and moving around these 'pairs' a challenge" (p. 3).

As a participant in a teacher research project, Richard questioned whether her own fear of manipulatives could be overcome (p. 2):

> I was drawn to this question because so many educators are now talking about manipulatives and hands-on activities. The theory is that if the student touches, feels, gets to play with, and uses more than just sight (book) and sound (lecture), then the student will learn and thus retain the material better. At first I was skeptical. My concerns were:
>
> - Why bother?
> - Okay, so what if this works for kindergartners: will this work with my adult education classes?
> - Will they be interested?
> - Will they like this?
> - Do I have enough time to let them play with these manipulatives if their goal is to get their GED as soon as possible?
>
> I wasn't sure.

Debra's starting point was to change her approach to teaching measurement equivalencies. Her previous approach had always been to ask GED students to memorize a given table of equivalencies, which she admitted was never successful. This time, she would ask students to fill out their own table of equivalencies. She wrote (p. 5):

> After collecting gallon jugs, yardsticks, a clock, a calendar, a cup, a pint, and a quart, off I went to begin. I set up three centers: weight and length, time, and volume. Students were separated into three groups. We all agreed that the group had to move along as a whole and could not move to the next center unless everyone in that group understood what was to be explored. The students agreed to try to teach each other.
>
> Students took water and found that eight ounces were a cup. They poured water into a pint and found that one pint was two cups. They looked at clocks, calendars, and rulers. They explored. It was a little messy, and it took more time than just saying, "Here it is, folks: memorize it." But did it work?

She reported on the students' as well as her own reactions to these adjustments (p. 5):

> The students told me after class that for the first time they could see what a pint was, that they understood measuring and could picture the equivalencies in their heads. They felt that this idea was good and wanted to do more of this learning.
>
> One of my students, Roberta, came in the next day and told me that she went home and cried about something that she had learned the day before. I felt pretty awful, and asked her what she meant. She said, "I never realized that when I gave my son eight ounces of formula that he was drinking a whole cup of formula." Aha! She made a connection that my students in the past had not.
>
> It was powerful for me that many students said that they never quite understood this before. One 16-year-old boy, who has lived in foster care for most of his life, said that he had never touched a measuring cup until now!

Richard reflected on her change in perspective (p. 10):

> The gaps in the adult learners' math backgrounds need to be addressed. How could anyone picture in their minds something that they have never seen?
>
> Manipulatives increased the number of connections the students made to tools, routine tasks, and basically, to their lives. They began to see that mathematics is in their lives.
>
> During this experience, many questions surfaced, and I began to try to think of better ways to do lessons. For instance, instead of introducing manipulatives with something unfamiliar, like pattern blocks, begin with something familiar. If you're teaching measuring, ask students to bring in their own measuring tools and go from there.

The final paragraphs of her story emphasize how emotional this experience proved to be for learner and teacher.

> At our "graduation" ceremony at the end of this school year, I called each student from my classes up to the front of the parish hall, one by one, to receive a carnation and a certificate of achievement. When I called Roberta, whom I spoke of earlier, and announced to everyone in our program that she got her GED, she walked up slowly and gave me a hug. She backed away saying that she wanted to tell everyone a story. How surprised I was to hear what she had to say (p. 11):
>
> > "My first day in Debra's math class, we tried to find how many ounces were in a cup. I went home and told my husband how stupid I was and cried all night. John [her husband] set up bottles and measuring cups, and we poured water all night until I understood.... I never liked math before, but I know that I can do it now."

The Role of Inquiry as a Motivator

Esther Leonelli (personal communication, 2 January, 2001) recounted an experience from her biweekly ABE/pre-GED Math, Science, and Writing summer school class. She asked her students, "What do you want to learn in science?" They chose the themes. Over the course of a few weeks, some of the students researched and refined their questions using the Internet.

> My main goal for the class was to have students come up with a genuine question for inquiry that could be tested in class. The prompt for student inquiry came up unexpectedly in the hands-on activity where we were exploring what a kilogram was using a balance, [with] standard and non-standard weights, to figure out how much it weighed. In the student workbook was the following text (Franco 1999, p. 6):
>
> > There is a joke that goes: "Which has greater mass: a kilogram of feathers or a kilogram of stones?" Explain the joke. Also, make sure to tell it to someone you know outside of your class.
>
> Some folks got the joke. Some didn't. Willie tried to explain it to Cora:
>
> "It's because they both weigh the same," he said. "What you're talking about is the weight."
>
> Cora said, "What do you mean? I know that stones are heavier than feathers."
>
> "Yeah," said Willie, "but even though a stone looks like it weighs more and the feathers are airy, a kilogram is like a pound, and if you have a pound of stones and a pound of feathers, they are the same."
>
> Further probing by me did not quite help make the connection between the weight, space, and density. However, the joke provided the genuine context for student science inquiry using math tools.
>
> Maria asked, "Why is it that a bag of magazines is heavier before you shred them than after you shred them?"
>
> There was much discussion about this. Willie was doubtful. I asked Maria why she thought that. She insisted that she spoke from her experience. I asked her questions like, "Do you mean that if you put unshredded magazines in a bag, that you could fit more in the bag and that it is heavier?"
>
> Willie asked, "Well, did the magazines fit in one bag after shredding or did you need another bag?"
>
> She said, "No! The magazines were definitely lighter after shredding them."
>
> Cora said, "Well, why don't we do an experiment and see?"
>
> The students enthusiastically embraced the idea. They planned how they would do it for the next class. I recorded their plan on the board. The students framed the question. We made a list of materials we would need and assigned people to bring them. On the day of the experiment, the pile of magazines kept slipping off the scale. To address this unexpected glitch, we weighed a class member on a bathroom scale. We then weighed her holding the stack of magazines. The students subtracted her weight from the total and used that for the weight of the magazines. Some of the students had to shred some magazines by hand while some used the machine. We placed the shredded pieces in the trash bags. Students were careful to collect pieces that fell on the floor. Then we weighed the bag(s). The students

found no difference between the two weights. All but Willie appeared a little disappointed that there wasn't a difference since it seemed to disprove their original hypothesis that the non-shredded magazines were much heavier.

Some factors they offered as to why it didn't follow their expectations: some pieces were shredded by hand so were larger than the machine-shredded pieces; the scale was off. We didn't have time to conduct more trials. In some ways the question was not really settled for them.

If there had been more time, I would have asked them to conduct the experiment a few more times with different amounts of magazines and asked them to describe how the shredded magazines differed from the non-shredded magazines in each case. I might have suggested that they include their descriptions of the volume (using the capacity of the bags as units) and record their findings and observations. Finally, I might have asked them what relationship or pattern they saw and how this would change or support the answer to their original question.

The important points in this activity for me were that, by the end of the five week class, the students posed their own question, devised an experiment to test their question, came in with their own tools, and carried out the activity.

Reflections on the Two Vignettes

Both vignettes exemplify some of the dilemmas and successes that we encounter in our adult education mathematics classes when we explore mathematics together. The environment, pedagogy, strategies, and mathematics interact in ways that are at times unexpected, and we and our students learn much more than just the mathematics.

Environment, Pedagogy, and Strategies

The first episode highlights the effect of the classroom environment on what people learn and feel. When Richard introduced common measurement tools—measuring cups, clocks, and rulers—along with "messy" exploration, she set the stage for "seeing" and "making connections that my students in the past had not." This process is in stark contrast to what she had encouraged before—memorization. Her purposeful adjustments to the classroom environment immediately connected the classroom to learners' lives in two distinct ways: Roberta and her husband carried over class activities to pouring and measuring at home; a young man assessed his past experience, realizing his lack of experience with even the most mundane measuring tools, like a measuring cup. Both teacher and students had an emotional response. Richard's student, Roberta, learned more than the fact that eight ounces is the same as one cup: she figured out that "I can do math!"

Leonelli waited to hear her students' questions. She, too, was purposeful about setting the stage to encourage students' questions. Leonelli identified the joke about whether a kilogram of feathers is heavier than a kilogram of stones as the puzzle that provided the "context for students' inquiry." We wonder about other factors as well. Might mucking around with scales and standard and nonstandard weights have evoked experiences outside the classroom, providing students with an opportunity to articulate their own contexts? Would all the learners have understood the joke without that exploration?

In both classes the lines between street and school blurred. The freedom to explore seemed to inspire students. One student in Leonelli's class brought her shredder to class, and a student in Richard's class took the class activity to her kitchen. The motivation appears to have come from the learners.

These two vignettes also leave us with important questions about both the environment and measurement. Richard's class, in which the result of using real tools was a surprise for the teacher as well as for the learners, begs the question: why don't we use real tools more often and more purposefully? And, having established that this is a viable resource, what experiences might we need to structure over time to deepen their knowledge of measurement?

In adult education, it is important to connect to what people know (Knowles 1984). We cannot emphasize enough how important this is yet how randomly this happens. Both scenarios suggest that the adult education classroom environment can capitalize on adults' motivation, experiences, and informal strategies.

Students' Mathematical Understandings

It is important for us to grapple with adults' understandings and formulation of their understandings about measurement and about mathematics. From both vignettes, we learn and have questions about students' understandings of measurement. In Leonelli's vignette, did students perceive that a given amount of shredded paper takes up a lot more volume than the same amount unshredded? Maria's strong belief that "the magazines were definitely lighter after shredding them" provided Leonelli and the rest of the class a difficult challenge: to give Maria a way to verify—or disprove—her claim. Leonelli needed a way to help her see that the paper's density decreases when it is shredded. The experiment was the vehicle, but Maria's basic understandings and beliefs contradicted the evidence.

We further wonder what common perceptions led Maria, and others in the class, to an erroneous conclusion. If they were recalling picking up the bags, for example, do they realize that the different pressures exerted by them when picking up the bags affects their perception of its weight? That is, a bag of shredded magazines has the weight spread out, so they pick it up with a larger area of their body, therefore perceiving that it is lighter. Distinguishing between weight and density is perhaps not something that students had thought about. Their ideas about, and experiences with, carrying a bag of magazines versus a bag of shredded paper weighed more heavily than the experiment they did in class.

Concerning Roberta, we note that she had been making her baby's bottle, and we imagine that she measured the contents carefully while mixing the formula, but she didn't connect that amount to any equivalence. She understood the number of ounces that went into the bottle, but the relation between that and a cup was not part of her mathematical vocabulary. In her practice, she functioned well and likely had a clear concept of the amount of liquid that filled the bottle, but in Richard's mathematics classroom, her life experience collided with a conventional measurement unit and she made a connection that astounded her. Having made that one connection, it is likely that other mathematical understandings will have shifted on the basis of this new insight. In situations like this where a learner and teacher experience an "aha" moment, after the emotion passes, staying on guard for opportunities to view a learner's mathematical understanding is crucial. So powerful were the understanding and making of the connection that Roberta ultimately was able to articulate her belief that she could now do math. Who knows what lies ahead?

MATHEMATICS CONTENT IN ADULT EDUCATION

Where do adult educators who are committed to building a bridge between the classroom and their learners' lives turn when they look for guidance on the mathematical content? For example, what aspects of measurement do adult education documents say are important for adults to be able to know and do? How do they compare to the school-based *Standards* of the National Council of Teachers of Mathematics (NCTM)?

Measurement is one of the Standards of *Principles and Standards for School Mathematics* (NCTM 2000), with emphasis on "understanding what a measurable attribute is and becoming familiar with the units and processes that are used in measuring attributes" (p. 44). We believe this is essential for adults as well; but because *Principles and Standards* was written for teachers of children, we examined policy documents that have implicitly or explicitly addressed this question for adult mathematics education.

During the 1990s and as this article was being prepared, efforts were under way to define and redefine the mathematical or numeracy content of adult basic education. We have identified seven salient documents and comment on their treatment of measurement.

Measurement for All?

The National Reporting System for Adult Education (NRS) is an outcome-based reporting system for state-administered, federally funded adult education programs. It holds ABE programs accountable for tracking students' progress (U.S. Department of Education 2000). An important principle of NCTM's *Standards*—that measurement be incorporated at all levels (prekindergarten to grade 12)—is not promoted by NRS. Across NRS's six levels, mathematical progress is primarily described by students' increasing ability to handle decontextualized computation of whole numbers and fractions. Measurement is mentioned explicitly only at the highest level and then is confined to making "mathematical estimates of time and space" and applying "principles of geometry to measure angles, lines, and surfaces." This is a limited view of measurement and its relevance in adults' lives.

In contrast to the NRS, *The Massachusetts Adult Basic Education Curriculum Framework* (Massachusetts Department of Education 1996) are more consistent with the NCTM *Standards.* At all levels of instruction, these frameworks encourage concrete and experiential approaches where "students engage in problem solving, communicating, reasoning, and connecting to: make and use exact and estimated measurements to describe and compare phenomena; select appropriate units and tools to measure to the degree of accuracy required; use systems of measurement" (p. 84). However, this document is limited in its examples of contextualization of the mathematical topics and skills.

A number of the documents are driven by realistic contexts that are important for adults. *Equipped for the Future Content Standards: What Adults Need to Know for the Twenty-first Century* (Stein 2000) and the earlier SCANS report (U.S. Department of Labor 1991) are grounded in data gathered from the workplace and from adults in their roles as workers, parents, and community members. Curricula developed within these frameworks do not emphasize specific mathematics skills as much as problem situations to which adults are expected to bring a full set of skills.

A Framework for Adult Numeracy Standards (Curry, Schmitt, and Waldron 1996) does not list a particular set of skills for measurement, either. Instead, this document combines geometry and measurement as it relates findings based on anecdotes of workers and employers on the use of measurement. Learners and stakeholders recognize that measurement skills can be crucially important at work. As one employer states, "In the workplace ... we measure everything. If someone is measuring something and they are taking samples and then see that it is going out of the acceptable margin[,] I mean, they need to stop everything and get someone to find out what is happening" (p. 49). This framework recommends that teachers encourage the use of exact and estimated measurements as well as address acceptable tolerances.

Adult Numeracy and Measurement: An International Perspective

In addition to what we glean from frameworks and standards written in the United States, we are aware of framework efforts in Australia and the United Kingdom that have taken a different approach. They have been successful in developing frameworks for adults because they focused on numeracy, which keeps context and mathematics content present at all times. The Australian *Certificates in General Education for Adults* is perhaps the most promising example (Kindler et al. 1996). It organizes numeracy by purposes rather than by mathematical skills.

National Standards for Adult Literacy and Numeracy in the United Kingdom (Qualifications and Curriculum Authority 2000) are organized by mathematical skills but give guidance and examples on where to use the skills. Contexts and use at each level include using the judgment of size for packing or storing things, using measures in cooking, weighing loose items that are sold by weight, making curtains or measuring to lay a carpet, and working out how many tiles are used to cover an area.

As this article was being prepared, the Australian and British standards' emphasis on contextualization is helping guide a revision of the previously mentioned Massachusetts curriculum framework.

Extending Mathematical Power to Adults

In preparing this article, we are committed to developing a curriculum that supports ABE teachers who share our classroom vision, with its regard for listening, discourse among students at all levels, and investigation based on realistic contexts. (Extending Mathematical Power [EMPower] is a four-year project [NSF grant number ESI-9911410] for the development of a mathematics curriculum for out-of-school youth and adults.) Because of its pervasiveness in everyday life, measurement easily lends itself to all the above. We primarily link measurement to geometry—for example, as students are asked to give their classroom a "fresh look," they measure as they make plans to paint the walls, retile the floors, or add trim to the top of the walls—but it also shows up in other units. For example, investigations about change over time lead to a discussion of units of measure, since they are crucial for determining scales on the *x*- and *y*-axes.

We place great emphasis on estimation, visualization, and relative size comparison. For example, in one piloted EMPower lesson, the teacher brought a cup with water and asked students what they might be able to measure. Working in small groups, they came up with descriptors and estimates for them: "distance around," water height, cup height, water temperature, cup weight, and "the measure of the cup's material." In addition, they talked about the different shapes they saw: circles of different sizes at the top and bottom of the cup; a cylinder; and, "depending on the angle from which you look at it (perspective), a triangle." All these ideas provided the teacher with an assessment. Rather than assume what they did and did not know, she listened as her students voiced their knowledge about measurable attributes and the units and processes for measuring those attributes, as NCTM suggests. The teacher stocked her classroom with rulers, string, measuring cups, measuring tapes, scales, and thermometers for students to use to check their estimates, as the Massachusetts ABE Math Standards suggests.

In order to welcome learners' experiences, roles, and goals from the start, the first EMPower lesson on geometry revolves around the items students bring in that they have made themselves. In one class, Dave brought the blueprints of a house he had built. He described all the details of the house, including the particulars about the architects' mistakes in the drawings; his classmates stood by, mesmerized. He pointed out errors in measurement and described what the actual construction entailed to correct those errors. However, when he made sketches of the shapes in the blueprint and described the objects, he faltered. He knew, for example, that the bricks were not squares, but he couldn't recall the name of their shape. When the word *rectangle* suddenly came to him, he could not spell it. With some help, he was able to spell it, and said, "Would you believe? I have built many $400,000 homes; I can measure really well. But I could never read or write." Did Dave begin to build a bridge between his experiential knowledge of measurement and the mathematics class version of measurement? Absolutely! Again the teacher is presented with the opportunity to capitalize on further insights that result from a newfound understanding of the connection between life and mathematics.

CONCLUSION

A comprehensive treatment of measurement for adults must address learners' informal knowledge base, connect school math to learners' out-of-school experiences, and increase the knowledge and skills of adults at all levels. When we listen, there is much for us to learn. Rich description, case studies, and classroom episodes provide us with valuable information that helps us begin to make sense of adults' understandings of mathematics and how those understandings develop. The two vignettes that we examined in this article provide us that type of information about adults working with measurement.

Measurement should be accessible at all levels of adult education, but—more important—it should appear in a variety of ways so that learners have concrete experiences with it. Real tools are powerful, as was seen in both vignettes. We advocate that documents like the National Reporting System, which is an accountability document, encourage this.

Connecting to adult contexts is crucial in the successful engagement of learners in their roles as workers, family members, and community members, and in their pursuit of further education. Everyday activities that involve measurement are resources that we can and should use in our classrooms.

Everyone has ideas about measurement. How do we develop or deepen those ideas? Given the opportunity to think about what they know, talk about it, and reflect on it, learners can connect what they know and understand to the mathematical conversation and what it is about. As teachers listen to and facilitate these conversations, the ideas about measurement—about the many and various forms it takes: actual measurement, temperature and other readings, the different dimensions of measurement (linear, two dimensional, and three dimensional)—can all enter into the classroom discourse in a more meaningful and investigative way. In all instances, learning is more powerful when it connects to learners' lives.

REFERENCES

Beach, King. "Transitions between School and Work: Some New Understandings and Questions about Adult Mathematics." In *Proceedings of the Seventh International Conference of Adults Learning Mathematics—a Research Forum*, edited by Katherine Safford-Ramus and Mary Jane Schmitt, pp. 8–12. Cambridge, Mass.: National Center for the Study of Adult Learning and Literacy, 2001.

Coben, Diana. "Mathematics or Common Sense? Researching 'Invisible' Mathematics through Adults' Mathematics Life Histories." In *Perspectives on Adults Learning Mathematics: Research and Practice*, edited by Diane Coben, John O'Donoghue and Gail Fitzsimons, pp. 47–53. Dordrecht, Netherlands: Kluwer Academic Publishers, 2000.

Curry, Donna, Mary Jane Schmitt, and Sally Waldron. *A Framework for Adult Numeracy Standards: The Mathematical Skills and Abilities Adults Need to Be Equipped for the Future.* Boston: World Education, 1996.

Franco, Betsy. *Key to Metric Measurement, Book 4: Metric Units for Mass, Capacity, Temperature, and Time.* Berkeley, Calif.: Key Curriculum Press, 1999.

Kindler, Jan R., Beth Marr Kenrick, Dave Tout, and L. Wignall. *Certificates in General Education for Adults.* Melbourne, Victoria, Australia: Adult, Community, and Further Education Board, 1996.

Knowles, Malcolm. "Introduction: The Art and Science of Helping Adults Learn." In *Andragogy in Action: Applying Principles of Adult Learning*, edited by Marlene Knowles et al, pp. 1–21. San Francisco: Jossey-Bass, 1984.

Lave, Jean. *Cognition in Practice: Mind, Mathematics and Culture in Every Day Life.* Cambridge: Cambridge University Press, 1988.

Massachusetts Department of Education. *The Massachusetts Mathematics Curriculum Framework: Achieving Mathematical Power.* Malden, Mass.: Massachusetts Department of Education, 1996.

Millroy, Wendy. *An Ethnographic Study of the Mathematical Ideas of a Group of Carpenters. Journal for Research in Mathematics Education* Monograph 5. Reston, Va.: National Council of Teachers of Mathematics, 1992.

National Council of Teachers of Mathematics (NCTM). *Principles and Standards for School Mathematics.* Reston, Va.: NCTM, 2000.

Nunes, Terezinha, Analúcia Schliemann, and David Carraher. *Street Mathematics and School Mathematics.* New York: Cambridge University Press, 1993.

Qualifications and Curriculum Authority. *National Standards for Adult Literacy and Numeracy.* Suffolk, England: QCA Publications, 2000.

Richard, Debra. "Can the Fear of Manipulatives Be Overcome?" In *Implementing the Massachusetts Adult Basic Education Math Standards, Vol. 2: Our Research Stories,* edited by Esther Leonelli, Martha Merson, and Mary Jane Schmitt, pp. 2–13. Holyoke, Mass.: Holyoke Community College/SABES Regional Center, 1994.

Schmitt, Mary Jane. "Developing Adults' Numerate Thinking: Getting Out from under the Workbooks." *Focus on Basics* 4 (September 2000): 1–5.

Stein, Sondra. *Equipped for the Future Content Standards: What Adults Need to Know and Be Able to Do in the Twenty-first Century.* ED Pubs document EX0099P. Washington, D.C.: National Institute for Literacy, 2000.

Sum, Andrew. *Literacy in the Labor Force: Results from the National Adult Literacy Survey.* Washington, D.C.: National Center for Education Statistics, 1999.

U.S. Department of Education. *National Reporting System for Adult Education: Implementation Guidelines.* Washington, D.C.: U.S. Department of Education, 2000.

U.S. Department of Labor. *What Work Requires of Schools: A SCANS Report for America 2000.* Washington, D.C.: Secretary's Commission on Achieving Necessary Skills, 1991.

Ward, Judith. "Arkansas GED Mathematics Instruction: History Repeating Itself." In *Proceedings of the Seventh International Conference of Adults Learning Mathematics—a Research Forum,* edited by Katherine Safford-Ramus and Mary Jane Schmitt, pp. 54–58. Cambridge, Mass.: National Center for the Study of Adult Learning and Literacy, 2001.